Mechanical Engineering: Principles and Practices

Mechanical Engineering: Principles and Practices

Edited by Rene Sava

CLANRYE INTERNATIONAL

www.clanryeinternational.com

Clanrye International,
750 Third Avenue, 9ᵗʰ Floor,
New York, NY 10017, USA

ISBN: 978-1-63240-580-7

Cataloging-in-publication Data

Mechanical engineering : principles and practices / edited by Rene Sava.
    p. cm.
Includes bibliographical references and index.
ISBN 978-1-63240-580-7
1. Mechanical engineering. 2. Engineering. I. Sava, Rene.
TJ153 .M43 2017
621.81--dc23

For information on all Clanrye International publications
visit our website at www.clanryeinternational.com

Printed in the United States of America.

Contents

Preface

The aim of this book is to present researches that have transformed the discipline of mechanical engineering and aided its advancement. This discipline studies the applications of engineering in manufacturing, designing and maintenance of mechanical systems. This book is a valuable compilation of topics, ranging from the basic to the most complex advancements in the field of mechanical engineering. It is compiled in such a manner, that it will provide in-depth knowledge about the theory and practice of this discipline. The text sheds light on the various principles and practical aspects of mechanical engineering. For all readers who are interested in this discipline, the case studies included in this book will serve as an excellent guide to develop a comprehensive understanding.

The world is advancing at a fast pace like never before. Therefore, the need is to keep up with the latest developments. This book was an idea that came to fruition when the specialists in the area realized the need to coordinate together and document essential themes in the subject. That's when I was requested to be the editor. Editing this book has been an honour as it brings together diverse authors researching on different streams of the field. The book collates essential materials contributed by veterans in the area which can be utilized by students and researchers alike.

Each chapter is a sole-standing publication that reflects each author's interpretation. Thus, the book displays a multi-facetted picture of our current understanding of applications and diverse aspects of the field. I would like to thank the contributors of this book and my family for their endless support.

Editor

Stress Analysis of a Rotating Body by Means of Photostress Method and Using Solidworks Programme

Ján Kostka[*], Peter Frankovský, František Trebuňa, Miroslav Pástor, František Šimčák

Technical University of Košice, Faculty of Mechanical Engineering, Košice, Slovakia
*Corresponding author: jan.kostka@tuke.sk

Abstract   The paper is dedicated to the examination of stress states of a rotating body which occur due to centrifugal forces. In order to determine stress and stain distribution, photoelastic PhotoStress method was applied and a measurement device – polariscope LF/Z-2 was used. The sample under examination was subjected to loads of various magnitudes and, subsequently, recorded with a photographic device. For objective evaluation of detected results, the object under examination was modelled and simulated in programme SolidWorks 2012.

Keywords: *photostress method, dynamic stress analysis, photoelasticity, isochromatic lines, solidworks*

1. Introduction

When carrying out a dynamic stress analysis, it is vital to be familiar with certain physical phenomena which occur in objects under examination. A body, in our case a body which is rotating around its fixed axis, has certain kinetic energy. It can be determined without any attempt to calculate performance which is necessary to put the body in rotating motion, i.e. directly through integration of kinetic energy values of individual material elements.

Kinetic energy of a material element can be examined via calculations of kinetic energy of a material point:

$$dE_k = \frac{1}{2} dm v^2 \tag{1}$$

where: dE_k - kinetic energy of the element,

dm - mass of the element under examination,

v- velocity of the element under examination.

Since velocities of individual material elements of the body are not identical, total kinetic energy of the rotating body under examination equals the sum of kinetic energy values of all individual elements of the body, i.e.:

$$E_k = \int \frac{1}{2} dm v^2 \tag{2}$$

The following applies to the velocity of an individual element of the body under examination:

$$v = \omega\, r, \tag{3}$$

where: ω - angular velocity of the rotating body,

r - distance of the relevant body from rotation axis.

Then, substituting for v we obtain:

$$E_k = \int \frac{1}{2} \omega^2 r^2 dm \tag{4}$$

Angular velocity ω is in given moment identical for all elements of the body, hence it can be placed before integral resulting in:

$$E_k = \frac{1}{2} I \omega^2, \tag{5}$$

where I (body moment of inertia) represents:

$$I = \int r^2 dm. \tag{6}$$

The rotating body is loaded by centrifugal forces having the nature of body forces. Stresses which occur during rotation are symmetrical with respect to rotation axis, hence can be expressed as a function of distance r from the rotation axis. Stresses in the rotating body of constant thickness h are evenly distributed along whole thickness of the rotating body, and stress σ_z equals zero in the direction of z-axis. Centrifugal force is defined as:

$$dF_o = dm\omega^2 r = \rho\, h\, r\, d\varphi\, dr \omega^2 r, \tag{7}$$

where: $d\varphi$ - angle delimiting the element of the rotating body,

ρ - specific mass of the material of the rotating body,

h - thickness of the material [5,6].

2. Photoelasticity

Photoelasticity is an experimental method which enables determination of deformations and stress states in a body. Measurement device which is used for the examination by means of this experimental method is known as polariscope. It is based on polarisation and temporary birefringence of a light beam. Temporary light birefringence occurs, provided that an optically isotropic material exhibits anisotropic characteristics after loading. To be more specific, the above-stated means that in the examined model there is a decomposition of one bean into two beams which vectors oscillate in different planes though perpendicularly to each other. These planes determine directions of principal stresses, and beams which oscillate in these planes are being distributed throughout the model under examination with a different speed due to the state of anisotropy. Phase shift occurs as a result of differences of the speed of individual beams. Phase shift has crucial influence on stress-state examination of the bodies [1].

2.1. Reflection Photoelasticity – PhotoStress

PhotoStress method represents reflection photoelasticity which is a technique of stress determination in structures, systems or components. It is as well applied when measuring surface deformations during dynamic load. It belongs to experimental methods and its principle has not considerably changed since the beginning of 1950's, although measurements are taken with rather advanced polariscopes.

A strain-optical layer is applied to the surface of the component which undergoes the measurement. When illuminated with polarised light and viewed through the polariscope, the coating on the component surface is seen as decomposition of strains in form of a colourful spectrum. These colourful patterns are known as isochromatic lines (isochromatics). These represent geometrical locations of points in which the difference of principal strains is constant, and immediately point at different magnitudes of strains with areas of maximum strains. The stress analysis can be performed thanks to an optical compensator attached to the polariscope. A video camera or a digital camera can be attached to the measurement device – the polariscope to record different values of loads [2,4].

3. Experimental Analysis of the Rotating Body

Figure 1. Reflection polariscope LF/Z-2 and stroboscopic light

The principle of PhotoStress method within the solution of dynamic effects lies in the scanning of phenomena which reoccur periodically during the same period of time. When examining the rotating structural element with photoelastic coating, the element under consideration is being illuminated for a short period of time and always in the same position in order to depict a static picture of

isochromatic lines. During circular movement of the component under consideration an impression of a static picture arises due to the source of stroboscopic light. Figure 1 depicts polariscope LF/Z-2 and stroboscopic light used in PhotoStress method [6].

The proposed sample with diameter 150 mm and a centre hole with diameter 5 mm was drawn in programme SolidWorks 2012. The sample was later attached to the rotating shaft through the centre hole with a nut. After drawing and after being stored in DWG format, the file was sent for cutting into a photoelastic material. Figure 2 depicts a drawing of the sample under examination with required size.

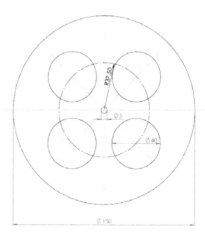

Figure 2. Drawing of the sample for experimental dynamic analysis

3.1. Material and Division of the Sample under Analysis

Right selection of material is an important criterion of rotating bodies in order to prevent its destruction at higher speeds. In our case, it was a photoelastic material PS-1 which was subjected to measurements. These measurements were transferred into initial solid plane plates after simulation drawing in programme SolidWorks and water-jet cutting. For the purposes of the experiment, the material PS-1 had been produced and delivered by American company Vishay Photoelastic Division of Measurements Group, Inc. The material exhibits high sensitivity and can be used in an elastic as well as an elastic-plastic area. Is is delivered together with a reflection layer and a protective temporary paper package. Listed in Tab. 1 are material characteristics of photoelastic material PS-1 [5].

Table 1. Material characteristics of photoelastic material PS-1

Strain-optical constant K (-)	0,150
Modulus of elasticity E (GPa)	2,5
Ductility (%)	5
Poisson's ratio μ (-)	0,38
Thickness (mm)	3,05
Tolerance (mm)	± 0,06
Sensitivity constant (°C)	150
Maximum temperature (°C)	150
Maximum elongation (%)	3

For verification of some material characteristics the solid plane plate PS-1 was subjected to load in point as shown in Figure 3. The plate was then viewed through reflection polariscope LF/Z-2.

Figure 3. Verification of material characteristics of PS-1

In order to measure and evaluate the object under examination and analysis objectively, it is very important to choose the most proper way of sample material separation. In our case, with the measured object made of a photoelastic material, we had to eliminate heat effect or residual stresses in the cutting area. The best and the easiest option was to separate the material with water jet. The sample under examination was carved out by Watting, s.r.o. with its seat in Prešov. The separation was done with the device type AquaCut 3001.20Wr.

3.2 Measurement Chain

The measurement chain, after set-up, included (position number in Figure 4): reflection polariscope LF/Z-2 (1),stroboscopic white light STROBEX Model 135M-11 (2), digital camera attachment (3), power supply (4), structure frame to attach the motor (5), signal generator (6), motor HSM 60 (7), photoelastic object to be examined (8), digital camera and portable computer. For determination of velocity of the rotating samples we used digital laser revolution counter Laser Tacho.

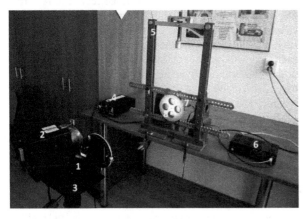

Figure 4. Measurement chain in dynamic stress analysis

3.3 Analysis of Isochromatic Fringes

When analysing isochromatic fringes, the polariscope LF/Z-2 was set to the position "MAGNITUGE". The laboratory was partially dimmed for better recording of colourful isochromatic lines. The rotating body under consideration was attached to the motor and, as revolutions were increasing in a clockwise direction, the body of a constant thickness was continually loaded by centrifugal forces. At very low revolutions of the body no

change of colourful lines occurred on the surface of the element under examination, i.e. the colour shade was grey as when the material PS-1 was not loaded. Colourful lines appeared with continuous increase of motor revolutions. These lines reoccurred when the whole colour spectrum was displayed. Black areas represented zero-stress areas. With every increase of rotational speed new pictures of the rotating model were taken. The examined components were not loaded to maximum capacity as there was a high risk of destruction, material breakdown and endangerment of persons carrying out the measurement. Figure 5 shows pictures of isochromatic lines (fringes) at continuous increase of revolutions of the light-weight disc during examination [2].

Figure 5. Isochromatic lines during continuous increase of rotations of the rotating body model under examination

Considering the pictures of the model, it is obvious that the most loaded parts were the narrowest parts of the component. In these so called critical areas, colourful patterns started to reoccur when gradually increasing the revolutions, while some fringes disappeared completely.

4. Evaluation and Comparison of Results

For more objective analysis and evaluation of the model, the object was modelled in programme SolidWorks 2012.

Object simulation was launched after input parameters and maximum speed of 7800 rev./min. were set – this maximum speed was set for the model rotating in clockwise direction – and model network had been created. After all calculations were completed, the model was depicted with all stresses, displacements and strains. Eight points were selected on the surface of the model which lead from the attachment centre to the edge of the model in order to recognise precise values. Differences of principal normal stresses, which were determined for the half of the object surface, were depicted in the programme as shown in Figure 6.

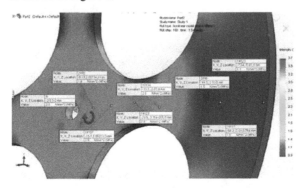

Figure 6. Differences of principal normal stresses along the half of the model

In the following step, a stress intensity diagram was created in which the behaviour of differences of principal normal stresses can clearly be identified. Whole sample was presented for better imagination of the above-mentioned processes (Figure 7). The sample is symmetrical, i.e. it is enough to take into consideration only one half of the model.

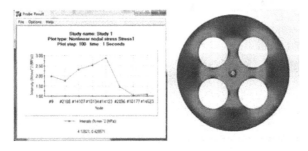

Figure 7. Stress intensity diagram and the model of whole rotating sample under examination

Closest to the centre hole the stress intensity equals 2 MPa. Later, it dropped to 1,8 MPa as a result of stress decomposition on a bigger surface, and started gradually increasing again up to node 14 123 where it reached its maximum value of 2,9 MPa. The given area can be considered critical. From this node the intensity started decreasing to 1 MPa, and at the end of the analysed model it slightly rose to 1,1 MPa.

In the most critical part of the model subject to examination, where stress intensity equals 2,9 MPa, stress intensity values were identified by means of "probe" function, i.e. along the edges of holes cut into the light-weight parts of the model as shown in Figure 8.

Differences of principal normal stresses showed identical values of 3,7 MPa. Considering the above-mentioned, it can be concluded that the direction of model rotation (clockwise or anti-clockwise rotation), or in our

case rotation of the sample examined by means of PhotoStress method at comparable speed of 7800 rev./min., is of no importance.

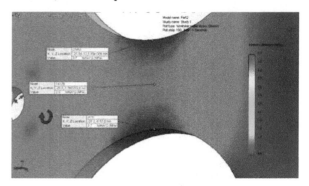

Figure 8. Differences of principal normal stresses in the narrowest area of the rotating disc model

5. Conclusion

In the experimental stress analysis and via subsequent verification by means of the finite element method carried out on the rotating body under consideration of centrifugal forces the most critical areas of the sample (model) were identified. These areas were the narrowest parts of the component subject to examination. Though not reported in this paper, there were other experimental examinations of a variety of other samples to identify stress distributions. These were radically changing due to different cut openings in the component. Not all forms of weight reduction were appropriate for the implementation into practical areas, since requirements of engineering and industry practice may be immensely challenging. Subsequently, the simulation was used as a verification tool to identify if graphical representations of stress distributions on the components were correct. The results of the analyses may have slightly varied due to potential inaccuracy of measurements, air humidity, temperature during measurement in the laboratory conditions or unstable network voltage, though it is relevant to sum up that this method of dynamic photoelasticity, which is newly implemented at the Department of Applied Mechanics and Mechatronics of the Faculty of Mechanical Engineering of the Technical university in Košice has a great future potential in the analysis of structural elements under consideration of centrifugal force effects.

Acknowledgement

This work was supported by projects VEGA 1/0937/12 and APVV-0091-11.

References

[1]   Trebuňa, F., Šimčák, F., Príručka experimentálnej mechaniky [Handbook of experimental mechanics]. TypoPress, Košice, 2007.

[2]   Trebuňa, F., Princípy, postupy, prístroje v metóde PhotoStress [Principles, procedures, devices in PhotoStress method]. TypoPress, Košice, 2006.

[3]   Trebuňa, F., Jadlovský, J., Frankovský, P., Pástor, M., Automatizácia v metóde Photostress [Automation in PhotoStress method]. 1. issue. Košice: TU – 2012.

[4]   Kostka, J., Optické metódy a ich uplatnenie v priemyselnej praxi [Optical methods and their utilisation in industrial practice]. Bachelor thesis, Košice, 2012.

[5]   Kostka, J.,Využitie metódy PhotoStress pri napäťovej analýze konštrukčných prvkov s uvážením vplyvu odstredivých síl [Application of PhotoStress method in stress analysis of structural elements under consideration of centrifugal force effect]. Diploma thesis, Košice, 2014.

[6]   Frankovský, P., Trebuňa, F., Application of photostress method in stress analysis of a rotating disc. Metalurgija 53.4 (2014): 541-544.

[7]   Kobayashi, A. S.,Handbook on Experimental Mechanics. Society for Experimental Mechanics, Seattle, 1993.

[8]   Trebuňa, F., Jadlovský, J., Frankovský, P., Bakšiová, Z., Kostelníková, A., Further Possibilities of Using Software PhotoStress for Separation of Principal Normal Stresses. Acta Mechanica,Slovakia.

Non-typical Hazards in Road Traffic

Marian Dudziak[1], Andrzej Lewandowski[2], Michał Śledziński[1,*]

[1]Poznan University of Technology, Chair of Basics of Machine Design, 60-365 Poznań, ul. Piotrowo 3, Poland
[2]Jan Sehn Institute of Forensic Research in Kraków, 31-033 Kraków, ul. Westerplatte 9, Poland
*Corresponding author: michal.sledzinski@put.poznan.pl

Abstract The authors describe a road traffic accident with an unusual cause which was a wrong defensive reaction of the driver on the road surface covered with spilled grain. The results of road tests are presented illustrating behaviour of the vehicle on the road surface in such conditions. Attention is paid to the psychological aspect of the drivers defensive reactions in the situation when they are taken aback with non-typical road surface conditions. There is a need to train drivers in right decision making skills (psychology) and in the ability to make effective defensive manoeuvres (driving technique) in order to improve road traffic safety.

Keywords: *road accident causes, driving psychology, road tests*

1. Introduction

According to the Main Police Headquarter data, 35,400 road traffic accidents occurred in Poland in 2013, resulting in almost 3,300 fatalities and 43,500 injured persons. Compared to the 2012 data a 4.5% drop in the number of accidents, almost 8% drop in the number of fatalities and over 5% in the number of the injured are noted [19]. While in 1997 in Poland, with the population of circa 38 million, 7,311 people died in road traffic accidents, in 2012 this number went down to 3,500 fatalities, despite a threefold increase in the number of motor vehicles, currently exceeding 24 million. The number of deaths on the Polish roads in relation to the increase in the number of motor vehicles is shown in Figure 1.

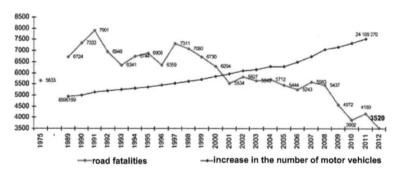

Figure 1. Number of road deaths in Poland in the population of ca. 38 million vs. increase in the number of motor vehicles [19]

Table 1. Structure of road traffic incidents caused by excess driving speed in Poland in 2011 [5]

Type of incident	Number	%	Fatalities	%	Injuries	%
Overturning of vehicle	2,012	21.5	169	13.7	2,679	20.5
Striking a tree	1,803	19.2	451	36.6	2,387	18.3
Rear-end collision	1,316	14.0	70	5.7	1,877	14.4
Head on collision	1,170	12.5	211	17.1	2,052	15.7
Side collision	730	7.8	109	8.8	1,101	8.4
Striking a post, traffic sign	554	5.9	66	5.4	703	5.4
Striking a pedestrian	439	4.7	98	8.0	432	3.3
Striking a safety barrier	270	2.9	27	2.2	356	2.7
Accident involving passenger	210	2.2	15	1.2	344	2.6
Striking a parked motor vehicle	174	1.9	10	0.8	223	1.7
Hitting a hole, pothole, bump	20	0.2	0	0.0	30	0.2
Striking an animal	9	0.1	0	0.0	14	0.1
Striking a railway gate	3	0.0	0	0.0	5	0.0
Other	469	7.1	6	0.5	628	6.7
Total:	9,179	100.0	1,232	100.0	12,831	100.0

Speeding was recognized as the main cause of road accidents. Table 1 describes results of driving with excessive speed in Poland in 2011 (according to the data of the National Council for the Road Traffic Safety).

Another category of causes of road traffic accidents includes factors such as: poor condition of vehicle (faulty vehicle lights, vehicle systems, tyres), condition of the road pavements [11,17] and insufficient or faulty lights.

Results of road traffic incidents can often be tragic, involving loss of human health and life, as well as property losses. Among the causes of road traffic accidents human factors is indicated as the primary one [1,16,18]. The most frequent causes of road incidents, that could be considered as classic ones, include improper driver behaviour in situations relating to: changing direction of vehicle movement (e.g., turning, making u-turns, backing up, merging into traffic), yielding the right of way, adjusting the speed to road and weather conditions and driving technique, keeping required distance to other vehicles, behaviour relating to pedestrians and other road users, overtaking, braking, stopping and parking, and finally, driving under the influence of alcohol, illegal drugs and psychotropic substances.

In turn, non-classic causes of road traffic accidents are associated with unusual and unexpected conditions encountered on the road, both in terms of the character of obstruction and the place of its occurrence, where most unusual obstacles, both in terms of shape and structure, are encountered on the carriageway in front of the moving vehicle. These may include, for instance, unguarded and unmarked potholes in the road pavement, unmarked obstacles within the right-of-way (e.g., unlit parked vehicles, unprotected road traffic accident sites); house animals and wildlife suddenly entering the road; big stones lying on the carriageway, lost elements of loads carried by other vehicles (such as: rolled hay, bulk or bagged grain or other objects spilled over the road); oversize farming equipment travelling on roads (e.g., combine harvesters, tractors). These are unusual and randomly occurring situations usually unexpected to drivers who usually are not alert to such hazards especially on road sections that they know very well. Note that when noticed too late any obstacle can become a serious safety hazard. Whether a driver manages to get out of trouble without any harm is mainly a matter of the psychological aspects of the situation (emotions, temperament, ability to assess the hazard, ability to come up with adequate defensive reactions). In such situations driver's personality traits, including intelligence understood as ability to draw on previous experience, effective controlling of cognitive processes, that is attention, perception and processing of information [20] become essential. Also psychological traits of the driver such are susceptibility to stress, his/ her actual psychophysical condition, ability to focus attention, driving efficiency and skills as well as external factors such as presence of other vehicles, music, or mobile phone are not without significance. It is known that driver's reaction to an obstacle is a process spread in time that encompasses a number of processes such as perception, recognition and identification, decision and motoric reaction. Also a time for the activation of a specific car system (braking, steering) needs to be allowed for. When in a typical situation that a driver can encounter on the road the statistical perception times range from 0 to 0.7 sec.), time of obstacle recognition and decision taking from 0.2 to 0.6 sec. and motoric reaction time is 0.25 sec. for braking and 0.2 sec. for turning, in a non-typical situation these times are considerably longer [12]. When reconstructing road traffic accidents the so-called total reaction time – meaning the time from the moment of coming up with a defensive reaction by the driver to till the moment of an appropriate car system activation - is usually adopted at the level of 1.0 sec. for daytime and 1.2 sec. after dark. It needs to be stressed that in a situation when the driver is surprised with an unusual obstacle he/ she has never encountered on the road before the reaction times may be significantly longer, which in an obvious way reduces probability of avoiding the hazard. In German literature on the subject, a term *Schrecksekunde* [15] is even mentioned, referring to the time in which the driver is unable to take any action due to being overwhelmed by terror.

Causes of road traffic accidents are investigated by institutions dealing with reconstruction of road traffic incidents [6,12,14]. In classic situations the road user is usually aware of the results of a failure to comply with the rules of the highway code. However the situation looks different when non-classic causes are involved.

An in-depth analysis of the causes and effects of such road incidents is fully justified. Especially, if they refer to non-typical, random incidents for which the driver is not mentally prepared [7,4,16]. Such incidents may include sudden changes of the type and condition of the road surface, weather conditions and condition of the vehicle. Taking the right decisions when performing defensive manoeuvres in order to minimize the potential effects of a dangerous situation requires knowledge and skills which an average driver usually does not possess. That was also the case in a road traffic accident involving a passenger car that run over a layer of grain spilled over the carriageway [2,8]. It was determined that the driver of the vehicle moving on the surface covered with grain lost control of the car as a result of a his defensive action. The vehicle left its lane and moved to the where it crashed into an oncoming car. The site of the road collision with the grain spilled over the pavement and the accident's results are shown in Figure 2 and Figure 3.

Figure 2. Grain spilled onto the road

Analysis of the causes of this accident involved, among other things, providing an answer to the question what psychomotoric possibilities were at the drivers disposal in order to avoid the crash [1,4,18]. The driver had to identify a non-typical hazard which required a decision how to react: whether to brake strongly or softly or to refrain from any additional manoeuvre, meaning going on driving with previous vehicle movement parameters.

Figure 3. Damaged car

Non-typical and unusual character of the situation and lack of knowledge on the car's behaviour on a pavement covered with a layer of grain or similar material made it necessary to conduct specialist road tests. The testing was necessary also due to insufficient data on practical instruction recommendations as to which potential defensive manoeuvre should be used by the driver to avoid the hazard. The effect of the change in the parameters of the vehicle movement on its stability in the existing conditions had to be determined.

2. Experimental Research

2.1. Test Conditions

The road testing was conducted on the runways of a disused military airfield. Two belts of grain, approximately 0.5 m wide each with a spacing corresponding to the vehicle wheels spacing were formed on the test road section. The grain layer was ca. 0.02 m thick. The test road section was ca. 70 m long. The experiment was carried out jointly by research teams from the Forensic Investigation Institute of Kraków and the Poznań University of Technology.

The purpose of the research was to analyze the traction characteristics of the vehicle and the stability of its movement [9] while driving on wet asphalt pavement covered with a layer of grain. The following parameters were measured in the course of the experiment: skid resistance on pavement surface covered with grain and on a control section, braking deceleration noted for the test vehicle on the test section of the carriageway covered with dry grain and on wet carriageway covered with grain. The experiment was carried out for vehicle speed range of 70-120 km/h.

2.2. Measuring Devices

The experiment was carried out using the following measuring instruments:

– Skid Resistance Tester SRT-3 developed at the Warsaw University of Technology,
– DATRON measuring system, for the measurement of vehicle dynamics parameters installed in the test car Ford Escort 1.6 16V, equipped with anti-lock braking system (ABS). The vehicle had Michelin Pilot HX 195/60R15 tyres,
– optical sensor V1 integrated with DLS processing module and AEP5 recorder for data acquisition (placed in the car) and RPM counters mounted on the vehicle's front wheels, photo eye sensor installed on the vehicle used for the recording device calibration, brake pedal sensor and the measuring computer.

The measuring system used allowed for the measurement of the following vehicle movement parameters: distance covered, time length of braking process, speed and deceleration of vehicle braking. Data sampling rate for the test was 50 Hz.

Weather conditions (ambient temperature and wind force) were monitored in the course of testing. The experiment was filmed using a video camera.

2.3. Test Procedure

The following measurements were taken during the experiment:
– skid resistance values m on wet asphalt pavement with no spilled grain and on carriageway covered with the spilled grain,
– relevant braking parameters of the test vehicle: time t and braking distance Sh on carriageway for different initial speeds u_0 and different pavement condition: uncontaminated wet pavement, pavement covered with dry grain and with wet grain.

It was assumed, that the specified and measured vehicle movement characteristics are the basic tool for an analysis and evaluation of the directional stability of the vehicle moving on a non-typical surface.

Figure 4 illustrates the brake test carried out on wet asphalt pavement covered with spilled grain.

Figure 4. Test car during braking test on wet asphalt pavement covered with spilled grain

3. Measurement Results

3.1. Skid Resistance

Skid resistance value m was measured with skid resistance tester SRT-3, used to determine road pavement quality [11]. Twenty three measurements were taken in the course of the experiment. The skid resistance values were obtained from 14 tests carried out for wet ashpalt pavement (with no grain spilled on it) and from 9 tests (carried out on the pavement with the grain spilled on it). The measurements were taken in two modes: with application of water (a) onto the road surface directly in front of the wheel whose parameters were being measured and with no additional application of water (b) in front of the wheel. The following mean skid resistance values were obtained: for wet asphalt pavement not covered with grain 0.46 (a) and 0.54 (b), and for wet asphalt pavement covered with grain 0.30 (a) and 0.35 (b).

The results of the measurements skid resistance value for wet asphalt pavement with no spilled grain on it agree with the values published in the literature, where for this type of pavement values falling within the range $\mu = 0.40$–0.55 [15] are given. For wet asphalt pavement covered with grain a significant drop in the skid resistance level was noted, with the value in the range 0.3–0.35.

3.2. Braking on Wet Asphalt Pavement

In the course of the experiment 8 braking tests were carried out for the car moving with various initial speeds, within the range from 70 to 110 km/h. Based on (momentary) vehicle movement parameters, recorded by the measuring system, mean car braking deceleration was determined in the function of the speed of motion and

braking distance, measured from the moment of brake application till the moment when the car comes to a stop.

Next the authors determined a Mean Fully Developed Deceleration (MFDD) reached by the vehicle following the braking process stabilization, excluding the initial and final stage. In this case the procedure adopted in European Standards for requirements imposed on brakes (ECE Regulation No 13) [3] was applied.

The obtained values of braking deceleration fall within the range 6.6–7.3 m/s^2 (with $a_m = 7.0$ m/s^2 mean value and 0.22 m/s^2 standard deviation of). In turn MFDD fell within the range 7.4–8.1 m/s^2 (with 7.9 m/s^2 mean value and 0.21 m/s^2 standard deviation).

3.3. Braking on Wet Asphalt Surface Covered with Grain

The authors carried out seventeen (7) brake tests on wet asphalt pavement contaminated with grain for various initial speed values υ_0 of the test car motion (within the speed range from 80 to 120 km/h). For each test, based on momentary parameters values recorded by the measuring system the diagrams of the following characteristics were made: car speed $u = f(S_h)$, braking deceleration $a = f(S_h)$ and relative wheel skid of $p = f(S_h)$. An example of the relevant car motion parameters record is shown in Figure 5.

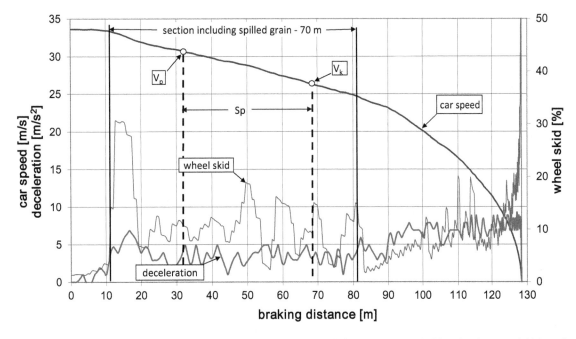

Figure 5. Characteristics of the change of car motion parameters during braking on asphalt pavement covered with grain, where: υ_0 – initial speed, υ_k – final speed of car in motion, S_P – test section

An analysis of the relative skid characteristics revealed a dominating change, at the time when the car was outside of the area which was covered with the spilled grain. The test section on which the stabilized operation of the ABS was analyzed, was a section of 35.0 m ±0.5 m length. The initial speed u_0 and final speed u_k of the vehicle, as well as the length of the test section S_P were determined for the test section defined in this way. Based on these data values of braking deceleration obtained by the vehicle on the test section and braking deceleration as follows $a = f(u_p, u_k, S_P)$ were determined. The procedure of the test section set-up is illustrated in Figure 5.

The authors carried out eleven (11) brake tests on dry asphalt pavement covered with grain. A mean value of braking deceleration from these tests was 3.3 m/s^2 (with the standard deviation of 0.24 m/s^2). Further six (6) brake tests were carried out on wet asphalt pavement covered with grain, obtaining a mean braking deceleration value of 2.9 m/s^2 (with the standard deviation of 0.22 m/s^2).

Thus brake test carried out on asphalt pavement covered with a layer of grain demonstrated that contamination has a significant effect on the traits and values of the deceleration characteristics during car braking. On the prepared test section, with wet asphalt pavement covered with grain the test car obtained mean breaking deceleration values that were significantly lower that on wet asphalt pavement not covered with grain. If the mean car breaking deceleration obtained on the test section on wet asphalt pavement is to be used as a reference (benchmark), then for pavement contaminated with dry grain mean deceleration on the test section was lower by 58% and for wet grain by 63%. The investigated cases exhibited no problems with controlling the direction of motion or any tendency to spontaneous change of the vehicle's path during abrupt braking of a vehicle equipped with anti-lock braking system (ABS).

In the course of the experiment a number of passes were also made with the car with speeds up to 120 km/h. During these passes the brake was not applied. The purpose of these tests was to find out if when driving on the road section covered with spilled grain any adverse phenomena leading to loss of motion stability occur. No directional stability loss of the vehicle was observed during any of the passes.

Contamination of the surface of the carriageway with grain had a significant adverse effect on the test vehicle's braking deceleration values. The car's initial speed had no effect on the value of braking deceleration obtained by the vehicle equipped with anti-lock braking system, regardless of the road surface condition (both for dry and wet asphalt pavements). No signs of directional stability loss of the vehicle was observed while driving on pavement covered with spilled grain within the speed range under study.

4. Final Remarks

Driving a car in non-typical road conditions which can be suddenly encountered on a carriageway (such as, for instance, spilled grain, mud or ice build-up, lost load such as boxes, rolled hay and straw, bottles, small food containers), poses a serious hazard to the road traffic safety. Drivers when taken aback by a change in the pavement condition or by an obstacle on the road usually react with panic resorting to various manoeuvres which may result in a road crash. Only adequately developed reflex motoric may prevent road accidents. Modern drivers often have uncritical trust in electronic equipment, driving support and both active and passive vehicle safety systems. However, one should not forget that it is the driver who is the most unreliable element in the driver-vehicle-road system. It needs to be stressed that in non-typical situations psychomotoric aspect becomes decisive, as driver's adequate defensive reaction may save his/ her health and life. The driver should know when it is right to carry out a manoeuvre, such as braking or passing around an obstacle in reaction to hazard, and when it is right to refrain from any counteraction.

The investigation carried out by the authors demonstrated that if the driver participating in the road traffic incident under analysis refrained from any defensive action or just applied the brakes the accident would not happen. Hence an obvious conclusion that the driver in the case under analysis must have had to ignore the hazard related to the contamination of the carriageway and went on driving in accelerated motion. In combination with the vehicle's rear drive (without traction control system) this fact lead to spinning of the wheels and loss of

the vehicle's directional stability, which in turn led to a head-on crash with a vehicle approaching from the opposite direction. The results of the investigation turned out to be different and contrary from the previous knowledge on how the driver should behave when driving in such non-typical conditions. A vehicle advancing in uniform motion through a carriageway section covered with spilled grain was not losing directional stability due to sufficient grip of wheels to the surface. In this particular case lack of the driver's defensive reaction would have guaranteed the vehicle's safe passage since the wheels' grip to the pavement contaminated with grain was similar to the grip of the wheels to compacted snow [6]. The investigation also confirmed that intensive braking on such surface with a car equipped with anti-lock braking system guaranteed maintaining of directional stability of motion while braking.

Carrying out of physical test of the motion of vehicles on non-typical surfaces makes it possible to develop recommendations and instructions for drivers on how to behave in situations when they are taken aback for instance by a sudden change in road surface condition. Investigation of driving reactions of motorists and their effect on road safety is also a subject of psychological analysis. These issues are a domain of scientific interest in disciplines such as road traffic psychology or driving psychology. Research on vehicle behaviour in real-life road conditions, especially on non-typical surfaces along with adequate psychological research on driver reactions may contribute to a significant improvement in road traffic safety.

Use of the results of research on vehicle dynamics in close combination with investigation of the driver's psychomotoric traits and the his/ her reacting to various hazards on the road should allow for developing of effective car, truck, motorcycle and bicycle driving techniques and methods and for preparing instructions for proper and safe use of roads by unprotected road users, that is pedestrians (in particular children).

The authors of this article conduct a wide range of research on the parameters of vehicle motion in various road conditions in terms of their influence on road traffic safety. The proposed approach to the subject may be a proposal of interdisciplinary co-operation between scientists dealing with road traffic psychology and experts in studying physical phenomena characteristic for dynamic behaviour of vehicles in various road traffic situations.

Preventive measures in the form of driver participation in practical training organized by driver improvement schools [4,10,16] in difficult road conditions and on non-typical road surfaces on dedicated driving tracks could minimize adverse effects of incidents as the one analyzed in this paper or those that happen in similar situations that every motorist may encounter on the road.

References

[1]   Bąk-Gajda, D., "Psychologiczne czynniki bezpieczeństwa ruchu drogowego," *Eksploatacja i Niezawodność*, 3, 2008.

[2]   Dudziak, M. (ed.), *Proces hamowania samochodu a bezpieczeństwo w ruchu drogowym*, Wydawnictwo i Zakład Poligrafii Instytutu Technologii i Eksploatacji, Poznań-Radom, 2002.

[3]   ECE Regulation No. 13.

[4]   Handbook of traffic psychology, Porter B.E. (ed.), *Elsevier*, 2011.

[5]   http://www.krbrd.gov.pl [Accessed Dec. 2, 2013].

[6]   Hugemann, W., *Unfall-rekonstruktion*, Autorenteam GbR, Schoenbach-Druck GmbH, Erzhausen, 2007.

[7]   Jenenkova, O., "Personal Characteristics of Aggressive Drivers in the Perception of Drivers and Road Traffic Inspectors," *Psychological Thought*, 7 (1), 2014.

[8]   Lewandowski, A., Anioła, M., Kurek, J. and Warszczyński J., "Hamowanie samochodu osobowego warunkach zagrożenia bezpieczeństwa ruchu drogowego, wywołanego zmiana stanu nawierzchni," in *VIII Konferencja Rekonstrukcji Wypadków Drogowych*, Kraków, 2000.

[9]   Litwinow, A., *Kierowalność i stateczność samochodu*, Wydawnictwo Komunikacji i Łączności, Warszawa, 1975.

[10] Nyberg, A., "The potential of driver education to reduce traffic crashes involving young drivers," Dissertation Linkoeping University. Faculty of Health Science, Linkoeping, 2007.

[11] Pokorski, J., Szwabik, B., "Dynamiczne charakterystyki przyczepności na nawierzchniach drogowych opon samochodowych," *Czasopismo Techniczne*, 6, 1998.

[12] Praca zbiorowa, *Wypadki drogowe vademecum biegłego sądowego*, Wydawnictwo Instytutu Ekspertyz Sądowych, Kraków, 2006.

[13] Prochowski, L., *Mechanika ruchu*, Wydawnictwo Komunikacji i Łączności, Warszawa, 2005.

[14] Prochowski, L., Unarski, J., Wach, W. and Wicher, J., *Podstawy rekonstrukcji wypadków drogowych*, WKŁ, Warszawa, 2008.

[15] Roehring, L., *Glueck finden*. Verlag Books on Demand GmbH, Norderstedt, 2010.

[16] Shirar, D., *Traffic Safety and Human Behavior*. Elsevier, Amsterdam, 2007.

[17] Szczepanik, C., *Podstawy modelowania systemu człowiek-pojazd-otoczenie*, PWN, Warszawa–Łódź, 1999.

[18] Waszkowska, M., Merecz, D., "Psychological Effects of Participation in Road Accidents – A Challenge for Public Health," *Medycyna Pracy*, 57, 2006.

[19] www.policja.pl [Accessed Dec. 1, 2013].

[20] www.biblioteka.psychelab.pl [Accessed Dec. 1, 2013].

3

Flow of a Maxwell Fluid in a Porous Orthogonal Rheometer under the Effect of a Magnetic Field

H. Volkan Ersoy*

Department of Mechanical Engineering, Yildiz Technical University, Istanbul, Turkey
*Corresponding author: hversoy@yildiz.edu.tr

Abstract The steady flow of a Maxwell fluid in a porous orthogonal rheometer with the application of a magnetic field is studied. It is shown that there exists an exact solution for the problem. The effects of the parameters controlling the flow are analyzed by plotting graphs. It is observed that the effects of the Deborah number, the suction/injection velocity parameter, and the Reynolds number in the absence of a magnetic field are similar to those in the presence of a magnetic field. It is displayed that the Hartmann number that is based on the applied magnetic field reveals the tendency to slow down the flow.

Keywords: *magnetohydrodynamics, Maxwell fluid, orthogonal rheometer, porous disk, exact solution*

1. Introduction

Magnetohydrodynamics (MHD) is the study of electrically conducting fluids in the presence of magnetic and electric fields and examines the phenomena associated with electro-fluid-mechanical energy conversion. Its applications in engineering are MHD generators, pumps, bearings, flow meters, plasma jets, fusion machines for power and space power generators. The MHD flow of non-Newtonian fluids has been a subject of great interest due to its widely spread applications in the design of cooling systems with liquid metals, purification of crude oil and polymer technology.

In nature, there are many fluids whose behavior cannot be described by the classical Navier-Stokes theory. The inadequacy of the theory of Newtonian fluids in predicting the behavior of some fluids, especially those with high molecular weight, has led to the development of non-Newtonian fluid mechanics. The constitutive equations of non-Newtonian fluids such as polymer solutions, greases, melts, muds, emulsions, paints, jams, soaps, shampoos and certain oils are of higher order and much more complicated than the Navier-Stokes equations. The mechanical behavior of non-Newtonian fluids cannot be described by a single constitutive equation. Therefore, many constitutive equations have been suggested in the literature. One of the most popular models is the Maxwell model. The importance of the Maxwell model lies in the fact that it can show the relaxation effects. The properties of polymeric fluids can be explored by the Maxwell model for small relaxation time. However, in some more concentrated polymeric fluids, the Maxwell model is also useful for large relaxation time. Maxwell fluids include

glycerin, toluene, crude oil, flour dough and dilute polymeric solutions.

Maxwell and Chartoff [1] devised an instrument called the orthogonal rheometer consisting of two parallel disks rotating about non-coincident axes with the same angular velocity. They pointed out that it is possible to determine the complex dynamic viscosity of a viscoelastic fluid by using it. Abbott and Walters [2] were the first to obtain an exact solution of a Newtonian fluid in the orthogonal rheometer. They also carried out a perturbation analysis in the case of a viscoelastic fluid. Rajagopal [3] showed that the velocity field corresponding to the motion in an orthogonal rheometer is a motion with constant principal relative stretch history. He further demonstrated that the adherence boundary condition is sufficient for obtaining a determinate problem since the flow of any homogeneous incompressible simple fluid which undergoes a flow characterized by this velocity field results in a second-order partial differential equation. The reader may also consult [4,5,6] for the studies that deal with the flow in an orthogonal rheometer.

The flow in the orthogonal rheometer in the presence of a magnetic field has been studied by a number of researchers. Mohanty [7] obtained an exact solution of the MHD flow in the case of a Newtonian fluid. Rao and Rao [8] investigated the flow in this geometry for a second grade fluid under the effect of a magnetic field. Kasiviswanathan and Rao [9] investigated the unsteady MHD flow of a Newtonian fluid in the same geometry when the disks are subjected to non-torsional oscillations. Kasiviswanathan and Gandhi [10] studied the flow of a micropolar fluid in the presence of a magnetic field. Ersoy [11] and Siddiqui et al. [12] investigated the MHD flow for an Oldroyd-B fluid and a Burger's fluid, respectively. Guria et al. [13] examined both the hydromagnetic flow and the heat transfer phenomenon when the disks are porous. Recently, Das et al. [14] studied unsteady hydromagnetic flow due to concentric rotation of eccentric disks.

The MHD flow due to the pull of disks in an orthogonal rheometer has also been one of the subjects which has attracted attention. After Ersoy's pioneering work [15], Asghar et al. [16] and Siddiqui et al. [17] investigated the influence of Hall current and heat transfer on the steady flow of an Oldroyd-B fluid and a Burgers' fluid, respectively. Hayat et al. [18] and Hayat et al. [19] studied the Hall current and heat transfer effects on the steady flow in the case of a generalized Burgers' fluid and of a Sisko fluid, respectively. Ersoy [20] obtained the velocity field and examined the distribution of force applied by the fluid in the same geometry for a second order/grade fluid.

We refer the readers to some papers by Hsiao [21,22,23,24] for flows of non-Newtonian fluids in the presence of a magnetic field.

The aim of this paper is to extend the study considered in [25] to the magnetohydrodynamic flow. The results obtained for the velocity field are presented graphically in terms of the Deborah number, the suction/injection velocity parameter, the Reynolds number, and the Hartmann number. It is shown that the presence of magnetic field results in the deceleration of flow.

2. Basic Equations and Solution

Let us consider a Maxwell fluid between two porous disks rotating with the same angular velocity Ω about two parallel and distinct axes perpendicular to the disks. The top and bottom disks are located at $z = h$ and $z = -h$, respectively. The distance between the axes of rotation placed in the plane $x = 0$ is 2ℓ (Figure 1). A uniform magnetic induction of strength B_0 is applied to the insulated disks in the z - direction. The induced magnetic field is neglected under the assumption of a small magnetic Reynolds number.

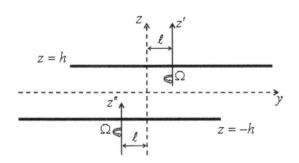

Figure 1. Flow geometry

The Cauchy stress $\mathbf{T}$ for a Maxwell fluid is related to the fluid motion in the form

$$\mathbf{T} = -p\mathbf{I} + \mathbf{S} \tag{1}$$

$$\mathbf{S} + \lambda\left(\dot{\mathbf{S}} - \mathbf{LS} - \mathbf{SL}^T\right) = \mu\mathbf{A}_1 \tag{2}$$

$$\mathbf{A}_1 = \mathbf{L} + \mathbf{L}^T, \mathbf{L} = \nabla\mathbf{v} \tag{3}$$

where p denotes the pressure, $\mathbf{I}$ the unit tensor, $\mathbf{S}$ the extra stress tensor, λ the relaxation time, $\mathbf{L}$ the velocity gradient, $\mathbf{L}^T$ the transpose of $\mathbf{L}$, μ the dynamic viscosity, $\mathbf{A}_1$ the first Rivlin-Ericksen tensor, and $\mathbf{v}$ the velocity vector. The dot represents the material time derivative. When $\lambda = 0$, the model (2) reduces to the classical linearly viscous model.

The governing equations are

$$\rho\dot{\mathbf{v}} = \nabla \cdot \mathbf{T} + \mathbf{J} \times \mathbf{B} \tag{4}$$

$$\nabla \cdot \mathbf{v} = 0 \tag{5}$$

$$\nabla \cdot \mathbf{B} = 0 \tag{6}$$

$$\nabla \times \mathbf{B} = \mu_m \mathbf{J} \tag{7}$$

$$\nabla \times \mathbf{E} = \mathbf{0} \tag{8}$$

$$\mathbf{J} = \sigma\left(\mathbf{E} + \mathbf{v} \times \mathbf{B}\right) \tag{9}$$

where ρ is the density, $\mathbf{J}$ the current density, $\mathbf{B}$ the magnetic induction, μ_m the magnetic permeability, $\mathbf{E}$ the electric field, and σ is the electrical conductivity of the fluid.

It seems reasonable to assume the following form for the velocity field

$$u = -\Omega y + f(z), v = \Omega x + g(z), w = w_0 \tag{10}$$

where u, v, and w are the x, y, and z - components of the velocity vector, respectively, and w is a constant as a result of (5). The direction of w is upward, so w_0 is taken to be positive.

The appropriate boundary conditions for the velocity field are

$$u = -\Omega(y - \ell), v = \Omega x, w = w_0 \text{ at } z = h \tag{11}$$

$$u = -\Omega(y + \ell), v = \Omega x, w = w_0 \text{ at } z = -h \tag{12}$$

$$u = -\Omega y, v = \Omega x, w = w_0 \text{ at } z = 0 \tag{13}$$

The boundary conditions for $f(z)$ and $g(z)$ from Eqs. (10)-(13) are

$$f(\pm h) = \pm\Omega\ell, \ g(\pm h) = 0$$
$$f(0) = 0, \ g(0) = 0 \tag{14}$$

We shall suppose that the extra stress tensor $\mathbf{S}$ depends only on z (see [25]). Using Eqs. (2), (3) and (10), we obtain the following equations:

$$S_{xx} + \lambda\left(w_0 S'_{xx} + 2\Omega S_{xy} - 2f'S_{xz}\right) = 0 \tag{15}$$

$$S_{xy} + \lambda\left(w_0 S'_{xy} + \Omega\left(S_{yy} - S_{xx}\right) - f'S_{yz} - g'S_{xz}\right) = 0 \tag{16}$$

$$S_{xz} + \lambda\left(w_0 S'_{xz} + \Omega S_{yz} - f'S_{zz}\right) = \mu f' \tag{17}$$

$$S_{yy} + \lambda\left(w_0 S'_{yy} - 2\Omega S_{xy} - 2g'S_{yz}\right) = 0 \tag{18}$$

$$S_{yz} + \lambda\left(w_0 S'_{yz} - \Omega S_{xz} - g'S_{zz}\right) = \mu g' \tag{19}$$

$$S_{zz} + \lambda w_0 S'_{zz} = 0 \tag{20}$$

where a prime denotes differentiation with respect to z. The solution of (20) gives

$$S_{zz} = C\exp\left(-z/(\lambda w_0)\right) \tag{21}$$

where C is a constant. We shall investigate the possibility of a solution to the steady problem in which $C=0$ [25,26,27,28].

We obtain the following equation by using Eqs. (17) and (19)

$$\lambda w_0 G' + (1-i\lambda\Omega)G = \mu F' \tag{22}$$

where $i = \sqrt{-1}$, $G(z) = S_{xz} + iS_{yz}$ and $F(z) = f + ig$.

Using Eqs. (4), (6) and (10), we get

$$\frac{\partial p}{\partial x} = \rho\Omega(\Omega x + g) - \rho w_0 f' + S'_{xz} + B_0 J_y \tag{23}$$

$$\frac{\partial p}{\partial y} = \rho\Omega(\Omega y - f) - \rho w_0 g' + S'_{yz} - B_0 J_x \tag{24}$$

$$\frac{\partial p}{\partial z} = 0 \tag{25}$$

Using Eq. (9), we have

$$J_x = \sigma\left(E_x + vB_0\right), J_y = \sigma\left(E_y - uB_0\right), J_z = \sigma E_z \tag{26}$$

Since the disks are insulated, we get $J_z = 0$ and $E_z = 0$. From Eq. (8), we have $\partial E_x/\partial z = 0$ and $\partial E_y/\partial z = 0$. We obtain the following equations by cross-differentiating Eqs. (23)-(25)

$$\rho(\Omega g - w_0 f') + S'_{xz} - \sigma B_0^2 f = \text{constant} \tag{27}$$

$$\rho(\Omega f + w_0 g') - S'_{yz} + \sigma B_0^2 g = \text{constant} \tag{28}$$

When we use Eqs. (27)-(28), it is obtained

$$G' - \rho w_0 F' - \left(\sigma B_0^2 + i\rho\Omega\right)F = \text{constant} \tag{29}$$

Combining Eqs. (22) and (29), we have

$$\left(\mu - \rho\lambda w_0^2\right)F'' - w_0\left(\rho + \lambda\sigma B_0^2\right)F' \tag{30}$$
$$- (1-i\lambda\Omega)\left(\sigma B_0^2 + i\rho\Omega\right)F = E$$

where E is a constant that can be determined from the boundary conditions. Let us use the following non-dimensional quantities:

$$\Gamma = \frac{F}{\Omega\ell}, \zeta = \frac{z}{h}, D = \lambda\Omega, \alpha = \frac{w_0}{\sqrt{\Omega\mu/\rho}}, \tag{31}$$
$$R = \frac{\rho\Omega h^2}{\mu}, M = \sqrt{\frac{\sigma}{\mu}} B_0 h$$

where D is the Deborah number, α the suction/injection parameter, R the Reynolds number, and M is the Hartmann number. The solution of Eq. (30) with the conditions $\Gamma(\pm 1) = \pm 1$ and $\Gamma(0) = 0$ is

$$\Gamma = \frac{Z_1\exp(A\zeta) + Z_2\exp(B\zeta) + Z_3}{Z_4} \tag{32}$$

where

$$Z_1 = 1 - \cosh(B), Z_2 = \cosh(A) - 1,$$
$$Z_3 = \cosh(B) - \cosh(A),$$
$$Z_4 = \sinh(A) - \sinh(B) - \sinh(A-B),$$
$$A = \frac{a + \sqrt{a^2 + bc + ibd}}{e}, \tag{33}$$
$$B = \frac{a - \sqrt{a^2 + bc + ibd}}{e},$$
$$a = \alpha\left(R + DM^2\right), b = 4R(1 - D\alpha^2),$$
$$c = M^2 + DR, d = R - DM^2, e = 2\sqrt{R}\left(1 - D\alpha^2\right)$$

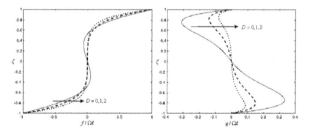

Figure 2. Variation of $f/\Omega\ell$ and $g/\Omega\ell$ with ζ ($R = 20$, $\alpha = 0.1$, $M = 0.5$, $D = 0,1,2$)

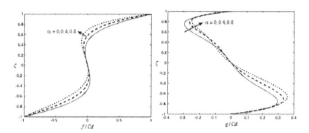

Figure 3. Variation of $f/\Omega\ell$ and $g/\Omega\ell$ with ζ ($R = 20$, $D = 0.1$, $M = 0.5$, $\alpha = 0,0.4,0.8$)

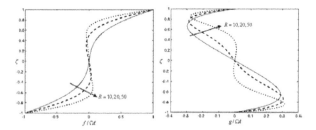

Figure 4. Variation of $f/\Omega\ell$ and $g/\Omega\ell$ with ζ ($D = 0.1$, $\alpha = 0.1$, $M = 0.5$, $R = 10,20,50$)

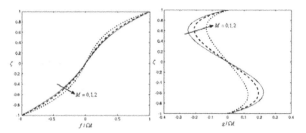

Figure 5. Variation of $f/\Omega\ell$ and $g/\Omega\ell$ with ζ ($D = 0.1$, $\alpha = 0.2$, $R = 5$, $M = 0,1,2$)

3. Results and Discussions

The problem that reflects the steady flow of a Maxwell fluid in a porous orthogonal rheometer has an exact solution of the velocity field even in the presence of a magnetic field. However, the components S_{xz} and S_{yz} of stress tensor, which are related to the x - and y - components of the force per unit area exerted by the fluid on the disks, cannot be obtained since the constitutive equations corresponding to the Maxwell fluid in the case of porous disks are of higher order than those corresponding to a classical viscous fluid including the same conditions.

The velocity field is entirely determined when the translational velocity components $f(z)$ and $g(z)$ are found. The effects of embedded parameters on the translational velocity components are examined through plots. It is obvious from Figure 2 that the curves in the main part of fluid belonging to the translational velocity components get closer to the z - axis when the Deborah number that is based on the relaxation time increases. An examination of Figure 3 shows that the curves start to approach the top disk but move away the bottom disk with the increase of the suction/injection velocity parameter that represents the upward axial velocity. It may be inferred from Figure 4 that the curves in the core region come near to the z - axis for large Reynolds numbers. The influence of the Hartmann number that is based on the applied magnetic field can be seen in Figure 5. An increase in the Hartmann number discloses that the curves tend to become flatter.

It should also be mentioned that the results obtained by Ersoy [25] can be obtained as the special case of the present analysis by taking the Hartmann number to be zero.

4. Conclusions

In this paper, the steady flow of a Maxwell fluid between two porous disks rotating about two distinct axes under the influence of a magnetic field is investigated. The conclusions which are drawn from this analysis can be listed as follows:

- The main part of fluid having the constant axial velocity tends to rotate about the z - axis for large values of the Deborah number.
- The thickness of the boundary layer adjacent to the top disk becomes thinner when the suction/injection velocity parameter increases, whereas an opposite behavior is observed for that near the bottom disk.
- The well-developed boundary layers exist near both the disks and are separated by a core region with increasing the Reynolds number. The main body of fluid both rotates about the z -axis and has a constant axial velocity.
- It is observed that the applied magnetic field reacts back on the flow as in the case of a Newtonian fluid. In other words, a Lorentz force which decelerates the flow is produced. As a result of this, the curves about which the fluid layers having the constant axial velocity w_0 rotate are closer to the z - axis.

References

[1] Maxwell, B. and Chartoff, R.P., "Studies of a polymer melt in an orthogonal rheometer," *Transactions of the Society of Rheology*, 9 (1), 41-52, 1965.

[2] Abbott, T.N.G. and Walters, K., "Rheometrical flow systems: Part2. Theory for the orthogonal rheometer, including an exact solution of the Navier-Stokes equations," *Journal of Fluid Mechanics*, 40 (1), 205-213, 1970.

[3] Rajagopal, K.R., "On the flow of a simple fluid in an orthogonal rheometer," *Archive for Rational Mechanics and Analysis*, 79 (1) 39-47, 1982.

[4] Rajagopal, K.R., "Flow of viscoelastic fluids between rotating disks," *Theoretical and Computational Fluid Dynamics*, 3 (4), 185-206, 1992.

[5] Srinivasa, A.R., "Flow characteristics of a multiconfigurational, shear thinning viscoelastic fluid with particular reference to the orthogonal rheometer," *Theoretical and Computational Fluid Dynamics*, 13 (5), 305-325, 2000.

[6] Ersoy, H.V., "On the locus of stagnation points for a Maxwell fluid in an orthogonal rheometer," *International Review of Mechanical Engineering*, 3 (5), 660-664, 2009.

[7] Mohanty, H.K., "Hydromagnetic flow between two rotating disks with noncoincident parallel axes of rotation," *Physics of Fluids*, 15 (8), 1456-1458, 1972.

[8] Rao, R. and Rao, P.R., "MHD flow of a second grade fluid in an orthogonal rheometer," *International Journal of Engineering Science*, 23 (12), 1387-1395, 1985.

[9] Kasiviswanathan, S.R. and Rao, A.R., "On exact solutions of unsteady MHD flow between eccentrically rotating disks," *Archiwum Mechaniki Stosowanej*, 39 (4), 411-418, 1987.

[10] Kasiviswanathan, S.R. and Gandhi, M.V., "A class of exact solutions for the magnetohydrodynamic flow of a micropolar fluid," *International Journal of Engineering Science*, 30 (4), 409-417, 1992.

[11] Ersoy, H.V., "MHD flow of an Oldroyd-B fluid between eccentric rotating disks," *International Journal of Engineering Science*, 37 (15), 1973-1984, 1999.

[12] Siddiqui, A.M., Rana, M.A. and Ahmed, N., "Magnetohydrodynamics flow of a Burgers' fluid in an orthogonal rheometer," *Applied Mathematical Modelling*, 34 (10), 2881-2892, 2010.

[13] Guria, M., Das, B.K., Jana, R.N. and Imrak, C.E., "Hydromagnetic flow between two porous disks rotating about non-coincident axes," *Acta Mechanica Sinica*, 24 (5), 489-496, 2008.

[14] Das, S., Jana, M. and Jana, R.N, "Unsteady hydromagnetic flow due to concentric rotation of eccentric disks," *Journal of Mechanics*, 29 (1), 169-176, 2013.

[15] Ersoy, H.V., "Unsteady flow due to a sudden pull of eccentric rotating disks," *International Journal of Engineering Science*, 39 (3), 343-354, 2001.

[16] Asghar, S., Mohyuddin, M.R. and Hayat, T., "Effects of Hall current and heat transfer on flow due to a pull of eccentric rotating disks," *International Journal of Heat and Mass Transfer*, 48 (3-4), 599-607, 2005.

[17] Siddiqui, A.M., Rana, M.A. and Ahmed, N., "Effects of Hall current and heat transfer on MHD flow of a Burgers' fluid due to a pull of eccentric rotating disks," *Communications in Nonlinear Science and Numerical Simulation*, 13 (8), 1554-1570, 2008.

[18] Hayat, T., Maqbool, K. and Khan, M., "Hall and heat transfer effects on the steady flow of a generalized Burgers' fluid induced by a sudden pull of eccentric rotating disks," *Nonlinear Dynamics*, 51 (1-2), 267-276, 2008.

[19] Hayat, T., Maqbool, K. and Asghar, S., "Hall and heat transfer effects on the steady flow of a Sisko fluid", *Zeitschrift für Naturforschung A*, 64 (12), 769-782, 2009.

[20] Ersoy, H.V., "Flow of a second order/grade fluid induced by a pull of disks in an orthogonal rheometer under the effect of a magnetic field," *International Review of Mechanical Engineering*, 6 (5), 966-971, 2012.

[21] Hsiao, K.-L., "MHD mixed convection for viscoelastic fluid past a porous wedge," *International Journal of Non-Linear Mechanics*, 46 (1), 1-8, 2011.

[22] Hsiao, K.-L., "MHD stagnation point viscoelastic fluid flow and heat transfer on a thermal forming stretching sheet with viscous dissipation," *The Canadian Journal of Chemical Engineering*, 89 (5), 1228-1235, 2011.

[23] Hsiao, K.-L., "Viscoelastic fluid over a stretching sheet with electromagnetic effects and non-uniform heat source/sink," *Mathematical Problems in Engineering*, 2010 (Article ID 740943), 1-14, 2010.

[24] Hsiao, K.-L., "Conjugate heat transfer for mixed convection and Maxwell fluid on a stagnation point," *Arabian Journal for Science and Engineering*, 39 (6), 4325-4332, 2014.

[25] Ersoy, H.V, "Flow of a Maxwell fluid between two porous disks rotating about noncoincident axes," *Advances in Mechanical Engineering*, 2014 (Article ID 347196), 1-7, 2014.

[26] Rajagopal, K.R. and Bhatnagar, R.K, "Exact solutions for some simple flows of an Oldroyd-B fluid," *Acta Mechanica*, 113 (1-4), 233-239, 1995.

[27] Hayat, T., Nadeem, S. and Asghar, S., "Hydromagnetic Couette flow of an Oldroyd-B fluid in a rotating system," *International Journal of Engineering Science*, 42 (1), 65-78, 2004.

[28] Hayat, T., Khan, M. and Ayub, M., "Exact solutions of flow problems of an Oldroyd-B fluid," *Applied Mathematics and Computation*, 151 (1), 105-119, 2004.

4

The Measurement of Standing Wave Patterns by using High-speed Digital Image Correlation

Róbert Huňady[*]**, Martin Hagara, František Trebuňa**

Technical University of Košice, Faculty of Mechanical Engineering, Košice, Slovakia
*Corresponding author: robert.hunady@tuke.sk

Abstract Paper deals with the measurement of standing wave patterns of a square plate by using Digital Image Correlation method that allows perform measurements of 3D displacements and strains. The experiment described in the paper had two phases. In the first phase, the natural frequencies of the plate were determined. The corresponding mode shapes of vibration were measured in the second phase. For the purpose of capturing a deformation of vibrating surface the correlation system Q-450 with two high-speed cameras had to be used for measurement.

Keywords: Chladni patterns, mode shapes of vibration, digital image correlation

1. Introduction

All objects have tendency to vibrate naturally at certain frequencies. These frequencies are known as the natural frequencies or eigen-frequencies of a structure. If the amplitudes of the vibrations are large enough and if natural frequency is within the human frequency range, then the vibrating object will produce sound waves that are audible. Each of the natural frequencies at which an object vibrates is associated with a standing wave pattern. When an object is forced into resonance vibrations at one of its natural frequencies, it vibrates in such a manner that a standing wave is formed within the object. Standing wave pattern can be described as a vibrational pattern created within a structure when the vibrational frequency of a source causes that waves reflected from border of the structure interfere with waves traveled from the source [1]. The result of the interference is that specific points appear to be standing still while other points vibrated back and forth. The points in the pattern that are standing still are referred to as nodal points or nodal lines. These lines occur as the result of the destructive interference of incident and reflected waves. In general, the higher natural frequency, the more complicated pattern occurs. At frequency other than a natural frequency, the interference of reflected and incident waves results in a disturbance of the vibration that is irregular and non-repeating. Standing wave patterns represent the lowest energy vibrational modes of the object. The mode shapes (patterns) of vibration are those that result in the highest amplitude vibrations with the least input of energy. Objects are most easily forced into resonance vibrations when they are excited at frequencies associated with their natural frequencies.

In 1787, a German physicist Ernst Chladni described a technique that allows show the mode shapes of vibration of a solid surface. Chladni repeated the pioneering experiments of Robert Hooke who, in 1680, had observed the different patterns associated with the vibrations of glass plates. Hooke ran a violin bow along the edge of a plate covered with flour and saw how these patterns emerge. Chladni's technique consisted of drawing a bow over a piece of metal whose surface was lightly covered with sand. The plate was bowed until it reached resonance, when the vibration causes the sand to move and concentrate along the nodal lines where the surface is still. The patterns formed by these lines are what are now called Chladni patterns or Chladni figures. Variations of this technique can be still used e.g. in the design and construction of acoustic instruments [2].

Natural frequencies and mode shapes of vibration are the key structural properties that together with a damping describe a dynamic behavior of a structure. These parameters are uniformly referred to as modal parameters and depend on the geometry, material properties and boundary conditions of the system. The process of determining (or estimating) the modal parameters is called Experimental Modal Analysis (EMA). This method is based on the investigation of the relationship between the excitation and the response of a structure. Its aim is a decomposition of vibration into modal contributions so-called modes, each of which is characterized by its modal parameters [3]. Modal analysis is the youngest field of dynamics whose beginnings date back to the forties of the last century. The invention of the fast Fourier Transformation (FFT) algorithm by Cooly and Tukey in 1965 finally paved the way for rapid and prevalent application of experimental modal analysis. Over the last fifty years, the different measurement techniques and the estimation algorithms have been developed and are being used successfully [4].

Accelerometers are most often used to measurement of vibration responses. They are very popular due to easy application and technical parameters. Accelerometers have wide dynamic and frequency range and relatively low weight. On the other hand, one accelerometer can measure a response at only one point. In addition, transducer applied on a structure influences the structural properties of an object at the place of its application. Progress achieved in computer technology and optics led to an improvement of unconventional optical methods that allow perform full-field contactless measurements with high spatial and time resolution. Thanks to this, it is possible to obtain response from any point of object without affecting the structure. The most famous optical techniques used in vibration analysis are Laser Doppler Vibro-Scanners (LDVS) and systems working on principle of Digital Image Correlation (DIC) [5].

2. Digital Image Correlation

The digital image correlation method is a modern contactless optic method, which uses digital image registration technique to precise measurement of plane and spatial deformations. Thanks to cameras with a high resolution, correlation system is able to observe a wide range of an object's surface points with variable contrast. It allows visualize measured quantities in the whole observed area. In the case, when the high-speed digital cameras are used, the correlation system can be applied to solve various dynamic problems in mechanics. The flexible area of measurement (mm^2 to m^2), material and geometry independence and spatial visualization of measured parameters are the characteristic features of DIC method resulting from its concept. This is the reason why digital imagine correlation has been successful applied for many areas of science and survey. It has been applied for testing of components in engineering and microelectronic, determination of material properties, analyzing of vibration and modal parameters, FEA validation and so on.

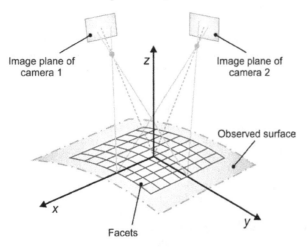

Figure 1. Basic principle of 3D digital image correlation method with a stereoscopic configuration of cameras

The principle of method is based on observation of stochastic speckle pattern, which is created on the surface of the tested object. In the 3D digital image correlation technique, a specimen surface is observed by two cameras in a stereoscopic configuration. A surface observed by cameras is divided into smaller subareas called facets (Figure 2) in such a way, that every one of them contains characteristic part of the pattern. Surface speckles represent material points of an object. Speckles copy deformations of a surface and they move together with an object.

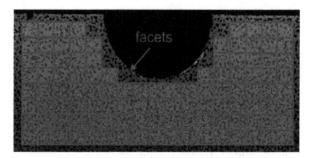

Figure 2. Facets determined on observed area of an obejct

With knowledge of the imaging parameters for each camera and the orientations of the cameras with respect to each other, the position of each object point in three dimensions is calculated. In order to evaluate surface displacements and strains on the object surface, a series of measurements is taken, while the specimen surface is moved due to a loading. The correlation algorithm tracks the observed gray value patterns for each camera and transforms corresponding facet positions in both cameras into 3D coordinates for each step, resulting in a track of each surface facet in 3D space. As the surface deformation is measured pointwise, displacements of individual surface points and subsequently surface strains can be evaluated.

3. Experimental measurement

The aim of an experiment was to capture standing wave patterns by using 3D DIC correlation system Q-450 produced by Dantec Dynamics company. This system uses two high-speed digital cameras Speed Sense Phantom with 1.3 Mpx CCD sensors. When the full resolution of sensor is used, the maximal acquisition rate of cameras is 3340 fps.

The object of measurement was a plate of square shape with dimensions 250 x 250 mm. The plate was made of poly methyl methacrylate sheet (PMMA) of thickness of approx. 2.0 mm. In the middle, the plate was attached to the drive rod of electromagnetic exciter. The exciter was connected to amplifier that amplifies the harmonic signal from signal generator.

3.1. Determination of Natural Frequencies

The first step of the experiment was to determine natural frequencies of the plate. For that purpose, a periodic swept sine signal was used as the excitation signal with a frequency varying from 15 Hz to 750 Hz linearly during the 60 seconds. The vibration responses of the plate were measured three times, at three different points, by one-point laser Doppler vibrometer PDV-100. The experimental setup is shown in Figure 3.

Figure 3. Experimental measurement of natural frequencies of the plate

The approximate values of natural frequencies were determined in time domain. If a frequency of excitation signal corresponds to arbitrary natural frequency of the plate, a resonance occurs and responses of the plate rise significantly (i.e. amplitudes of vibration are maximal). There is a relatively simple to identify these moments in the time course of the measured response. This is the way how to obtain a rough estimate of natural frequencies because there is a difficult to distinguish the frequencies of the coupled and closed modes of the plate. Figure 4 shows the velocity response measured at one of analyzed points.

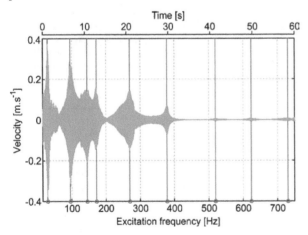

Figure 4. Time course of measured response of the plate

The values of natural frequencies of the plate that were determined by the mentioned way are listed in Table 1.

Table 1. Natural frequencies of the plate

f [Hz]	29	96	145	172	270	378	520	623	730

3.2. Measurement of Standing Wave Patterns

To capture standing wave patterns, the harmonic sine signal was fed to the input of the exciter. The plate was excited by its natural frequencies from Table 1, individually. The cameras recorded an oscillating surface of the plate with sampling frequency of 2000 fps. Figure 5 shows the configuration of the correlation system Q-450 during the measurement.

Figure 5. Experimental configuration of system Q-450 during the measurement of standing wave patterns of the plate

Of course, before the measurement, there was necessary to create a stochastic speckle pattern on the top side of the plate. We used a self-adhesive vinyl foil with pre-printed pattern. The view from the camera 1 and camera 2 is shown in Figure 6.

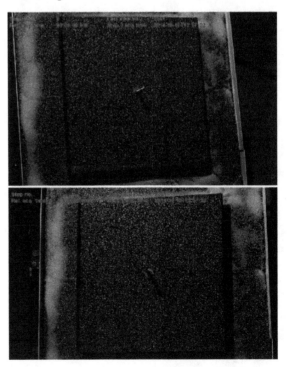

Figure 6. View from the cameras to a measured surface of the plate

Considering that the resultant relative displacements are being determined in a correlation process when digital images recorded in the individual time steps are being compared to each other, the images capturing the stationary and undeformed plate had to be used as reference.

In order to be possible to visualize the standing wave patterns (similarly as in the case of Chladni patterns), the resultant deformation of the plate had to be plotted in the form of absolute displacements. In addition, there was used a function that removes the movements of rigid body. The individual standing wave patterns of the plate are given in Table 2.

was similar as in the previous measurement, i.e. the plate was excited in successive steps by harmonic signal with the frequency corresponding to natural frequencies of the plate. In this case, the plate was covered by salt that moved to nodal place of the vibrating plate. Some of the Chladni patterns obtained by this way are given in Table 3.

Table 2. Standing wave patterns of the analyzed plate

29 Hz	96 Hz
145 Hz	172 Hz
270 Hz	378 Hz
520 Hz	623 Hz
730 Hz	

Table 3. Chosen Chladni patterns of the analyzed plate

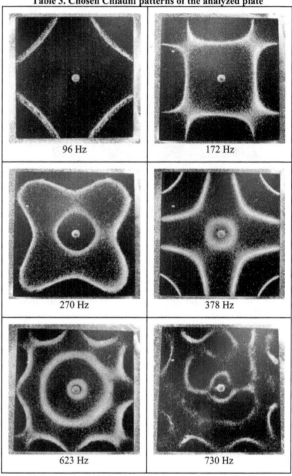

96 Hz	172 Hz
270 Hz	378 Hz
623 Hz	730 Hz

4. Conclusions

The paper presents the methodology of measurement the standing wave patterns by using digital image correlation method. The accuracy of measurement depends on several factors. The one of them is the knowledge of the accurate values of the natural frequencies. The best way to obtain these values is experimental modal analysis. The second factor is quality of the excitation. If the intensity of excitation is too low, some modes will not be adequately excited. This is seen in the case of modes with frequency 520 Hz and 730 Hz. The last factor is the sampling frequency of cameras. The higher time resolution will be the more accurate mode shapes of vibration can be captured.

Acknowledgement

The work has been accomplished under the projects VEGA 1/0937/12, ITMS 26220220182, KEGA 021TUKE-4/2013.

3.3. Chladni Patterns

For the purpose of verification, we performed also modified Chladni's experiment. Experimental procedure

References

[1] Behrman A., *Speech and Voice Science,* Plural Pub., 2007, 450 pp.

[2] Sperry W.Ch., *The Structure of Musical Sound*, iUniverse, 2010, 372 pp.

[3] Trebuňa F., Šimčák F., Huňady R., *Vibration and Modal Analysis of Mechanical Systems (in Slovak)*, TU Košice, 2012.

[4] Brown D.L., Allemang R.J., "The Modern Era of Experimental Modal Analysis: One Historical Perspective", *In: Sound and Vibration,* 40th Anniversary Issue, January 2007, p. 16-25.

[5] Huňady R., Trebuňa F., Hagara M., Schrotter M., "The use of modan 3D in experimental modal analysis", *In: Applied Mechanics and Materials*, Vol. 486, 2013, p. 36-41.

5

Comparison between Theoretical, Analytical & Experimental Vibration Analysis Method & Different Circular Tile Cutter by Using FFT&ANSYS

Hirave S.H.[1,*], **Jadhav S.M.**[1], **Joshi M.M.**[2], **Gajjal S.Y.**[2]

[1]Department of Mechanical Design Engineering, NBN Sinhgad School of Engineering, Pune
[2]Department of Mechanical Engineering, NBN Sinhgad School of Engineering, Pune
*Corresponding author: sunilhrv@gmail.com

Abstract Circular cutters with uniform radial cracks are extensively used in the cutting processes. Unwanted noise, vibration & accidental failure associated with the cutting process have become an important economic and technological problem in the industry. The knowledge of natural frequencies of components is of great interest in the study of response of structures to various excitations. Hence, it is important to study a circular tile cutter with central hole, which fixed at inner edge and free at outer edge with its dynamic response. In this study some efforts are taken for analyzing vibration characteristics of tile cutter used in tile cutting industries with free boundary condition but having different (enlargement of stress concentration holes, slot end hole diameter ,numbers of cutting teeth's, aspect ratio, effect of number and length of cracks, variable radial slit, circular concentric slit, thickness).Results From the FEA and FFT it is found that, this study is needful for design data preparation to avoid resonance and vibration in tile cutting industries from which finding out natural frequency and vibration w.r.t. different changes in tile cutter. Thus Theoretical, ANSYS & FFT results obtained are to be compared.

Keywords: *circular tile cutter, aspect ratio, teeth's, radial slit, circular concentric slit, different cracks, vibration ,FFT & ANSYS*

1. Introduction

Unwanted noise, vibration & accidental failure associated with the cutting process has become an important economic and technological problem in the industry that can be solved by this dissertation work. Also natural frequency of theoretical, analytical & experimental methods are to be compared by using some theoretical formulae, FFT and ANSYS. Comparison results are tabulated in table and comparison graphs plotted in bellow. Natural frequency determine by 3 vibration method as discussed in short as bellow phases.

Figure 1. Sample of circular Tile-Cutter

Test Specimens:-Spring steel are chosen with same b/a ratio i.e. aspect ratio (Inner to outer radius ratio).Following are the material properties for the specimen tile cutter.

Young's modulus (E) $= 2.1 \times 10^{11} \, \text{N/m}^2$
Poisson's ratio (γ) $= 0.3$
Density of material (ρ) $= 7800 \, \text{kg/m}^3$
Dia $=110$ mm, t$= 1.85$mm, N$=14500$ rpm.

Different changes in Geometry:-Specimens with variable increase in no. radial crack, increase length of radial crack, slot end hole diameter, geometry of cutter tooth, enlargement of stress concentration holes, different thickness, aspect ratios , different no of cutting teeth are selected. These specimen sizes are chosen to facilitate the measurements by using the same fixture for all the specimens. As boundary conditions for cutter are inner edge fixed and outer edge free.

2. Different Phases

2.1. Theoretical Method

W. T. Norris and J. E. T. Penny have given values of $(\beta a)^2$ for various ratios of inner radius to outer radius and different mode characteristics. First six harmonics are given in order for aspect ratio (b/a =0.1). $(\beta a)^2$ Is non dimensional frequency parameter used to calculate natural frequencies of annular tile cutter.

$$(\beta\alpha)^2 = \omega a^2 \sqrt{\frac{3\rho(1-\upsilon^2)}{Eh^2}}$$

Where, h = half thickness in mm
E = Young's modulus N/m^2
ρ = Density of the tile cutter material in kg/m^3
ω = frequency in rad/s.
υ = Poisson's ratio taken as 0.3
a = Outer radius of tile cutter in mm

By putting all theoretical values as mentioned in test specimen w.r.t. different mode given bellow results are as tabulated follows. We will compare only aspect ratio and thickness these two parameter with theoretical, FFT & ANSYS results. Beside from that, remaining parameter will get from comparison in between different individual values of results. In that FFT as an experimental method & ANSYS as an analytical method.

2.2. Experimental Method

Procedure-Experimental work is done by using FFT analyzer. Natural frequencies are detected by hitting the tile cutter with impact hammer which was mounted on vice jaw. The response at a point of a plate is measured by using an accelerometer. FFT analyzer analyzed the output of accelerometer. Analysis is done experimentally with the help of FFT analyzer, accelerometer, impact hammer.

Text fixture- Clamping was obtained by using two 20mm diameter nuts and one bolt with washers are fastened below the tile cutter. Bolt is used to restrict the movement at inner edge of annular tile cutter in x and y and z direction. The fixture was hold in vice, which is rigidly fixed on table in which concrete foundation.

2.3. Analytical Method

ANSYS is universal software, which is used on simulation of the interactions in physics structures, vibration, fluid dynamics, thermal transfer and electro mechanics for engineers. We can simulate with ANSYS structures and then test them in the virtual environment. ANSYS can import CAD data and sketch of the geometry. ANSYS Workbench is a platform, which integrates simulation technologies and parametric CAD systems with unique automation and performance. some steps for analyzing as geometry Creation, Material Data selection, Boundary Condition, Mesh of Finite Elements and Modal Analysis.

3. Comparitive Results

3.1. Aspect Ratio

Sr.No	Natural frequency(Hz) by Aspect ratio(R)(CUTTER 1,6,&7)								
(Mode)	a =110 mm,R=0.182 (CUTTER 1)			a =108 mm,R=0.185 (CUTTER 6)			a =105mm,R=0.190 (CUTTER 7)		
	Theoretical	Analytical	FFT	Theoretical	Analytical	FFT	Theoretical	Analytical	FFT
1	563.60	570.31	566.4	589.74	593.05	634.8	633.82	630.6	595.7
2	616.69	659.2	664.1	643.64	679.31	732.4	687.73	716.13	683.6
3	774.88	1035.5	1084	807.76	1060.3	1093.8	861.31	1091.7	1093.8

3.2. Thickness

Natural frequency by thickness (CUTTER 1&8)					
a =110 mm, R=0.182 t=1.85 mm(CUTTER 1)			a =110 mm, R=0.182 t=1.2 mm(CUTTER 8)		
Theoretical	Analytical	FFT	Theoretical	Analytical	FFT
563.60	570.31	566.4	450.84	468.96	488.3
616.69	659.2	664.1	493.37	560.11	585.9
774.88	1035.5	1084	619.93	871.43	878.9

3.3. Other Parameter

Making stress concentration holes (cutter 2) and enlargement them (cutter 3)-Compare cutter 1,2 &3

Sr.No.	Natural frequency (ω in Hz)					
Mode	Cutter 1 (without hole)		Cutter 2 (making hole)		Cutter 3(enlarge hole)	
	FFT	Analytical	FFT	Analytical	FFT	Analytical
1	605.5	570.31	576.2	564.41	546.9	562.09
2	664.1	659.2	664.1	645.97	654.3	645.95
3	1084.0	1035.5	1054.7	1012	1044.9	1013.8

Increasing slot end diameter (cutter 4) and Number of teeth (cutter 5)-Compare cutter 1,4 &5

Sr.No.	Natural Frequency (ω in Hz)					
Mode	Cutter 1 (Original 9 teeth 4mm slot end dia.)		Cutter 4 (9 teeth & Increasing slot end diameter i.e.6 mm)		Cutter 5 (Increasing slot end 6 mm with10 teeth)	
	FFT	Analytical	FFT	Analytical	FFT	Analytical
1	605.5	570.31	546.9	545.65	556.6	548.52
2	664.1	659.2	654.3	647.9	654.3	648.59
3	1084.0	1035.5	1005.9	997.8	986.3	987.91

Variable radial slit (cutter 9) and circular concentric slit (cutter 10)-Compare cutter 1, 9 & 10

Sr No.	Natural Frequency (ω in Hz)					
Mode	Cutter 1 (Original)		Cutter 9 (3 cracks of length 24.5 mm with 2 mm dia.end circle)		Cutter 10 (3 cracks of length 24.5 mm with 2 mm circular end crack)	
	FFT	Analytical	FFT	Analytical	FFT	Analytical
1	605.5	570.31	556.6	547.59	566.4	551.4
2	664.1	659.2	644.5	640.56	654.3	644.16
3	1084.0	1035.5	1015.6	1007.1	1025.4	1018.9

Increasing number of cracks of same length 24.5 mm from centre with same teeth and outer dia.- Compare cutter 11,12 & 13.

Sr No.	Natural Frequency (ω in Hz)(cracks from same length 24.5 mm)					
Mode	Cutter 11 (3 cracks)		Cutter 12 (6 cracks)		Cutter 13 (9 cracks)	
	FFT	Analytical	FFT	Analytical	FFT	Analytical
1	537.1	554.29	517.6	531.76	507.8	**518.36**
2	644.5	648.27	644.5	632.3	654.3	**633.99**
3	1044.9	1027	1035.2	1007.1	1044.9	**1022.5**

Increasing length of cracks up to 30 mm from centre with same teeth and outer dia.-Compare cutter 11,12,14 & 15

Sr.No.	Natural Frequency (ω in Hz)							
Mode	(3 cracks)				(6 cracks)			
	Cutter 11 (24.5 mm crack length)		Cutter 14 (30 mm crack length)		Cutter 12 (24.5 mm crack length)		Cutter 15 (30 mm crack length)	
	FFT	Analytical	FFT	Analytical	FFT	Analytical	FFT	Analytical
1	537.1	554.29	527.3	543.5	517.6	531.76	498.0	507.42
2	644.5	648.27	556.6	644.28	644.5	632.3	517.6	628.08
3	1044.9	1027	1044.9	1017.1	1035.2	1007.1	996.1	995.55

4. Comparitive Graphs

4.1. Aspect Ratio

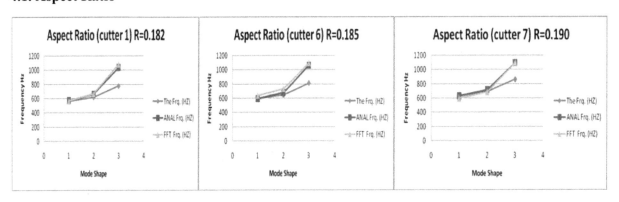

Graph (a.1). (1,6,7)-Natural frequency for same cutter with different theoretical ,analytical & FFT values

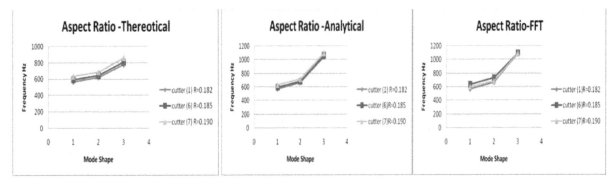

Graph (a.2). (1,6,7)-Natural frequency for different cutter with same theoretical ,analytical & FFT values

4.2. Thickness

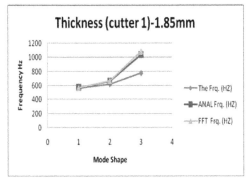
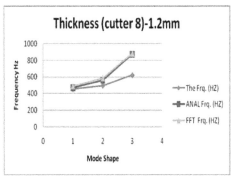

Graph (b.1). (1,8)-Natural frequency for same cutter with different theoretical ,analytical & FFT values

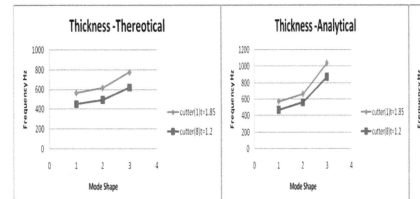

Graph (b.2). (1,8)-Natural frequency for different cutter with same theoretical ,analytical & FFT values

4.3. Other Parameter

Making stress concentration holes (cutter 2) and enlargement them (cutter 3)-Compare cutter 1,2 &3

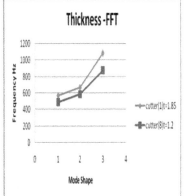

Graph (c.1.a). (1,2,3)-Natural frequency for same cutter with different analytical & FFT values

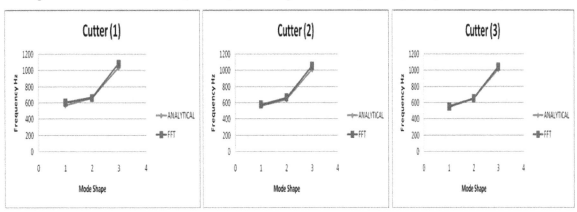

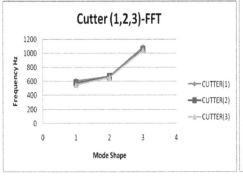
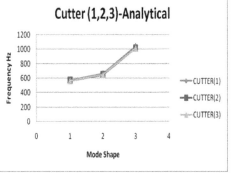

Graph (c.1.b). (1,2,3)-Natural frequency for different cutter with same theoretical ,analytical & FFT values

Increasing slot end diameter (cutter 4) and Number of teeth (cutter 5)-Compare cutter 4&5

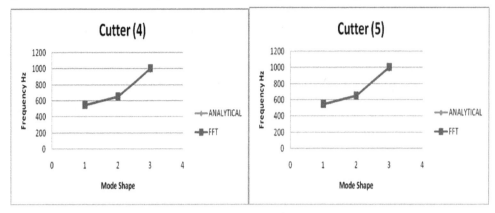

Graph (c.2.a). (1,4,5)-Natural frequency for same cutter with different analytical & FFT values

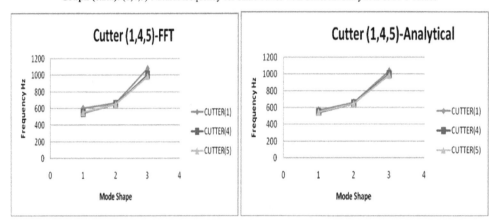

Graph (c.2.b). (1,4,5)-Natural frequency for different cutter with same analytical & FFT values

Variable radial slit (cutter 9) and circular concentric slit (cutter 10) –Compare cutter 1, 9 & 10

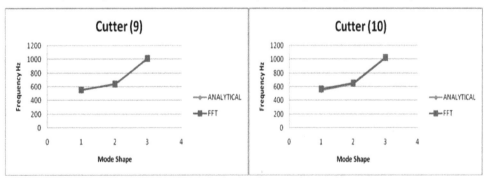

Graph (c.3.a). (1,9,10)-Natural frequency for same cutter with different analytical & FFT values

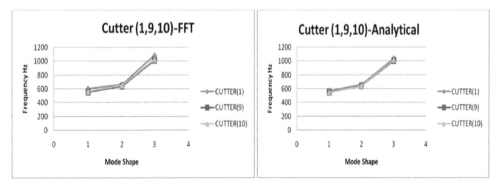

Graph (c.3.b). (1,9,10)-Natural frequency for different cutter with same analytical & FFT values.

Increasing number of cracks of same length 24.5 mm from centre with same teeth and outer dia.- Compare cutter 11,12 & 13

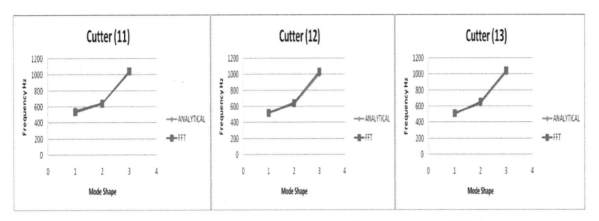

Graph (c.4.a). (11,12,13)-Natural frequency for same cutter with different analytical & FFT values

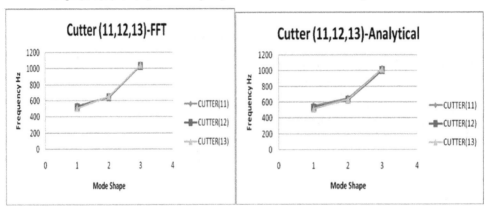

Graph (c.4.b). (11,12,13)-Natural frequency for different cutter with same analytical & FFT values.

Increasing length of cracks up to 30 mm from centre with same teeth and outer dia.-Compare cutter 11,12,14 & 15

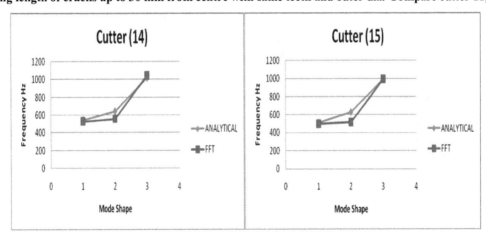

Graph (c.5.a). (11,12,14,15) -Natural frequency for same cutter with different analytical & FFT values

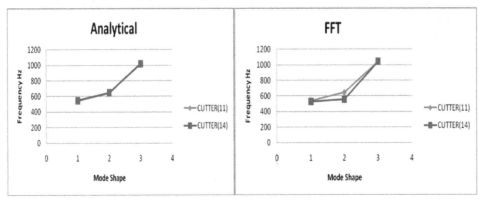

Graph (c.5.b). (11,14)-Natural frequency for different cutter with same analytical & FFT values

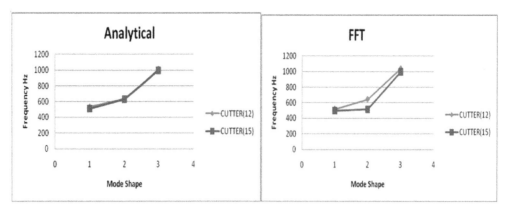

Graph (c.5.c). (12,15)-Natural frequency for different cutter with same analytical & FFT values.

5. Conclusion

Thus vibration analysis of circular tile cutter is done by using some theoretical, FFT & ANSYS. After prediction of natural frequency it will gives that future design scope to avoid resonance and vibration in tile cutting industries. Vibration described in terms of deformation in ANSYS analytical method and in terms of acceleration in FFT experimental method. From overall comparing all graph result it can gives that as,

a) Aspect ratio increases natural frequency increase. Also vibration will minimum in high aspect ratio tile cutter.(cutter 1,6& 7)

b) Thickness of tile cutter reduced then natural frequency also reduced. But vibration will more in reduced thickness tile cutter.(cutter 1 &8)

c) By making and enlargement of stress concentration holes natural frequency reduced. But enlargement of dia. of hole may cause high vibration (cutter 3) than only making of holes (cutter 2). So selected only making of mini. dia of holes causes less vibration.

d) Increasing slot end dia. natural frequency will minimum and as vibration also minimum.(cutter 1 &4)

e) Increase in number of teeth natural frequency will increases and vibration also increases.(cutter 4& 5)

f) By making 3 cracks at end of crack there is small hole having low natural frequency and high vibration than same no of crack but at end of crack there is small circular crack. (cutter 9 & 10)

g) Increasing no of cracks from 3 to 6 of same crack length (24.5 mm) natural frequency decreases with increase in vibrations. Next 6 to 9 cracks natural frequency and vibration both will increased. So select tile cutter with minimum no of cracks, up to 6 cracks will be fine w.r.t. mini. vibration.(cutter 11,12&13)

h) Increasing length of cracks (30 mm) with same no of cracks (3 or 6) natural frequency will decreases but vibration will more. So select minimum length of crack and minimum no of crack for minimum vibration than original cutter (1) (cutter 11,12, 14 & 15).

References

[1]   W.M. Lee a, J.T.Chen, "Eigen solutions of a circular flexural plate with multiple circular holes using the direct BIEM and addition theorem", Journal of Engineering Analysis with Boundary Elements, Vol.2 Issue.4, July 2010, Pgs:1064-1071.

[2]   L. Cheng,Y.Y. Li, L.H. Yam ,"Vibration analysis of annular-like plates ",Journal of Sound and Vibration ,2003, Pgs: 1153-1170.

[3]   Sasank Sekhar Hot ,Payodhar Padhi ,"Vibration of plates with arbitrary shapes of cutouts", Journal of Sound and Vibration, 2007, Pgs: 1030-1036.

[4]   D.V.Bambill, S.La.Malfa, C.A.Rossit, P.A.A.Laura, "Analytical and Analytical Investigation on Transverse Vibration of Solid, Circular and Annular Tile cutters Carrying a Concentrated Mass at an Arbitrary Position with Marine Applications", Journal of ocean Engineering, vol31, Pgs: 127-138.

[5]   K. Ramesh, D.P.S. Chauhan and A.K. Mallik, "Free Vibration of Annular Tile cutter with Periodic Radial Cracks", Journal of sound and vibration, 1997, Pgs: 266-274.

[6]   J.C. Bae and J.A. Wickert, "Free Vibration of Coupled Tile cutter-Hat Structures", Journal of sound and vibration, 2000, Pgs: 117-132.

[7]   M. Haterbouch, R. Benamar "Geometrically nonlinear free vibrations of simply supported isotropic thin circular plates" Journal of Sound and Vibration, 2005, Pgs:903-924.

[8]   P.A. Lauraa, U. Masia, D.R. Avalosb, "Small amplitude, transverse vibrations of circular plates elastically restrained against rotation with an eccentric circular perforation with a free edge" Journal of Sound and Vibration, 2006, Pgs: 1004-1010.

[9]   W.T. Norris and J.E.T Penny, "Computation of the Resonant Frequencies of an Annular Tile cutter Encysted at its Inner Edge and Free at its Outer Edge", Aston University, Dec 2002.

[10] Chi-Hung Huang, "Vibration of Cracked Circular Tile cutters at Resonance Frequency"Journal of Sound and vibration, 2000, Pgs: 637-656.

6

Modelling and Design of an Auto Street Light Generation Speed Breaker Mechanism

O.A. Olugboji, M.S. Abolarin, I.E. Ohiemi, K.C. Ajani[*]

Department of Mechanical Engineering, School of Engineering and Engineering Technology, Federal University of Technology, Minna, Nigeria
*Corresponding author: clemajan@gmail.com

Abstract Energy usage is inevitable to Man's daily living. For any task to be achieved in life energy must be converted from one form to another. This method of power generation is one of the most recent forms of energy generation concepts. The mechanism converts the kinetic energy of vehicles at speed bump into electric energy. A preliminary modelling of the speed breaker system was developed. The component members were designed and modelled on SolidWorks software based on the properties of the selected materials. A static and fatigue analysis of the spring using SolidWorks reveals a yield stress of 172.3 Mpa. The computational fluid dynamics of the air reveals an average total pressure and velocity of 29.74 bar and 1.018 m/srespectively.The conceptual design was carried out and resulted to the selection of materials needed for each component of the design. A prototype of the speed breaker mechanism was constructed and tested. The mechanism generated 6V D.C which validates the principle. The light sensor automatically opens the valve to on the light when it is dark and closes it to off the light at dawn.

Keywords: *development, speed breaker, electrical energy, compression, light sensor*

1. Introduction

A speed breaker or speed bump is a traffic calming device that uses vertical deflection to slow otor-vehicle traffic in order to improve safety conditions on roads. The use of vertical deflection devices is widespread around the world and they are most commonly found where vehicle speeds are statutorily mandated to be low so as to guarantee safety to road users.A speed breaker mechanism is a device that converts the impulse energy produced by the passage of the vehicle over a speed bump into the kinetic energy needed to rotate the rotor of an alternator to generate electrical Energy.

There has been a wide attention of researchers in design and development of renewable energy conversion systems. Electrical energy is normally generated through thermal Energy, Mechanical Energy nuclear Energy and chemical Energy. Researchers over the years have devised a new method that involves mechanical vibrations. There are three categories of devices that generate electricity by vibration. These are piezoelectric generation device, electrostatic generation device [1] and the Work function energy devices.

Piezoelectricity is the electric charge that accumulates in certain solid materials in response to applied mechanical stress [2]. Piezoelectric energy generation still remains an emerging technology, though several investigation has been carried out on it since the late 1990s [3,4,5,6]. Elementary principles of vibration theory, reveals that the resonance of a system can be tuned by changing the stiffness or mass [7,8,9,10,11].

Different from previous research on generating energy from translational base excitation, Gu and Livermore [12] focused on rotational motion. As identified by Aswathaman [12], three different mechanisms are currently being used in power generation via speed breakers. These are: Roller type mechanism [14,15], the Rack- Pinion mechanism [12,16], Crank-shaft mechanism [17]. When vehicles move on the road, the piezoelectric materials under the road are vibrated due to vehicle suspension in the tires that force the road and produces electricity in large amount [18]. The energy of water in the hosepipe is used to drive the turbine. As the water rushes through the pipe it turns the blades of the small turbine to generate electricity [19].

Epileptic power supply is a common problem prevalent in developing countries. Anytime there is power failure especially at night, the streets are turned into darkness which in turns affects the social economic life. It is against this backdrop that this research was conceived.

The work aims to develop a speed breaker mechanism for generating electrical energy by compression method with auto light sensing system. The speed breaker mechanism will be developedusing a static analysis of the spring and computational fluid dynamics (CFD) analysis of the air flow using SolidWorks software. A prototype of a speed breaker mechanism is also developed.

2. Materials and Method

2.1. Materials and Equipment.

The conceptual design of the speed breaker system include essential component such as: The speed breaker where the force is applied, Helical spring to return the breaker to its original position when the load is removed, Air pump /compressor, Receiver tank, Mini turbine, the automatic light sensor system which include transistor TIP31, Light emitting diodes, resistors, relays, wires, gears, 12v battery, vero board, capacitors, 7808 regulator, toggle switches and white junction box. The operations were performed using the following instruments: measuring tape, scriber, hack saw and the electric arc welding machine.

2.2. Methods

2.2.1. Speed Breaker

The speed breaker is required to have adequate mechanical strength to withstand continuous loading and unloading. Therefore, Mild steel was selected because of its cost and its good machinability. The speed breaker was cut and machined and welded based on the specification.

2.2.2. Spring Selection

The primary function of the spring in this application is to withstand twist and pull resulting from the applied force. Based on the strength, functional requirements and the properties of the candidate materials as earlier stated, Stainless steel was selected as the most suitable material for its excellent ability to withstand high stress, good fatigue resistance and long endurance for shock and impact loads. The spring stiffness and spring deflection for the chosen spring are 17500 N/m and 0.01M.

2.2.3. Cylinder and Piston Assembly

The Cylinder Piston Assembly serves as a unit that compresses air into the receiver tank. The reciprocating compressor is selected since its application requires the compression of air at a very high pressure.

2.2.4. Receiver Tank

Receiver tanks are used as air storage facility for mechanical units powered by compressed air. Mild steel is selected, as the ideal material due to its cost and its ability to be turned into different shape and form. The receiver tank was equally fabricated to specification.

2.2.5. Alternator Selection

Electric generator or Alternator is a device that converts mechanical energy to electrical energy. Direct current (D.C) alternator of different ratings and Alternating Current (AC) alternator of different ratings were used. A 6V DC alternator was used to power at least three LED bulbs of 1.8 Volts each.

2.2.6. Mini Turbine

The Turbine unit and its components were installed on the machine by drilling and fastening of screws, bolts and nuts using the drilling machine, screw driver and spanner respectively. All the components fabricated were assembled on the supporting frame of the speed breaker mechanism by welding, filing, drilling and fastening operations. The air compressor is installed between the speed bump and a DC alternator. The weight of the vehicle on the speed breaker pushes the speed breaker downward. The movement of the speed breaker pushes the piston through the barrel thereby compressing the air into an air receiver tank. The stored air is then released by the help of a valve to rotate a coil through an electromagnetic field to generate electric current.

2.2.7. Supporting Frame Work, Accessories and Assembly

The supporting frame was made from a square mile steel pipe (30 mm x 30 mm 0.5 mm).The operation was carried out by measurement, marking out, cutting and welding. The cylinder liners was made from a steel pipe with an inner bore diameter of 55 mm and outer diameter of 70 mm the sleeve was made smooth by turning the bore size to 60 mm using a lathe machine. The resulting cylinder was welded on a 3 mm mild steel sheet to form a dead end. The metal sheet was drilled using a power drilling machine to provide an orifice for the air that will be compressed out of the cylinder. The piston rod was welded to a circular connector where the piston can be joined to the piston rod. The piston rings were mounted on it and using the piston pins, the piston was connected to the piston rod. The air receiver tank was installed on the supporting frame by placing it on the seat provided for it at its base and by welding the cylinder holder at the top to the body of the supporting frame. The speed breaker was constructed from a 3 mm thick mild steel sheet. The road surface was made from plywood.

2.2.8. Automatic Light Sensor Unit

In the circuit as shown in Figure 1 below, SW2 and SW3 are switches connected to the battery to form a series and parallel connection. When the two 6Vbatteries are connected in series they yield 12V at the discharging mode which is connected to the entire circuit. While when connected in parallel, the total emf of the 6V batteries at the charging mode is connected to the incoming 6V from the turbine connected to D5.The SW2 is connected to the positive side of the B2and SW3 is connected to GND of B1. The diode D5 allows voltage coming in from the turbine to charge the battery thereby preventing back emf coming from the battery and over charging. The SW1 prevents excessive flow of current from the D7 to D13 by changing the value of resistors connected in series with the light emitting diodes D7 to D13. The LDR (light dependent resistor sensor R2) in series with the variable resistor (VR2) controls the base of Q1 to switch RL1 (relay switch) during the day and night. R1is controlled by RL1 (day and night switching) to bias D2 or D1 to switch the bases of Q2 or Q3 depending on the intensity of light or darkness on the LDR. Q2 and Q3 are connected to forward and reverse polarity of the motor MT1 which in turn closes or opens the release valve to stop or start the turbine.

2.2.9. Computational Fluid Dynamics (CFD) of the Air Flow Using Solid Works

Several steps were taken in order to perfectly execute the fluid flow analysis. A project file in SolidWorks Flow

Simulation containing all the settings, results of a problem and each project that is associated with a SolidWorks configuration was created. The analysis was then Ran and the Vectors, Contours, Isolines, Cut Plots, Surface, Flow Trajectories, Isosurfaces were plotted. The Results for XY Plots, Goals was presented using Microsoft Excel, while Surface Parameters Point Parameters was Reported using Microsoft Word.

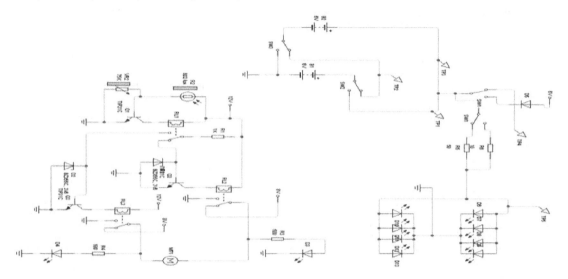

Figure 1. Schematic circuit diagram showing the light sensing unit

2.2.10. Testing of the Speed Breaker Mechanism

The speed bump was subjected to a continuous loading and unloading force for a period of five minutes. Air was compressed continuously into the receiver tank by the to and fro movement of the piston in the cylinder liners. The air outlet valve was then opened to release the compressed air from the receiver tank into the blade of the mini-turbine. The turbine was turned for a period of two second and 3 volt was generated as read from the connected voltmeter. The above procedure was repeated for a period of 10, 15, 20, 25, 30 and 35 minutes. The results is as shown in Table 1.

3. Design Analysis and Calculations

3.1. Theoretical Analysis

3.1.1. Solid Simulation (Static and Fatigue) Analysis of the Spring Using Solid Work.

Finite-element analysis (FEA) is used to predict the response of the speed breaker mechanism model to the influence of forces and torques. FEA analyzes large or complicated models where analytical solutions are impossible. FEA software breaks the model into thousands of small tetrahedral elements and solves for them numerically.

3.1.2. Static Analysis

Static analysis computes the effects of static loading on the model. It is the determination of stress, strain, displacement, and the factor of safety at each segment of a model. In Simulating the model, the location and magnitude of each load is specified, as well as where and how the model is supported. Identification of the areas with high and low stress quickly shows where the model can be improved on either by the addition of supports or by the removal of excess material. The part model used for this simulation is the spring and must be specified.

To specify the material, the Feature Manager Design tree on the left-side panel was selected. An icon named "Material <not specified>" as shown in Figure 2 was Right-clicked and "Edit material" was selected on the Materials window. Furthermore, the material was selected by clicking on the folder named SolidWork materials and Steel Alloy on the dropdown menu selected. Finally, Stainless Steel alloy was selected from the array of steel alloy.

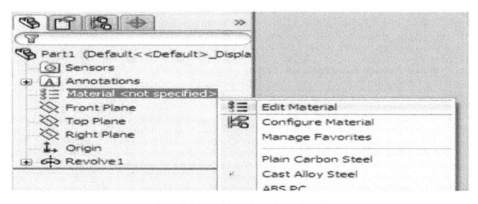

Figure 2. Material Selection using Solidwork

3.2. Design calculation

3.2.1. Speed Bump Design

The Speed bump is meant to bear the load of the input force. The material for this component is mild steel. The Tensile Strength of mild steel is 407 MPa

Yield Stress of Mild Steel, $\sigma_{Yield} = 210 MPa$

Allowable Stress,

$$\sigma_{all} = \frac{\sigma_{yield}}{\text{Factor of Safety}} \quad [20] \tag{1}$$

Factor of safety, F.O.S = 1.5

$$\sigma_{all} = \frac{210}{1.5} = 140 MPa$$

But,

$$\sigma_{all} = \frac{\text{Max. Load}, F_{max}}{\text{Cross Sectional Area of Bump}, A_b} \tag{2}$$

Designing for a cross sectional Area of 200 mm × 60 mm, therefore, the Allowable Maximum Load that can be bored by the cross section

$$F_{max} = \sigma_{all} \times A_b = 140 \times 0.2 \times 0.06 = 168 kN \tag{3}$$

Therefore the speed bump should not be loaded beyond 168 KN force. The Bump is supported at both ends by spring, which results to reactions F_A and F_B, due to the application of load, F as shown in Figure 3.

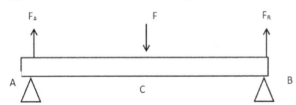

Figure 3. Schematic diagram of the bump under load

Sum of all vertical forces, $\sum F_v = 0 \uparrow +$

$$F_A + F_B = F \tag{4}$$

Sum of moment about Support A, $\sum M_A = 0$ (Clockwise Positive)

$$F \times \frac{1}{2} - F_B \times 1 = 0$$

$$F_B = \frac{F \times 1}{21} = \frac{F}{2} \tag{5}$$

From equation 4 above, making F_A, subject of the formulae

$$F_A = F - F_B = F - \frac{F}{2}$$

$$F_A = \frac{F}{2}$$

The shear force at A is equal to $R_A = 0.5 F$, where x = 0 and continues on a straight line to the mid-point of the beam where x = L/2; beyond which it continues to

decrease uniformly to -0.5 Fl at C where it remained constant till point B where x = L.Also the bending moment is zero at A and B (when x = 0, x = L) and increase in a triangular form to a point C. The Bending Moment is maximum where shear force is zero. Taking Moment about C,

$$M_c = \frac{F \times 1}{4} = \frac{350 \times 0.2}{4} = 17.5 Nm \tag{6}$$

Let b= Length of the bump, d=Thickness of the bump

Also, considering a rectangular section cut from the speed bump ABCD as shown in Figure 4 as a segment of the bump

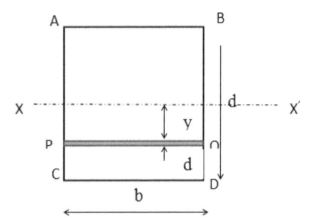

Figure 4. A Cut segment of the speed bump

Considering a strip PQ of thickness dy parallel to X-X' axis and at a distance y.

$$\text{Area of Strip} = b.dy \tag{7}$$

Moment of inertia of strip about X-X' axis

$$= b.y^2 dy \tag{8}$$

Moment of inertia of the whole section is the integration equation 8 from $-\frac{d}{2}$ to $+\frac{d}{2}$

$$I_{xx} = \int_{-d/2}^{d/2} b.y^2 . dy = b \int_{-d/2}^{d/2} y^2 . dy \quad [18] \tag{9}$$

$$= b \left[\frac{y^3}{3} \right]_{-d/2}^{d/2} = \frac{bd^3}{12} \tag{10}$$

From bending moment equation,

$$\frac{M_c}{I_{xx}} = \frac{\sigma_{all}}{y} \tag{11}$$

Also from Figure 4, $y = \frac{d}{2}$

Substituting for y and I_{xx} in equation 11

$$\frac{12 M_c}{bd^3} = \frac{2\sigma_{all}}{d}$$

$$d = \sqrt{\frac{6 M_c}{b.\sigma_{all}}} \tag{12}$$

Minimum Thickness of the bump material

$$sd = \sqrt{\frac{6 \times 350 \times 50}{200 \times 140}} = 1.94 \text{mm}.$$

3.2.2. Spring Design

Consider a helical spring shown in Figure 5 below under a load, F

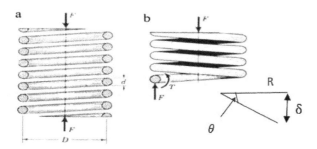

Figure 5. Schematic Diagram of a Helical Spring under Load

Consider a helical spring under a load, F, Where, D = Diameter of coil; d = diameter of the coil wire; R = Radius of Coil; δ = Deflection of the coil under load; C = Modulus of Rigidity of the Spring Material; n = Number of coils or turns θ = Angle of Twist , l = $2\pi Rn$ = Length of wire, τ = Shear Stress .

From Hooke's Law,

$$F = k\delta \tag{13}$$

Calculating for a minimum deflection (δ) of 0.01 m and applying force F_A as calculated from equation 2 above

$$k = \frac{F}{\delta} \tag{14}$$

$$\frac{175}{0.01} = 17500 \text{N} / \text{m}$$

Also Allowable Stress for mild steel $\sigma_{all} = 140$MPa

Force acting on spring, $F_A = F_B = 175$ N

Cross sectional area of spring,

$$A = \frac{F_A}{\sigma_{all}} = \frac{\pi d^2}{4} \tag{15}$$

Minimum spring wire diameter,

$$d = \sqrt{\frac{4F_A}{\pi . \sigma_{all}}} = \sqrt{\frac{4 \times 175}{3.142 \times 140}} = 1.26 \text{mm} \tag{16}$$

Also the minimum deflection as a result of 175 N load

$$\delta = \frac{64 F_1 R^3}{Cd^4} \text{ [20]} \tag{17}$$

Making the minimum mean radius subject of the formula from equation 17

$$R = \sqrt[3]{\frac{\delta \times Cd^4}{64F_1}} = \sqrt[3]{\frac{10 \times 80000 \times 1.26^4}{64 \times 175}} = 5.647 \text{mm} \tag{18}$$

But mean diameter of spring, D

$$D = 2R = 2 \times 5.647 = 11.29 \text{mm}$$

$$n = \frac{Cd^4}{64R^3 k} = \frac{800 \times 1.26^4}{64 \times 5.647^3 \times 10} = 7 \tag{19}$$

The Length of the wire

$$l = 2\pi Rn = 2 \times \frac{22}{7} \times 5.647 \times 7 = 248.47 \text{mm} \tag{20}$$

3.2.3. Cylinder and Piston Assembly Design

3.2.3.1. Load Bearing Capacity of Piston Rod

The application of load on one face of the piston leads to pressure, since pressure is a function of Surface area. The other face of the piston moves to and fro a fixed wall. Failure or breakage of piston rod will only occur due to excessive compressive stress developed in the piston rod.

$$\text{Compressive Stress } \sigma = \frac{\text{Force, F}}{\text{Area, A}} \tag{21}$$

$$A = \pi r^2$$

Where,
r = radius of the piston rod, σ = stress, A = cross section area of the piston rod.

The compressive yield stress of mild steel (σ) = 152 N/mm^2

Applying a factor of Safety of 1.5,
The allowable Compressive stress

$$\sigma_{all} = \frac{\sigma_{yield}}{1.5} = \frac{152}{1.5} = 101.33 \text{MPa}$$

$$\text{Force = Stress} \times \text{Area}$$
$$= 101.33 \times 3.142 \times 5^2 = 7957.5 \text{ N} \tag{22}$$

Therefore the maximum Load the piston can bear is 7957.5 Newton. The load of 8 kN on the buckling chart for cylinder corresponds to a piston rod diameter of 12 mm, the cylinder bore size diameter is 60 mm, Length of the piston and piston rod assembly is 300 mm, the set pressure of 24.93 bars by interpolating between 7 kN and 10 kN

3.2.3.2. The Maximum inside Pressure of Barrel

The maximum inside pressure in the barrel will be due to the effect of the maximum Load 7957.5 N

$$\text{Pressure} = \frac{\text{Force}}{\text{Cross Sectional Area of Piston Rod}} \tag{23}$$

Force on the piston =7957.5 N
Radius of Piston = 30 mm (determined from buckling chart for cylinders in appendix B)
Cross sectional area of piston = 2828.57 mm^2

$$\text{Pressure} = \frac{7957.5}{2828.57}$$

$$\text{Pressure} = 2.8133 \text{N} / \text{mm}^2$$

This means that pressure of 2.8133 MPa acts on the walls of the cylinder barrel when the air cylinder is loaded with 7957.5 N force.

3.2.3.3. Thickness of the Barrel

The internal diameter of the barrel is 60 mm. The outer diameter of the barrel will be calculated for as follows.

Inner radius, r_1

$$r_1 = \frac{D_i}{2} = 30mm \tag{24}$$

The maximum tensile yield strength for mild steel is 250 MPa. Applying a factor of safety of 1.5, the working tensile stress becomes 166.7 MPa

The circumferential stress σ_c at the inner radius r_i is 166.7 MPa,

$$166.7 = \frac{b}{(30)^2} + a \tag{25}$$

Also, since pressure at the inner surface of the cylinder is calculated as 2.257 MPa, it implies that radial stress σ_r at inner radius is 2.8133 N/mm^2, Equation 17 becomes

$$\sigma_r = \frac{b}{r^2} - a$$

$$2.8133 = \frac{b}{(30)^2} - a \tag{26}$$

Adding equation 26 and 27 we get,

$$2.8133 + 166.7 = \frac{b}{(30)^2} + a + \left(\frac{b}{(30)^2} - a\right)$$

$$169.5133 = \frac{2b}{(30)^2}$$

$$b = 76280.99$$

Therefore, from equation 26, replacing 'b' results to [20]

$$\sigma_r = \frac{b}{r^2} - a$$

$$a = \frac{b}{r^2} - \sigma_r = \frac{76280.99}{(30)^2} - 2.8133 = 81.9434$$

Therefore, Lamen's equation for this case become

$$\sigma_r = \frac{76280.99}{r^2} - 81.94335 \tag{27}$$

$$\sigma_c = \frac{76280.99}{r^2} + 81.94335 \tag{28}$$

Now the air cylinder barrel must be strong enough to absolve all the stress such that the stress at the outer surface of the barrel must be zero. That is, the radial stress at outside radius $r_o, \sigma_r = 0$

Therefore, equation 28 becomes

$$\sigma_r = \frac{76280.99}{r_o^2} - 81.94335 = 0$$

$$r_o = \sqrt{\frac{76280.99}{81.94335}} = 34.7mm$$

Barrel Minimum wall thickness (t) = Outer radius - Inner radius

$$t = r_o - r_i = 34.7 - 30 = 4.7mm \tag{29}$$

3.2.4. Receiver Tank Sizing

Receiver tanks are found in most modern compressed air systems. Hence, one of the most common design tasks in sizing of a compressor system is the sizing of receiver tanks. To size the receiver tanks appropriately, we need to identify the key function of the receiver tank. In this design, the receiver tank is serving as a buffer. Therefore, the Volume of the tank, taking into consideration the free air delivery from the air compression unit, is given by equation below. The height of the tank is 350 mm which is 0.35 m.

Therefore, Volume of Receiver Tank

$$V_R = \pi r^2 h \tag{30}$$

The radius of the Tank (r) is given by

$$r = \sqrt{\frac{V_R}{\pi h}} = \sqrt{\frac{0.01447}{3.142 * 0.35}} = 0.013158m \tag{31}$$

4. Results

Table 1. Speed beaker prototype test Results

Loading time(minute)	Time to power turbine(sec.)	Voltmeter Reading(V)
5.0	2.0	3.0
10.0	8.0	4.2
15.0	9.0	4.5
20.0	10.0	4.5
25.0	10.0	6.0
30.0	10.0	6.0
35.0	10.0	6.0

Table 2. Dimensions of the Air Receiver Tank

Description of Parameter	Value	Unit
Diameter of the Base	0.263	M
Height of the Tank	0.35	M
Minimum Wall Thickness	0.0023	M
Volume of the Tank	0.01447	m^3

Table 3 below gives a summary of the results obtained from the analysis of the fluid flow simulation. The solid work fatique analysis of the spring is as presented in Figure 6 with the various colors showing the probable regions of fatique.

Figure 6. Fatique Analysis

Table 3. Summary of Flow Simulations

Name	Unit	Value	Averaged Value	Minimum Value	Maximum Value	Progress	Use In Convergence	Dalta	Cniteria
GG Av Total Pressure 1	[bar]	29.74	28.07	27.16	29.74	28%	Yes	1.61	0.45
GG Av Velocity (X) 1	[m/s]	1.018	1.112	1.018	1.159	100%	Yes	0.024	0.731
GG Av Velocity (Y) 1	[m/s]	0.045	0.024	7.547e-004	0.048	100%	Yes	0.008	0.130

Figure 7 shows a graph of the global goal average pressure against the total pressure obtained from the CFD analysis. The total pressure increases as the goal average pressure increases. The graph shows the adequacy of the air pressure for effective performance of the mechanism.

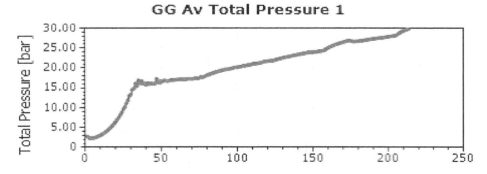

Figure 7. Total Pressure against GG Av Total Pressure 1

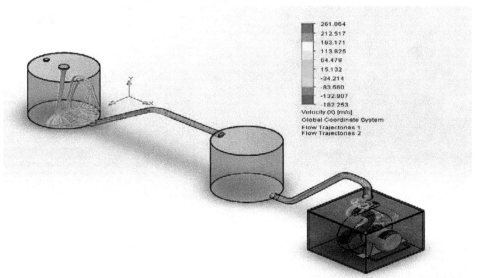

Figure 8. Fluid flow simulations showing the velocity of air flow

5. Discussions

Table 1 reveals the results of the testing of the speed breaker mechanism. A peak voltage of 6V was generated with a 10 seconds time taken to power the turbine. Table 2 presents the capacity of the air receiver tank. The minimum Volume of the air receiver tank was evaluated as 0.01447 m³. Also a minimum wall thickness of 0.0023 m was obtained. The results of the solid simulation of the spring were presented in Figure 6. This involves the results obtained from the static and fatique analysis of the spring. The solid simulation was carried out by subjecting the system to repeated loading and unloading. This action weakens the affected part over time even when the induced stresses are considerably less than the allowable stress limits. This phenomenon is known as fatigue. Each cycle of stress fluctuation in the system, weakens the mechanism to some extent. After a number of cycles, the object becomes so weak that it fails. Figure 8 shows the fluid flow simulation on solid work for the velocity of the air flow. This reveals that the air speed decreases with time.

The existing sources of energy such as oil, coal just to mention a few, may not be adequate to meet the ever increasing energy demands. Development of renewable energy sources has emerged in recent times as a vital component of the power industrial revolution. A large amount of energy is unutilized at speed breakers through friction, every time a vehicle passes over the speed breaker. Based on the results obtained in this work, the following conclusions were drawn. The speed breaker mechanism is a source of electricity generation. Thus it is concluded that stakeholders in the power sector should embrace the use of a speed breaker mechanism as an alternative power source for lighting street lights. This source of energy is pollution free and can be used to power street lights and traffic indicators on roads. The results obtained from the development design, modelling and simulation were used to construct a prototype of the Speed breaker mechanism and the test results revealed that the mechanism generated an average of 6 volts D.C. The result obtained from the constructed prototype is an indication that the speed breaker mechanism is a viable source of renewable energy and will function efficiently on a mega scale. The automatic light sensing system will also enhance an effective energy utilization.

6. Conclusion

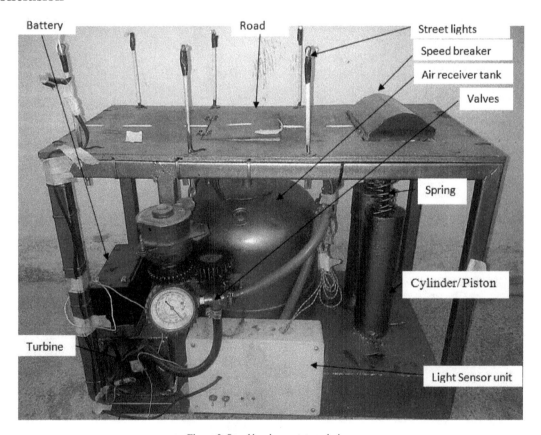

Figure 9. Speed breaker prototype design

References

[1]   G. Ankit, An experimental study of generation of electricity using speed breaker, Int. Academy of Sci., Engineering and Technology, Volume 2, Issue 2, 2012.

[2]   F.J. Holler, D.A. Skoog, S. R. Crouch, Principles of Instrumental Analysis, sixth ed., Cengage Learning, pp. 9., 2007.

[3]   N. M. White, P. Glynne-Jones, S. P. Beeby, A novel thick-film piezoelectric micro- generator. Smart Materials and Structures. Vol.10, No. 4 pp. 850-852, 2001.

[4]   A. Abbasi, Application of piezoelectric materials and piezoelectric network for smart Roads, Int. J. of Electrical and Computer Engineering (IJECE) Vol. 3, No.6, pp. 857-862, 2013.

[5] A. Manbachi, R. S. Cobbold, Development and application of piezoelectric materials for ultrasound generation and detection, Ultrasound 19 (4) pp. 187-196,2011.

[6] G. Gautschi, Piezoelectric Sensorics: force, strain, pressure, acceleration and acoustic emission sensors, Materials and Amplifiers Springer, 2002.

[7] E. S. Leland, P. K. Wright, Resonance tuning of piezoelectric vibration energy scavenging generators using compressive axial preload. Smart Mater Structure. 15, pp. 1413-1420.2006.

[8] C. Eichhorn, F. Goldschmidtboeing, P. Woias, A frequency tunable piezoelectric energy converter based on a cantilever beam, Proceedings of Power MEMS, pp. 309-312, 2008.

[9] Y. Hu, H. Xue, H. Hu, A piezoelectric power harvester with adjustable frequency through axial preloads. Smart Material Structure 16, pp. 1961-1966, 2007.

[10] J. Loverich, R. Geiger, J. Frank, Stiffness nonlinearity as a means for resonance frequency tuning and enhancing mechanical robustness of vibration power harvesters. Proc. SPIE, pp. 6928-6935, 2008.

[11] S. E. Jo, M. S. Kim, Y.J. Kim, Passive-self-tunable vibrational energy harvester. In proc. of 16th int. solid-state sensors and actuators, 2011.

[12] L. Gu,C. Livermore, Passive self-tuning energy harvester for extracting energy from rotational motion. Appl. Physicspp. 97, 2010.

[13] V. Aswathaman, M. Priyadharshini, Every Speed Breaker is now a Source of Power. International Conference on Biology, Environment and Chemistry, Volume 1, IACSIT Press, Singapore, 2011.

[14] S. Shakun, A. Ankit, Produce Electricity by the Use of Speed Breaker. Journal o Engineering Research and Studies, Article 30, Volume 2, 2011.

[15] G. Ankit, B. Meenu, Power Generation from Speed Breaker. Int. J. of Advance Research in Sci. and Engineering, Volume 2, Issue 2, 2013.

[16] F. Noor, M. Jiyaul, Production of Electricity by the Method of Road Power Generation. Int. J. of Advances in Electrical and Electronics Engineering, Volume 1, 2011.

[17] K. Gogoi, Generation of electricity from speed breaker using crank shaft mechanism,thesis submission www.scribd.com, 2010.

[18] A. Abbasi,. "Application of Piezoelectric Materials and Piezoelectric Network for Smart Roads." International Journal of Electrical and Computer Engineering (IJECE) Volume 3, No.6 pp. 857-862, 2013.

[19] A.L. Ling, Compressor Selection and Sizing (Engineering Design Guideline): KLM Technology Group, 2008.

[20] R.S. Khurmi, J.K. Gupta, A Textbook of Machine Design, Eurasia Publishing House Ltd., Ram Nagar, New Delhi, 2005.

7

Determination of Center of Gravity and Dynamic Stability Evaluation of a Cargo-type Tricycle

Ekuase Austin, Aduloju Sunday Christopher*, Ogenekaro Peter, Ebhota Williams Saturday, Dania David E.

National Engineering Design Development Institute, Nnewi, Nigeria
*Corresponding author: chrisaduloju@yahoo.com

Abstract Dynamic stability of vehicles is a major concern to vehicle manufacturers, as this determines how safe a vehicle will be on the road, to passengers and other road users. The location of centre of gravity (CG) on a vehicle determines its stability. The objective of this work is to evaluate the dynamic stability of a modeled cargo tricycle. The mass Properties capability of the SolidWorks software was used to determine the CG location on the tricycle. Result shows that the model will response to side load with a yaw motion and it's an oversteer vehicle. Therefore it is unstable at high speed above its critical speed. The rollover threshold (F_c) for the tricycle model is 0.32g.

Keywords: dynamic stability, center of gravity, cargo tricycle, tipping treshold, yaw motion, oversteer

1. Introduction

Dynamic stability of vehicles is a major concern to vehicle manufacturers, as this determines how safe a vehicle will be on the road, to passengers and other road users. The location of centre of gravity (CG) on a vehicle determines its stability. In an attempt to safely exploit the available capacities of both the highway system and motor transport fleet, the Canadian government had proposed regulatory principles and recommended limits. The regulatory principles were formed based on a research conducted to understand the influence of heavy vehicle weights and dimensions on the stability and controllability of the vehicles which are used in the highway system. One of the seven performance indicators for stability recommended is the rollover threshold [1]. Systems Technology Inc [2] studied the rollover accident experience of small cars as a function of the rollover potential. Result shows that rollover accident rate (fatal accident per 100,000 new car years) decreases with increasing rollover threshold. Robertson and Kelly [3] conducted a methodical analysis of rollover accident experience for passenger cars and utility vehicles. In their work, they considered a wider range of vehicle and rollover was taken as the 'first harmful event' in accidents per 100,000 vehicle-years. Their findings reveal a direct relationship between rollover threshold and accident rates, and the inclusion of utility vehicle increased the accident rates. The high involvement of the utility vehicles made the Federal Motor Vehicle Safety Standard (FMVSS) to propose a minimum rollover threshold of 1.2 for new utility vehicle [4].

In Asian countries like India, Bangladesh, Thailand and Nepal, three wheeled vehicle TWV is commonly used as means of public transportation, utility vehicles and carrying freight. The use of TWV also known as tricycle in Africa is expected to increase over the next decades. Because of their low cost and maneuverability in small turn radii and crowded roads, three wheelers is preferred compared to 4-wheel vehicles, hence they are important means of transportation in developing countries [5]. Delise et al [6] however discussed the poor stability of the three-wheeled vehicle as an important factor for accidents. Relatively higher centre of gravity have being cited as contributing factor for rollover, therefore passenger safety is compromised. Nevertheless, Van Valkenburg et al [7], ascertain that it is possible to make a three wheel vehicle as stable as a 4-wheel car.

We have analyzed the body shape of tree wheelers for reduction in fuel consumption [8] but Patrick. J [9] finds it disturbing that most three wheeler designers do not know or follow the procedures for designing stability into these three wheelers even when it can be achieved. In order to guarantee the safety of drivers and road user, the dynamic stability of a modeled three wheel vehicle for manufacturing should be examined for better stability. Therefore the objective of this paper is to evaluate the dynamic stability of a modeled cargo tricycle.

2. Dynamic Stability Analysis of Tricycle

2.1. Lateral Stability

2.1.1. Vehicle Response to Side Loads at the CG: Static Margin

Figure 1 show three conditions of the bicycle model of a four wheeled vehicle. It is assumed that each end of the vehicle has single tire with double cornering stiffness of the actual tire in the bicycle model. The CG is assumed on the centerline of the vehicle. The horizontal forces which act upon the CG and the tire contact patches are of interest. Three centers of gravity positions are considered i.e CG, CG1 and CG2. A vehicle headed straight along path P,

travelling at steady velocity V is represented in figure 1(a). The velocities of the axle line above the front and rear tire contact patches are VF and VR, and have same magnitude and direction as V.

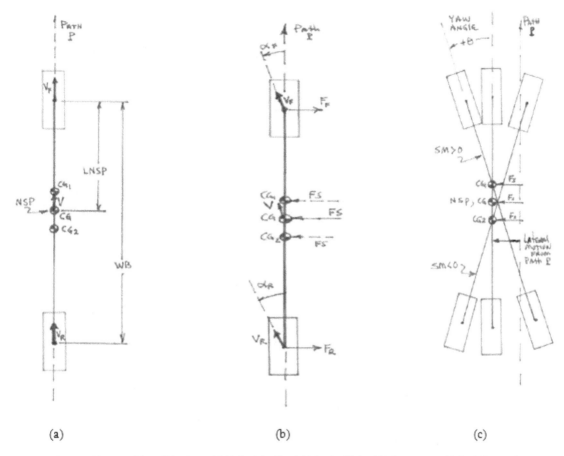

(a)   (b)   (c)

Figure 1. Three condition of bicycle model (a) Straight Ahead (b) Apply side load (c) Response to side load (Source: [7])

Figure 1b shows the vehicle traveling along path P as the side load is applied. Lateral forces at the tires, FF and FR, would develop to counteract the side load. Corresponding slip angles αF & αR would be present as the velocities veer off their original paths. Figure 1a shows three vehicle responses after the application of the side load. The response is a sideways movement of the CG off the path P an angular rotation about the CG, which is the yaw response and indicated by angle θ. The point at which a side load is applied and there is no yaw response is called the Neutral Steer Point (NSP). The distance from the front axle line to the NSP is:

$$LNSP = \left[\frac{\bar{C}_R}{\bar{C}_F + \bar{C}_R} \right] WB, ft \qquad (1)$$

Where C_F, C_R = cornering stiffness values of a single front and rear tire

$\bar{C}_F, \bar{C}_R$ = total cornering stiffness values for the four wheeler bicycle

$$\bar{C}_F = 2C_F$$

$$\bar{C}_R = 2C_R$$

WB = wheel base
LG = distance from front axle line to CG.

Static Margin

The character of the yaw response of a vehicle is determined by the location of the NSP relative to the CG. The distance from the CG rearward to the NSP divided by the wheelbase is termed "static margin" (SM).

$$SM = \left[\frac{\bar{C}_R}{\bar{C}_F + \bar{C}_R} - \frac{LG}{WB} \right]. \qquad (2)$$

If the same tires are used at each end, $\bar{C}_R \approx \bar{C}_F$, equation (2) can be written as

$$SM = \left[\frac{1}{2} - \frac{LG}{WB} \right]. \qquad (3)$$

The value of SM can be positive, negative or zero and it's an indicator of a vehicle yaw response.

For *SM = 0*, the center of mass is at one half of the wheelbase, *LG = WB/2*. In that case, a load at the CG will not cause any yaw response. The front and rear angle will make the vehicle sideslip. It is termed "***Neutral steer***".

For *SM = (+)*, the CG is ahead of the NSP, shown as CG1, then LG/WB is less than one half. Figure 4C shows the yaw response in which the front slip angle is larger than the rear and the vehicle is headed in the direction of applied force. This is considered "***stable***".

For $SM = (-)$, the CG is behind NSP, shown as CG2, then LG/WB is larger than one half. Figure 1C shows the rear slip angle is larger than the front and the vehicle is turned against the direction of the applied force. This is considered **"unstable"**.

2.1.2. Vehicle Response in Turn

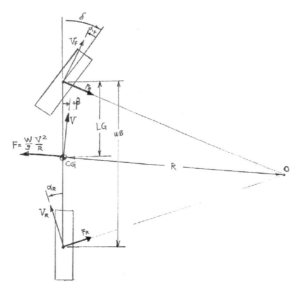

Figure 2. Vehicle response in a turn Source: [7]

The Figure 2 shows the stability of a vehicle in a turn. If a vehicle is negotiating a corner of radius R, travelling at a velocity V and with a steer angle of δ. The vehicle has a Front axle velocity VF and Rear axle velocity VR with their corresponding slip angles (α_F and α_R) as indicated in the figure above.

The side load is a centrifugal force F which acts on the CG and it increases with the square of the velocity V. Angle β is the angle between the vehicle centerline and the direction of the velocity of the CG. Hence the steering angle in degrees can be expressed as:

$$\delta = 53.7\frac{WB}{R} + \left(\alpha_F - \alpha_R\right) \qquad (4)$$

Equation (4) can be express in terms of CG location, tire cornering, stiffness and stability by writing torque and force balances for the vehicle of figure (2).

$$\delta = 53.7\frac{WB}{R} + \left[\frac{W_F}{\bar{C}_F} - \frac{W_R}{\bar{C}_R}\right]\frac{V^2}{gR}. \qquad (5)$$

The $\frac{V^2}{gR}$ term is lateral acceleration a_y in terms of g's.

Where v = velocity of CG along the path
R = radius of the CG path
G = acceleration of gravity

The term in the brackets has special significance and is called **"understeer gradient"** and denoted with symbol K

$$K = \left[\frac{W_F}{\bar{C}_F} - \frac{W_R}{\bar{C}_R}\right] \qquad (6)$$

Where W_F = weight on front wheels

W_R = weight on rear wheels

$\bar{C}_F$ = total front cornering stiffness

$\bar{C}_R$ = total rear cornering stiffness

Hence stability is analyzed in a turn as δ changes as the lateral acceleration a_y increases. δ is the sum of a constant term, plus an expression involving one of the following terms which could be positive, zero, or negative

From Equation (4) $\left(\alpha_F - \alpha_R\right)$

From Equations (6) $K = \left[\dfrac{W_F}{\bar{C}_F} - \dfrac{W_R}{\bar{C}_R}\right]$

From Equation (2) $SM = \left[\dfrac{C_R}{C_F + C_R} - \dfrac{LG}{WB}\right]$.

The condition that change the signs of these terms are equivalent: if $\alpha_F > \alpha_R$, then K > 0 and SM >0. The sign change of these terms determines stability and defines vehicle steer characteristics.

Neutral Steer: $\alpha_F = \alpha_R$, K = 0 and SM = 0

Steering angle δ remains constant at the values 57.2 WB/R, degrees. As the vehicle negotiates a radius R and slowly increases the velocity causing lateral acceleration to increase, then lateral forces at each end of the vehicle increases to cause slip angles αF and αR to increase. For a neutral steer vehicle, the vehicle centerline will slightly rotate toward the direction of velocity V, hence decreasing the side slip angle β and thus increase both front and rear slip angles the same amount as δ remain constant.

Understeer: $\alpha_F > \alpha_R$, K > 0 and SM > 0

As steering angle increases with speed, the front slip angle is larger than the rear. Though, it is self correcting. If the δ is not increase further, the positive yaw will turn the front of the vehicle toward the outside of the original path, thereby increasing the radius R and reducing the centrifugal force. Thus, a positive SM is termed **"stable"**.

Oversteer: $\alpha_F < \alpha_R$, K < 0 and SM < 0

The rear slip angle is larger than the front, i.e the rear is side slipping more than the front, and hence the steering angle is reduced from its neutral steer value. If there is no correction, the vehicle moves outward from the original path, decreasing the radius R and increasing the centrifugal force requiring a greater correction. The correction is a turn in the direction of skid which can lead to the vehicle spinning. Thus a negative SM is termed **"unstable"**.

2.1.3. Application of Static Margin and Understeer Gradient in Three Wheelers

For three wheelers with two wheels at the rear and one wheel at the front, the expression for K and SM can be written in terms of single tire stiffness values as:

$$K = \left[\frac{W_F}{\bar{C}_F} - \frac{W_R}{2\bar{C}_R}\right] \qquad (7)$$

$$SM = \left[\frac{C_R}{C_F + 2C_R} - \frac{LG}{WB}\right] \qquad (8)$$

If the same tires are used, then each tire would have the same stiffness values, $C_F = C_R$

2.1.4. Critical Speed

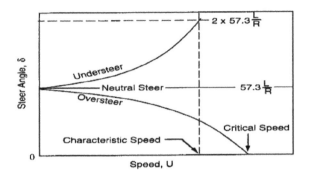

Figure 3. Change of steer angle with speed (Source: [4])

Characteristic speed is experience by understeer vehicle which is the speed at which the steer angle required to negotiate a turn is twice the ackerman angle. While critical speed is associated with oversteer vehicle. It is the speed at the steer angle of a vehicle is zero and above which the vehicle become directionally unstable as shown in Figure 3. It can be represented by the equation below:

$$V_{crit} = \sqrt{-57.3Lg / K} \qquad (9)$$

V_{crit} = critical speed, L = Wheelbase, g = acceleration due gravity, K = Understeer gradient

Rollover stability is another form of vehicle stability, which is the resistance to tipping over in a turn. The "quasi – static" model is used to examine the level of a_y, that causes the inside tire(s) to have zero vertical load when negotiating a turn. This level of lateral acceleration is term the "tipping threshold" and can be represented as (F_c).

2.2. Rollover Stability

The value of F_c for a three wheeler involves the longitudinal placement of CG as well as its height and the vehicle track. Figure 4 shows a three wheeled vehicle negotiating a right turn with vertical reactions at the tire contact point and the lateral friction coefficient multiplied by their vertical loads represent the lateral forces. The Lateral acceleration, ay, multiplied by the vehicle weight, W represents the side load at the CG. The subscript "i" denotes the inside front wheel and "o" denoted the outside front wheel. HG represents the height of the CG, TR represents the front track, and LG is the distance from the front axle rearward to the CG. The CG is assumed to be on the centerline of the vehicle. The vehicle tips over at some value of the side load; however this may not actually occur since the vehicle slides before tipping over. Hence the largest value of lateral acceleration (a_y) that the vehicle will experience is equal to the coefficient of lateral friction at the tires. The lateral forces are simply whatever is needed to keep the vehicle from sliding, for tipping to be examined. This will occur when the inside front wheel leaves the ground or when $WF_i = 0$. At this point, the vehicle is supported by the outside front and rear wheel. This can be expressed as:

$$a_y \geq \frac{TR(WB - LG)}{2(WB)(HG)} = F_c \qquad (6)$$

The expression shows that increasing the front track TR, decreasing LG and lowering the CG all contribute to a larger F_c value.

Table 1. Rollover Threshold for vehicle types

Vehicle Type	CG Height (inches)	Tread (inches)	Rollover Threshold
Sports Car	18-20	50-60	1.2-1.7
Compact car	20-23	50-60	1.1-1.5
Luxury Car	20-24	60-65	1.2-1.6
Pickup Truck	30-34	65-70	0.9-1.1
Passenger Van	30-40	65-70	0.8-1.1
Medium Truck	45-55	65-75	0.6-0.8
Heavy Truck	60-85	70-72	0.4-0.6

Source: [5]

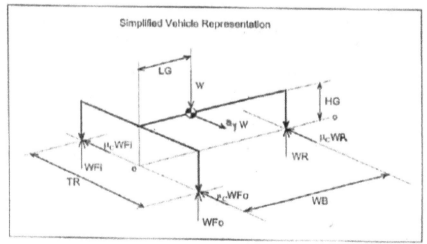

Figure 4. Force diagram for tipping analysis, Right hard turn (Source: [7])

3. Simulation Procedures

A tricycle was drawn and modeled with Solidworks software and its components assembled to form the complete model. The engine, chassis, cabin, tires and bucket in the assembled model were configured using Solidworks material library to their appropriate material for manufacturing. This was to achieve a close estimate of the tricycle weight in reality to enhance efficient weight distribution on the tires.

Table 2. Modeled Tricycle mass Properties

Parameter	Value	Unit
Mass	63676060.99	G
Volume	146456892.21	mm^3
Surface Area	29331589.12	mm^2
Center of Mass (X, Y, Z)	9.75, -15.08, -263.90	Mm
Principal Axis of Inertia Ix	1.00,-0.03,0.03	
Principal Axis of Inertia Iy	0.04, 0.33,-0.94	
Principal Axis of Inertia Iz	0.02, 0.95,0.33	
Principal Moment of inertia Px	257660011537.85	g.mm^2
Principal Moment of inertia Py	636802732968.45	g.mm^2
Principal Moment of inertia Pz	666445377210.40	g.mm^2

The mass of the Tricycle was determined with the Mass Properties capability of the SolidWorks software. The mass properties and moments of inertia of the tricycle is shown in the table below. Also the position of the center of mass (gravity) was represented with the cursor appropriately located at the coordinates X9.75, Y-15.08 and Z -263.90 (Figure 5a&b).

4. Results

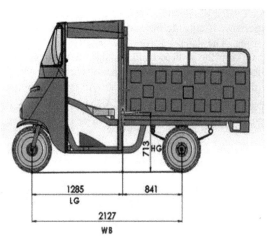
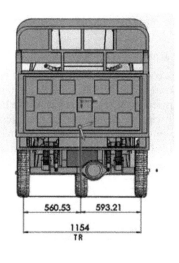

Figure 5. The position of the CG on the tricycle (a) Side View (b) Rear view

At the location of the center of gravity on the tricycle model, the following parameters value were determined in order to evaluate how stable the tricycle will be when manufactured. From Figure 5 above:

Table 3. Simulation results

Parameter	Value	Unit
Wheelbase distance (WB)	2127	mm
Distance from front axle to CG (LG)	1285	mm
LG to WB Ratio (LG/WB)	0.604	
Height of CG from the ground (HG)	713	mm
Track length	1154	mm
Mass of Tricycle	637	Kg
Weight on Front Wheel (W_f)	252.077	Kg
Weight of a Rear wheel	192.46	Kg
Weight on Rear Wheels (W_r)	384.92	Kg
Static Margin (SM)	-0.274; SM <0	
Understeer Gradient (K)	-0.929; K>0	
Lateral acceleration Fc	0.32	g
Critical Speed (V_{crit})	128	mph

5. Discussion

The tricycle remains stable at low speed below its critical speed. At its critical speed and beyond, because the CG of the vehicle falls behind the Neutral Steering Point (NSP) the rear wheel slip angle is larger than the front wheel which makes the Static margin of the tricycle negative. Hence the vehicle respond to side load with a yawing motion i.e. a sideways movement of the CG about its original path of travel is experienced. As a result the tricycle is unstable at speed above its critical speed.

From simulation results, analysis can be made when the tricycle negotiate a turn. At low speed cornering because the tricycle rear axle weight is greater than the front axle, the rear wheel slip angle is larger than the front wheel. Understeer gradient (K) is negative making the tricycle oversteer. This means as speed increases, the steering angle continually decreases until it become zero at the tricycle critical speed. Because tricycle is oversteer, it is unstable above its critical speed when negotiating a turn.

According to Patrick fenner [9], to achieved better rollover stability the CG height for a vehicle should be less than half the track length (HG < TR/2). The values from the tricycle model show that the CG height (HG) is larger than half of the track length (TR). Rollover threshold (F_c) for the tricycle model is 0.32g. Bigger value for F_c is better as it offer more stability.

For lateral and rollover stability to be improved on the tricycle model the value of LG has to be reduced and HG must be lowered which change the positioning of the center of gravity on the tricycle model. This can be achieved by reducing the total volume and weight of the model.

6. Conclusion

A design cargo tricycle model dynamic stability was evaluated. Solidworks software was used for solid modeling and stability analyses. The Mass Properties capability of the SolidWorks Software was used to determine the location of the center of mass on the model. The mass, CG height, track length, and wheelbase measurement were determined and used to calculate values for front & rear weight distribution, static margin, understeer gradient and lateral acceleration for the tricycle model. Result shows that the tricycle model will respond to side load with a yaw motion about its CG at high speed above the critical speed which makes it unstable. Also because the tricycle is an oversteer vehicle, it become unstable during cornering at speed above the critical speed. The tipping threshold for the tricycle model is 0.32g, but can be improve by reducing the value of LG and HG.

Acknowledgement

This project was supported by National Agency for Science and Engineering Infrastructure (NASENI) and National Engineering Design Development Institute (NEDDI).

References

[1] Pearson, J. R. (1989). Medium combination vehicle use in Canada. *Transportation planning and technology*, *14*(2), 125-135.

[2] Wade A. (1986), "Validation of Tires Side Force Coefficient and Dynamic Response Analysis Procedures: Field Test and Analysis Comparison of a Front Wheel Drive Subcompact," Systems Technology Inc., USA.

[3] Robertson, L.S., & Kelley, A.B. (1988). "Static Stability as a Predictor of Overturn in Fatal Motor Vehicle Crashes," *Journal of Trauma*, 29(3), 313-319.

[4] Thomas D.G, Fundamentals of vehicle dynamics. Warrendale, USA. Society of Automotive Engineers.

[5] Mukherjee, S., Mohan, D., & Gawade, T.R (2007). Three-wheeled scooter taxi: A safety analysis *Sadhana*, 32(4), 459-478.

[6] DeLisle, A., Laberge-Nadeau, C., & Brown, B. (1988). Characteristics of three- and four-wheeled all-terrain vehicle accidents in Quebec. *Accid. Anal. & Prev.* 20(5) 357-366.

[7] Van Valkenburgh, P.G., Klein, R.H., & Kanianthra, J. (1982) Three-wheel Passenger Vehicle Stability and Handling. *SAE Trans.* 820140: 604-627.

[8] Metu C., Aduloju S.C., Bolarinwa G.O., Olenyi J., Dania D. E. (2014). Vehicle Body Shape Analysis of Tricycles for Reduction in Fuel Consumption. *Innovative System Design and Engineering* 5(11), 91-99.

[9] Patrick F, (2010), On The Golden Rule of Trike Design. Retrieved from www.deferredprocrastination.co.uk.

[10] Patrick J. F, (2006), Designing Stable Three Wheeled Vehicles, With Application to Solar Powered Racing Cars. Retrieved from www.academia.edu/687820.

8

Investigation of Metal and Metal Oxide nanocoating on Fins in HPHE with Silver Water NanoFluid

Aysar A. Alamery[*], Zainab F. Mahdi, Hussein A. Jawad

Institute of laser for postgraduate studies, University of Baghdad, Iraq
*Corresponding author: aysar_Alamery@yahoo.com

Abstract An improvement of energy saving and the heat transfer characteristics via introducing nanocoating of fins is a new method in HVAC systems. ANSYS software is the preferred choice to design a model for nanocoating on fins in Thermosyphon Heat Exchangers. The investigation of thermal resistance and temperature distribution are presented. The influence of enhancing the working system is examined using different nanocoating metal and metal oxide materials of fins (Ag, Cu, AL, BeO, Al_2O_3 and CuO) in heat pipe. The effect of Nanocoating by metals is explored in which the maximum improvement occurred at the evaporator section in transient operation condition with 71.9585 for AL, 71.383 for Cu and 70.708 for Ag while, the effect of Nanocoating by metal oxide is (73.79 for CuO, 73,41 for AL_2O_3 and 71.96 for BeO). The obtained results showed that the best nanocoat material for this purpose is the aluminum for metals and copper oxide for metal oxides. The calculations give important indication that using the nanocoating just in the evaporator section to reduce the cost. The large amount of heat transfer occur in the evaporator section according to the result of thermal resistance in this section then large benefit of latent heat of the air is guiding to increase the energy saving when using HPHE in HVAC systems.

Keywords: *HVAC, HP (Thermosyphone) HE,ANSYS, Nanocoat, temperature distribution, Thermal resistance*

1. Introduction

Heat pipes have been utilized in heat transfer related applications for many years. Depending on their application area, they can operate over a wide range of temperatures with a high heat removal capability [1]. Heat pipes are heat transfer devices which use the principles of thermal conduction and latent heat of vaporization to transfer heat effectively at very fast rates [2]. It is essentially a passive heat transfer device with an extremely high effective thermal conductivity [3] and very efficient for the transport of heat with a small temperature difference via the phase change of the working fluid [4]. The two-phase heat transfer mechanism results in heat transfer capabilities from one hundred to several thousand times that of an equivalent piece of copper [3]. Key factors affecting on thermal performance of a HPHE are: velocity, relative humidity (RH) and dry-bulb temperature (DBT) of input air, type and filling ratio (FR). Heat pipe technology has found increasing applications in enhancing the thermal performance of heat exchangers in microelectronics, energy saving in HVACs [5]. It can be used for operating rooms, surgery centers, hotels, cleanrooms etc, temperature regulation systems for the human body and other industrial sectors and are passive components used to improve dehumidification by commercial forced-air HVAC systems. They are installed with one end upstream of the evaporator coil to pre-cool supply air and one downstream to re-heat supply air. This allows the system's cooling coil to operate at a lower temperature, increasing the system latent cooling capability. Heat rejected by the downstream coil reheats the supply air, eliminating the need for a dedicated reheat coil. Heat pipes can increase latent cooling by 25-50% depending upon the application. [6]. **P.G. Anjankar and Dr.R.B.Yarasu. (2012)** [7] studied the new design and thermal performance of thermosyphon. **W. Srimuang et al. (2012)** [8] presented the knowledge of two-phase closed thermosyphon (TPCT) as being used nowadays and the application of TPCT to air to air heat exchanger. **H.A. Mohammed et al. (2013)** [9] studied numerically the effect of using louvered strip inserts placed in a circular double pipe heat exchanger on the thermal and flow fields utilizing various types of nanofluids.

The major aim is to examine the combined internal (heat pipe) and external (fin surface) effects of HPHE when Nanocoating materials of metal and metal oxide are done.

2. Modeling of Heat Pipe Heat Exchanger

The simulation of the heat pipe is achieved from the real design [10]. Dimensions of the model for the heat pipe were constructed in ANSYS _Pre_ prossor package [11].

The top and bottom layers are represented by the nanocoating, Six nanocoating with metal and metal oxide materials (Ag, Cu, AL, BeO, Al$_2$O$_3$, CuO) of fins respectively with their thermophysical properties [12,13]. These nanoparticles are used in coating with the aim of improving the thermal properties of the base metal (aluminum).

3. Boundary Conditions

The Boundary condition of the tube is used as temperature distribution along HP which can be obtained by phase change properties from liquid (sliver water nano fluid) in evaporator section to the vapor in the condenser section) [14]. Boundary condition of the thermal model were specified as surface loads through ANSYS codes [11]

4. Results and Discussion

The results which approach from the Ansys program after modeling the heat pipe are shown in Table 1. For All tested fins, two fixed positions were taken in evaporator according to the direction of air flow.

Table 1. Temperature distribution results which approach from the Ansys program when the program in steady and transient conditions of operation runs after the modeling of the heat pipe with all selected materials

HPHE without Nanocoat (Steady State operation Condition)				
Total HPHE	Bottom of Evap	Top of Evap	Bottom of Cond	Bottom of Cond
HPHE without Nanocoat (Transient operation Condition)				
Metal nanocoating materials(Steady State operation Condition)				
NanoAg				
NanoCu				
NanoAl				
Metal nanocoating materials(Transient operation Condition)				
NanoAg				
NanoCu				
NanoAl				

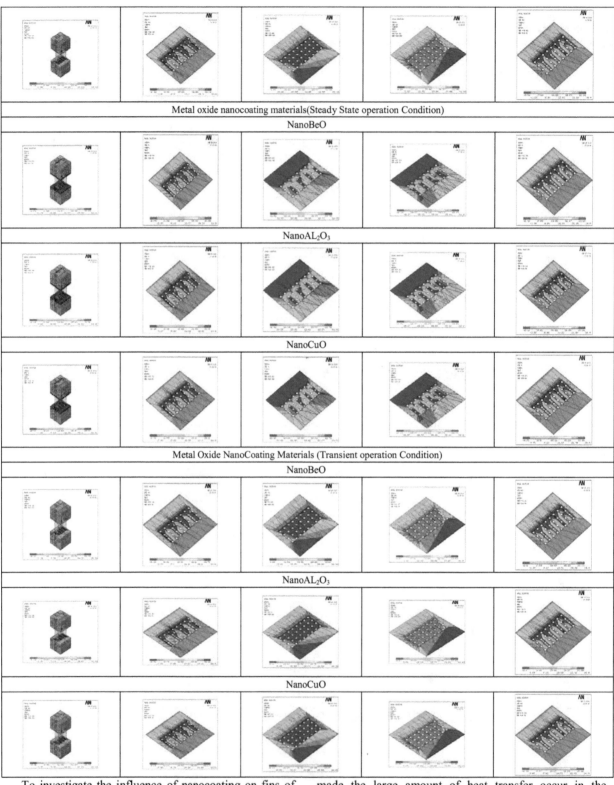

Metal oxide nanocoating materials(Steady State operation Condition)

NanoBeO

NanoAL$_2$O$_3$

NanoCuO

Metal Oxide NanoCoating Materials (Transient operation Condition)

NanoBeO

NanoAL$_2$O$_3$

NanoCuO

To investigate the influence of nanocoating on fins of heat pipe performance, all the thermal resistances were determined for the evaporator and condenser sections separately and the sum was considered as the total thermal resistance of the heat pipe. The respective average temperature of evaporator and condenser are recorded to calculate the system thermal resistance. Results in Figure 1 show the decrease of the thermal resistance along the evaporator section and it becomes zero in the adiabatic section after that a small increase occurs in the condenser section compared with the evaporator section this what made the large amount of heat transfer occur in the evaporator section compared with condenser.

Figure 2 Show temperature distribution of two selected positions on fins (with and without metal nanocoating materials) in the evaporator and condenser section with the and. From these figures the results observed the small effect of nanocoat in decrease thermal resistance in the evaporator section while the large effect was observed in the condenser section with the steady and transient operation conditions. Gradual decreasing For AL but fast decreasing of R for other types of nanocoating metals.

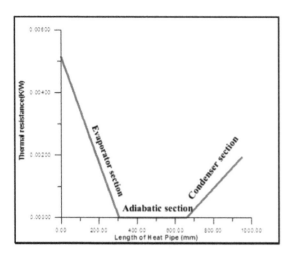

Figure 1. Thermal resistance a long heat pipe with metal and metal oxide nanocoating materials

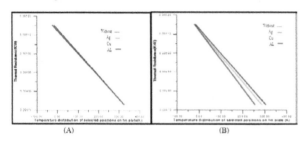

Figure 2. A,B Temperature distribution in two selected positions on fins (with and without metal nanocoating materials)in the evaporator section with the. A-Evaporator in steady state conditions. B- Evaporator in transient conditions

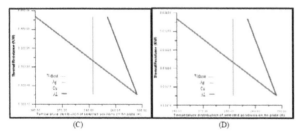

Figure 2. C,D Temperature distribution in two selected positions on fins (with and without metal nanocoating materials)in the condenser section with the. C-Condenser in Steady State Conditions. D-Condenser in Transient Conditions

After calculations of the percentages of thermal performance enhancement with nanocoating layers by using the equation ((Without Nanocoat-With Nanocoat)/(Without Nanocoat))*100 the results were arranged in Table 2. As observed from the values the using of nanocoat were successful in the evaporator section especially in transient case of operation which is the fact and the aluminum is the best metal.

Table 2. Percentages of thermal performance enhancement with metals nanocoating layers

Evaporator steady state			Evaporator Transient		
Ag	Cu	AL	Ag	Cu	AL
-0.0626	-0.066	-0.66	70.708	71.383	71.9585
-0.034	-0.0337	-0.0337	5.909	15.857	15.938
Condenser steady state			Condenser Transient		
Ag	Cu	AL	Ag	Cu	AL
5.5	-0.005	-0.5	0.84	0.92	1.03
1.94	9.41	9.37	9.39	9.39	9.35

Figure 3 Show temperature distribution of two selected positions on fins (with and without metal oxide nanocoating materials) in the evaporator and condenser section with the R_e and R_c. From these figures the results observed the negative effect of nanocoat in decrease thermal resistance in the evaporator section when the HPHE operate in the steady state conditions. The small effect in decrease thermal resistance was observed in the evaporator section with transient operation conditions. In condenser section for both conditions of operation, gradual and early decreasing For all types of nanocoating metal oxides compared with fins without nanocoat.

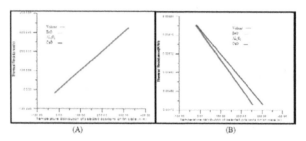

Figure 3. A,B Temperature distribution in two selected positions on fins (with and without metal oxide nanocoating materials) in the evaporator section with the R_e. A-Evaporator in steady state conditions. B-Evaporator in transient conditions

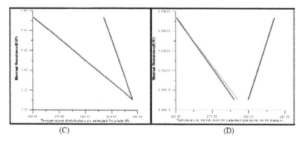

Figure 3. C,D Temperature distribution in two selected positions on fins (with and without metal oxide nanocoating materials) in the condenser section with the. C-Condenser in Steady State Conditions. D-Condenser in Transient Conditions

Table 3. Percentages of thermal performance enhancement with metal oxide nanocoating layers

Evaporator steady state			Evaporator Transient		
BeO	AL2O3	CuO	BeO	AL2O3	CuO
-0.66	-0.66	-0.66	71.96	73.41	73.79
-0.034	-0.034	-0.034	15.938	16.13	16.18
Condenser steady state			Condenser Transient		
BeO	AL2O3	CuO	BeO	AL2O3	CuO
9.385	9.385	9.29	9.37	9.28	9.28
9.385	9.385	9.29	9.37	9.28	9.28

After calculations of the percentages of thermal performance enhancement with nanocoating layers by using the equation ((Without Nanocoat-With Nanocoat)/(Without Nanocoat))*100 the results were arranged in Table 3. As observed from the values the using of metal oxide nanocoat were successful in the evaporator section especially in transient case of operation and copper oxide is the best one.

5. Conclusions

To improve the thermal conductivity of heat pipes coat fins by nano materials, The effective surface area of heat flux absorbent should be increased and increasing the effective thermal conductivity of the fluid. The thermal performance of fins coating by nano metal and metal oxide materials is investigated by ANSYS software, The effect of Nano coating on temperature distribution on the fins were explored in which the maximum enhancement occurred at the evaporator section in transient operation condition with 71.9585 for AL, 71.383 for Cu and 70.708 for Ag. While The effect of Nano coating by metal oxide on temperature distribution on the fins were explored in which the maximum enhancement occurred at the evaporator section in transient operation condition too with (73.79 for CuO, 73,41 for AL_2O_3 and 71.96 for BeO). The best nanocoat material for this purpose is the aluminum for metals and copper oxide for metal oxides especially in transient case of operation.

References

[1] M. Shafahi, V. Bianco, K. Vafai and O. Manca, "Thermal performance of flat-shaped heat pipes using nanofluids", International Journal of Heat and Mass Transfer 53, 1438-1445, (2010).

[2] S. Ravitej Raju, M. Balasubramani, B. Nitin Krishnan,K. Kesavan and M. Suresh, "Numerical Studies On The Performance Of Methanol Based Air To Air Heat Pipe Heat Exchanger", International Journal of ChemTech Research, Vol. 5, No. 2, pp. 925-934, April-June 2013.

[3] A. Khare, A. Paul and G.Selokar, "Design Development of Test-Rig to Evaluat performance of Heat Pipes in Cooling of Printed Circuit Boards", VSRD-MAP, Vol. 1 (2), pp. 65-79, 2011.

[4] M. R. SARMASTI EMAMI, S. H. NOIE AND M. KHOSHNOODI, "Effect of Aspect Ratio and Filling Ratio on Thermal Performance of an Inclined Two-Phase Closed Thermosyphone", Iranian Journal of Science & Technology,

Transaction B, Engineering, Printed in The Islamic Republic of Iran, Vol. 32, No. B1, pp. 39-51, 2008.

[5] E. Firouzfar, M. Soltanieh, S. H. Noie and M. H. Saidi, "Investigation of heat pipe heat exchanger effectiveness and energy saving in air conditioning systems using silver nanofluid", Int. J. Environ. Sci. Technol. 9: 587-594.

[6] E. Firouzfar and M. Attaran," A review of heat pipe heat exchangers activity in Asia", World Academy of Science, Engineering and Technology 47, 2008.

[7] P.G. Anjankar and R.B.Yarasu, "Experimental analysis of condenser length effect on the performance of thermosyphon", International Journal of Emerging Technology and Advanced Engineering, ISSN 2250-2459, Volume 2,, March 2012.

[8] W. Srimuang, P. Khantikomol and P. Amtachaya, "Two phase closed thermosyphon (TPCT) and Its application for an air-to-air heat exchanger", The Journal of KMUTNB., Vol. 22, No. 1, Jan.-Apr. 2012.

[9] H.A. Mohammed, H. A. Hasan and M.A. Wahid," Heat transfer enhancement of nanofluids in a double pipe heat exchanger with louvered strip inserts", International Communications in Heat and Mass Transfer 40, pp. 36-46,2013.

[10] Y.H. Yau, "Experimental thermal performance study of an inclined heat pipe heat exchanger operating in high humid tropical HVAC systems", International Journal of Refrigeration 1143-115230 (2007).

[11] A. A. Alamery, H. A. Jawad, H. A. Ameen and Z.F. Mahdi, "Effect of Hard material Nanocoating on Fins in (Thermosyphon) HPHE with Nano Working Fluid", International Journal of Engineering Research & Technology (IJERT), Vol. 3 Issue 6, June-2014.

[12] J. H.Lienhard IV and John H.Lienhard V, "A heat transfer textbook", Third edition, 2001.

[13] S. Zhang and N. Ali, "Nanocomposite thin films and coatings processing, properties and performance", Copyright by Imperial College Press, 2007.

[14] S.H. Noie, S. Zeinali Heris, M. Kahani and S.M. Nowee, "Heat transfer enhancement using Al2O3/water nanofluid in a two-phase closed thermosyphon", International Journal of Heat and Fluid Flow 30 pp. 700-705, 2009.

9

Computational Study of the Turbulent Flow inside a Waste Heat Recovery System with a 25° inclined Angle Diffuser

Sobhi FRIKHA, Zied DRISS[*], Mohamed Aymen Hagui

Laboratory of Electro-Mechanic Systems (LASEM), National School of Engineers of Sfax (ENIS), University of Sfax (US), B.P. 1173, Road Soukra km 3.5, 3038 Sfax, TUNISIA
*Corresponding author: Zied.Driss@enis.rnu.tn

Abstract In this paper, we are interested on the study of the turbulent flow inside a waste heat recovery system with a 25° inclined diffuser. For thus, we have developed a numerical simulation using a CFD code. Particularly, we are interested to visualize the temperature, the velocity, the total pressure, the dynamic pressure, the vorticity, the turbulent kinetic energy, the turbulent dissipation rate and the turbulent viscosity. The numerical model is based on the resolution of the Navier-Stokes equations in conjunction with the standard k-ε turbulence model. These equations were solved by a finite volume discretization method.

Keywords: waste heat recovery, inclined diffuser, heat exchanger, power generator, water heating

1. Introduction

Exhaust of the engines is a main source that a large amount of energy wastes through it.. Researchers confirm that more than 30–40% of fuel energy wastes from the exhaust and just 12–25% of the fuel energy convert to useful work [1,2,3]. The dumped thermal energy can be recovered and a heat exchanger is necessary to transmit the heat from hot gases to working fluid at excellent efficiency. A heat exchanger is thermal equipment, which is built for efficient heat transfer between two fluids of different temperatures. Heat exchangers are used mainly in industrial sectors (chemicals, petrochemicals, steel, food processing, energy production, etc) and transportation (automotive, aeronautics), but also in the residential sector and tertiary (heating, air conditioning, etc). The choice of a heat exchanger, for a given application depends on many parameters such as field temperature and pressure of fluids, and physical properties of these fluids, maintenance, cost and space. In the literature, we can find relevant studies on heat exchanger networks [4-9]. CFD numerical simulations have been used with success to better understand the flow and heat transfer behavior in particular configurations of heat exchangers and to improve heat exchanger designs. Deng et al. [10] designed two thermoelectric exchangers models shown by CFD simulation and used Wilcox k-ω model to discuss on different internal structures, lengths and materials on the exchangers performance, The same study has been performed by Kumar et al. [11] which modeled three exchangers (rectangular, triangular and hexagonal) by FLUENT software and experimentally produced and tested the best model. Zhang and Li [12] proposed a structure of two-stage-distribution and the numerical investigation shows that the flow distribution in plate-fin heat exchanger is more uniform if the ratio of outlet and inlet equivalent diameters for both headers is equal. Wen [13] employed CFD technique to simulate and analyze the performance of fluid flow distribution and pressure drop in the header of plate-fin heat exchanger. Wasewar [14] studied the flow distribution through a plate-fin heat exchanger by using FLUENT. A modified header is proposed and simulated using CFD. The modified header configuration has a more uniform flow distribution than the conventional header configuration. Hence, the efficiency of the modified heat exchanger is seen to be higher than that of the conventional heat exchanger. Gan et al. [15] performed a CFD simulation on the tubes of a heat exchanger used in closed-wet cooling towers. Pressure drop was found to depend on the tube configurations and water to air ratio. Predicted pressure loss coefficient was found inversely proportional to transverse pitch but was in direct relationship with water to air mass flow rate. Sheik Ismail et al. [16] numerically investigated the effects of nozzle and header orientation on the hydrodynamic performance of a plate-fin heat exchanger, their studies showed that the orientation of the header and nozzle plays a major role in the exchanger performance. In this context, we are interested in studying the turbulent flow inside a waste heat recovery system with a 25° inclined diffuser.

2. Numerical Model

The computational domain is shown in Figure 1. It is defined by the interior volume of the gas flow and by the two diffusers. It is limited by the inlet and the outlet of the water. In the inlet of the gas and the water, the mass flows are equals to 0.0335 kg.s^{-1} and 0.07 kg.s^{-1} respectively, and the temperature values are defined by T=526°C and T=10°C respectively. In the outlet, we consider the atmospheric pressure of 101325 Pa.

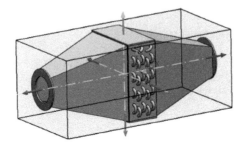

Figure 1. Computational domain

In this study, we have used the Software "Solid Works Flow Simulation" for calculation. The k-ε turbulence model has been considered for the analysis of turbulent flow. In fact, this model has been used in different anterior works and satisfactory results were obtained [17-23]. We have used the Local Initial Mesh option. This mesh option allows us to specify an initial mesh in a local region of the computational domain to better resolve the model geometry and flow particularities in this region. Local mesh settings are applied to all cells intersected by a component, face, edge, or a cell enclosing the selected vertex. We have used two local meshing for the entry diffuser and the heat exchanger and a basic meshing for the outlet meshing (Figure 2). In these conditions, the cells size for the local mesh is equal to 0.001 m and for the basic mesh is equal to 0.01 m.

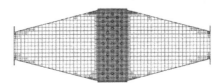

Figure 2. Meshing

3. Numerical Results

Two longitudinal planes defined by z=0 m, y=0 m and three transverse planes defined by x=-0.2 m, x=0 m, and x= 0.2 m are considered. Particularly, we are interested to visualize the temperature, the velocity, the total pressure, the dynamic pressure, the vorticity, the turbulent kinetic energy, the turbulent dissipation rate and the turbulent viscosity.

3.1. Temperature

Figure 3 presents the distribution of the temperature in the longitudinal planes defined by z=0 m and y=0 m. According to these results, it is clear that the temperature is at its maximum at the entry of the diffuser, which is the value of the inlet boundary condition. The temperature

shows a decrease at the sides of the first diffuser and through the heat exchanger. The decreased value of the temperatures is the quantity of heat transported to the water in the heat exchanger. After going out of the heat exchanger, it is clear that the temperature of the gas decreases but still represents important values.

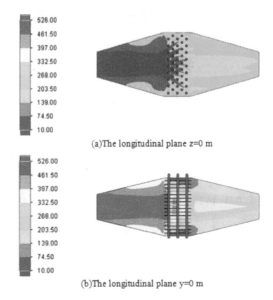

(a)The longitudinal plane z=0 m

(b)The longitudinal plane y=0 m

Figure 3. Distribution of the temperature in the longitudinal planes

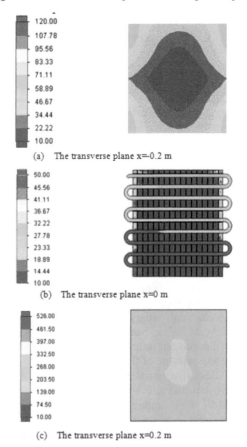

(a)   The transverse plane x=-0.2 m

(b)   The transverse plane x=0 m

(c)   The transverse plane x=0.2 m

Figure 4. Distribution of the temperature in the transverse planes

Figure 4 presents the distribution of the temperature in the transverse planes defined by x=-0.2 m, x=0 m and x=0.2 m. In the first plane x=-0.2 m, the temperature distribution is located in the upstream of the heat exchanger. However, in the third plane x=0.2 m, the temperature distribution is located in the downstream. According to these results, the high temperatures are concentrated in the middle of the diffusers and there is a decrease of the temperature along the sides of the diffusers. The row defined by the plane x=0 m presents the evolution of the water in the middle of the heat exchanger. In these conditions, the water reaches a temperature equal to T=50°C. The last transverse plane defined by x=0.2 m presents only the gas through the second diffuser. All these figures show a decrease in the value of the gas temperature. This fact is due to the heat loss through the heat exchanger. All the temperatures are between 200°C and 300°C. In the diffuser sides, it has been observed a heat loss more superior.

3.2. Magnitude Velocity

Figure 5 presents the distribution of the magnitude velocity in the longitudinal planes defined by z=0 m and y=0 m. According to these results, the maximum value of the velocity appears in the gas inlet which is imposed by the boundary conditions. The magnitude velocity shows a decrease in the first diffuser. Also, a decrease has been noted on the sides of the first diffuser which there is an important drop of the magnitude velocity. Out of the heat exchanger, the magnitude velocity value decreases in the second diffuser. At the end of the diffuser, an increase of the magnitude velocity value has been noted due to the reduction of the diffuser section. Indeed, it is clear that through the reduction of the size of the diffuser, the magnitude velocity value increases at the end of the diffuser.

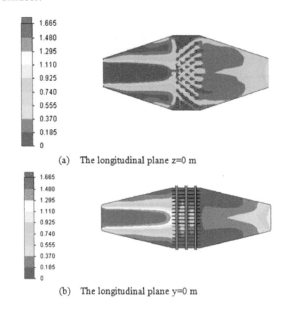

(a) The longitudinal plane z=0 m

(b) The longitudinal plane y=0 m

Figure 5. Distribution of the velocity in the longitudinal planes

Figure 6 presents the magnitude velocity in the transverse planes defined by x=-0.2 m, x=0 m and x=0.2 m. In the planes x=-0.2 m and x=0.2 m, the maximum value of the magnitude velocity is shown in the middle of the diffuser. A progressive decrease of the magnitude velocity is shown at the two sides of the diffuser. In the transverse plane defined by x=0 m, the magnitude velocity in the tubes is constant and do not changes through this plane.

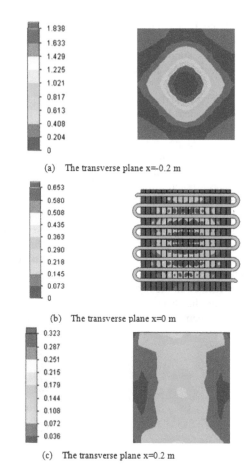

(a) The transverse plane x=-0.2 m

(b) The transverse plane x=0 m

(c) The transverse plane x=0.2 m

Figure 6. Distribution of the velocity in the transverse planes

3.3. Total Pressure

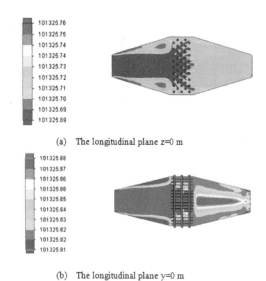

(a) The longitudinal plane z=0 m

(b) The longitudinal plane y=0 m

Figure 7. Distribution of the total pressure in the longitudinal planes

Figure 7 presents the distribution of the total pressure in the longitudinal planes defined by z=0 m and y=0 m. According to these results, a compression zone characteristic of the maximum value of the total pressure has been observed on the heat exchanger upstream and the middle of the diffuser. A progressive decrease has been observed on the sides of the first diffuser. A decrease of the total pressure has also been noted at the second diffuser and the heat exchanger downstream. The total pressure decreases out of the heat exchanger. In fact, the total pressure has approximately the same value and do not have a great value change in all the length of the diffuser.

Figure 8 presents the distribution of the total pressure in the transverse planes defined by x=-0.2 m, x=0 m and x=0.2 m. While examining these results, it can easily be noted that the total pressure is on its maximum in the intake and is globally uniform in the diffuser for the first plane defined by x=-0.2 m. A pressure drop of the total pressure has been noted in the sides of the diffuser. In the plane defined by x=0.2 m, the distribution of the total pressure shows that the maximum values of the total pressure are located in the middle of the diffuser, and decreases in the sides. In the transverse plane defined by x=-0.2 m and x=0 m, results confirm that the distribution of the total pressure is uniform in the tube and the difference is located in the heat exchanger. An increase of the total pressure has been observed in the middle. Along the sides, it has been noted a decrease of the total pressure.

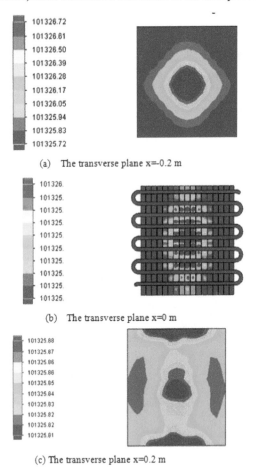

(a)   The transverse plane x=-0.2 m

(b)   The transverse plane x=0 m

(c) The transverse plane x=0.2 m

Figure 8. Distribution of the total pressure in the transverse planes

3.4. Dynamic Pressure

Figure 9 presents the distribution of the dynamic pressure in the longitudinal planes defined by z=0 m and y=0 m. These results show that the compression zone characteristics of the maximum values of the dynamic pressure are localized in the heat exchanger and the middle of the diffuser. A decrease of the dynamic pressures values appear in the sides of the diffuser and in

the second diffuser. Indeed, the dynamic pressure decreases out of the heat exchanger. In the second diffuser and through the gas flow, the dynamic pressure decreases also. This fact is due to the difference in the section and the size of the diffuser. In the end of the second diffuser, a progressive increase of the dynamic pressure has been observed.

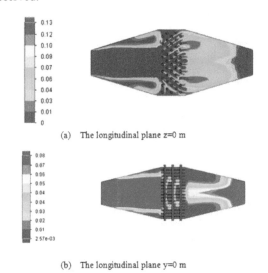

(a)   The longitudinal plane z=0 m

(b)   The longitudinal plane y=0 m

Figure 9. Distribution of the dynamic pressure in the longitudinal planes

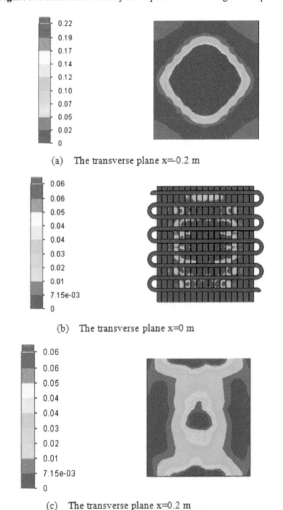

(a)   The transverse plane x=-0.2 m

(b)   The transverse plane x=0 m

(c)   The transverse plane x=0.2 m

Figure 10. Distribution of the dynamic pressure in the transverse planes

Figure 10 presents the distribution of the dynamic pressure in the transverses planes defined by x=-0.2 m, x=0 m and x=0.2 m. According to these results, it has been noted that the dynamic pressure is on its maximum in the intake and is globally uniform in the diffuser. A pressure drop has been noted in the sides of the diffuser. In the plane defined by x=0.2 m, it is clear that the depression zones are located in the middle and the superior sides of the diffuser. In the transverse planes defined by x=-0.2 m and x=0 m, the distribution of the dynamic pressure is uniform in the tube and there is a difference in the heat exchanger. Also, an increase of the dynamic pressure has been observed in the middle of the heat exchanger. However, along the sides, it has been noted a decrease of the dynamic pressure.

3.5. Vorticity

Figure 11 presents the distribution of the vorticity in the longitudinal planes defined by z=0 m and y=0 m. According to these results, the vorticity has been observed in the entry of the gas inlet. In the middle of the diffuser, the vorticity is low and does not presents an important value. The vorticity presents an increase in the middle of the heat exchanger near the obstacles like the fin plates. In the heat exchanger downstream, the vorticity importantly decreases. With the second diffuser an increase of the vorticity has been noted in the outlet of the diffuser.

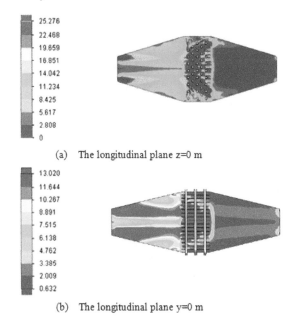

(a)   The longitudinal plane z=0 m

(b)   The longitudinal plane y=0 m

Figure 11. Distribution of the vorticity in the longitudinal planes

Figure 12 presents the distribution of the vorticity in the transverse planes defined by x=-0.2 m, x=0 m and x=0.2 m. According to these results, the wake characteristic of the maximum values of the vorticity has been observed on the sides of the diffuser. Far the side of the diffuser, the vorticity value decreases. In the plane x=0 m, the vorticity presents higher values in the elbow which can produce turbulence of the water. In the heat exchanger the vorticity is quite high in the middle and starts to decrease on the sides. In the plane x=0.2 m, the gas goes out of the heat exchanger. The vorticity is shown on the sides of the diffuser. Also, a progressive decrease has been observed in the middle of the diffuser.

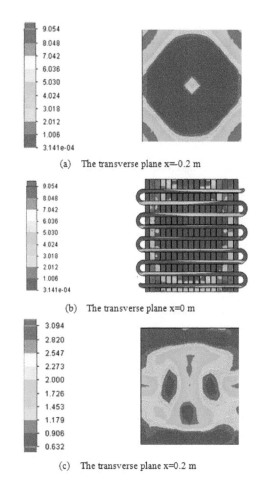

(a)   The transverse plane x=-0.2 m

(b)   The transverse plane x=0 m

(c)   The transverse plane x=0.2 m

Figure 12. Distribution of the vorticity in the transverse planes

3.6. Turbulent Kinetic Energy

Figure 13 presents the distribution of the turbulent kinetic energy in the longitudinal planes defined by z=0 m and y=0 m. The results show a wake characteristic of the maximum values of the turbulent kinetic energy in the diffuser. The results show also a decrease in the value of the turbulent kinetic energy almost at the end of the heat exchanger and at the entry of the second diffuser. After then, the turbulent kinetic energy decreases progressively.

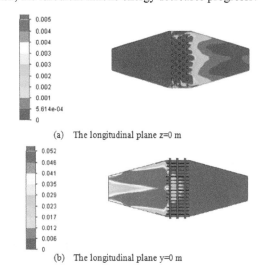

(a)   The longitudinal plane z=0 m

(b)   The longitudinal plane y=0 m

Figure 13. Distribution of the turbulent kinetic energy in the longitudinal planes

Figure 14 presents the distribution of the turbulent kinetic energy in the transverse planes defined by x=-0.2 m, x=0 m and x=0.2 m. The results, presented in the plane x=-0.2 m, show a wake characteristic of the maximum value of the turbulent kinetic energy in the sides of the diffuser. In the middle of the diffuser, the turbulent kinetic energy decreases. Also, a progressive decrease in the far sides of the diffusers has been noted. The plane x=0 m presents a wake characteristic of the maximum value of the turbulent kinetic energy. In the sides, a progressive decrease of the turbulent kinetic energy has been observed. The turbulent kinetic energy in the tubes presents low values. The plane x=0.2 m presents the turbulent kinetic energy in the second diffuser where the gas goes out of the heat exchanger. In this zone, the turbulent kinetic energy values are very low.

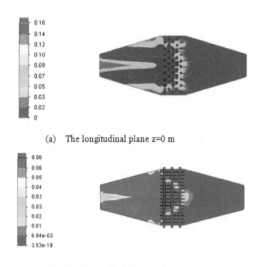

(a)   The longitudinal plane z=0 m

(b)   The longitudinal plane y=0 m

Figure 15. Distribution of the dissipation rate of the turbulent energy in the longitudinal planes

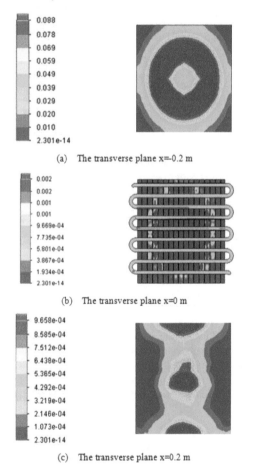

(a)   The transverse plane x=-0.2 m

(b)   The transverse plane x=0 m

(c)   The transverse plane x=0.2 m

Figure 14. Distribution of the turbulent energy in the transverse planes

3.7. Dissipation Rate of the Turbulent Kinetic Energy

Figure 15 presents the distribution of the dissipation rate of the turbulent kinetic energy in the longitudinal planes defined by z=0 m and y=0 m. The results show a wake characteristic of the maximum values of the dissipation rate of the turbulent kinetic energy in the sides of the diffuser. In the middle of the diffuser and in the gas inlet, the dissipation rate of the turbulent kinetic energy decreases. The wake extension has been observed until the gas leaves the heat exchanger where there is a great drop on the dissipation rate of the turbulent kinetic energy values.

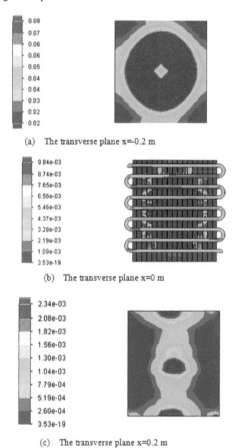

(a)   The transverse plane x=-0.2 m

(b)   The transverse plane x=0 m

(c)   The transverse plane x=0.2 m

Figure 16. Distribution of the dissipation rate of the turbulent energy in the transverse planes

Figure 16 presents the dissipation rate of the turbulent kinetic energy in the transverse planes x=-0.2 m, x=0 m and x=0.2 m. In the plane x=-0.2 m, the dissipation rate of the turbulent kinetic energy has been observed in the sides of the diffusers. After that, the dissipation rate of the turbulent kinetic energy decreases progressively until they reach the limit of the diffusers. In the middle of the diffuser, the values of the dissipation rate of the turbulent kinetic energy are low. The plane x=0 m shows a wake characteristic of the maximum values of the dissipation rate of the turbulent kinetic energy in the middle of the planes. The dissipation rate of the turbulent kinetic energy decreases in the other sides of the diffuser. The dissipation rate of the turbulent kinetic energy in the tubes is very low and does not presents a significant effect. The plane x=0.2 m presents the dissipation rate of the turbulent kinetic energy in the second diffuser in the exit of the gas from the heat exchanger. These figures present very low values of dissipation rate of the turbulent kinetic energy. According to these results, it is clear that the values of the dissipation rate of the turbulent kinetic energy increases in the middle of the heat exchanger and then it starts to decrease progressively until the sides of the diffuser.

3.8. Turbulent Viscosity

Figure 17 presents the distribution of the turbulent viscosity in the longitudinal planes defined by z=0 m and y=0 m. According to these results, a wake characteristic of the maximum values of the turbulent viscosity has been observed in the second diffuser in the outlet of the heat exchanger. A decrease in the value of the turbulent viscosity has been noted in the gas inlet for the first diffuser. Indeed, the turbulent viscosity values decrease importantly in the heat exchanger. The results show an increase in the middle of the diffuser and a decrease in the sides of the diffuser.

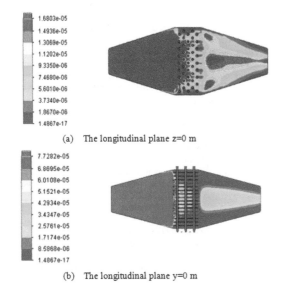

(a)   The longitudinal plane z=0 m

(b)   The longitudinal plane y=0 m

Figure 17. Distribution of the turbulent viscosity in the longitudinal planes

Figure 18 presents the distribution of the turbulent viscosity in the transverse planes defined by x=-0.2 m, x=0 m and x=0.2 m. While examining these results, a wake characteristic of the maximum values of the turbulent viscosity has been observed in the middle of the diffuser. In the sides of the diffuser, the turbulent viscosity decreases. According to the plane defined by x=0 m, an increase in the middle of the heat exchanger has been noted and a progressive decrease has been observed in the sides. The plane defined by x=0.2 m presents the distribution of the turbulent viscosity in the second diffuser after the heat exchanger outlet. The results show an increase of the values of the turbulent viscosity in the middle of the heat exchanger and present a progressive decrease in the sides.

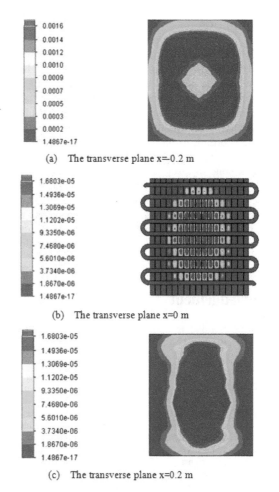

(a)   The transverse plane x=-0.2 m

(b)   The transverse plane x=0 m

(c)   The transverse plane x=0.2 m

Figure 18. Distribution of the turbulent viscosity in the transverse planes

4. Comparison with Analytical Results

Figure 19 presents the evolution of the water temperature profiles through the tube for the inclined diffuser. These profiles present the same evolution of the curve. In fact these profiles present a progressive increase of the water temperature through the tube. The comparison with the analytic method shows a good agreement and confirms the validity of the numerical method.

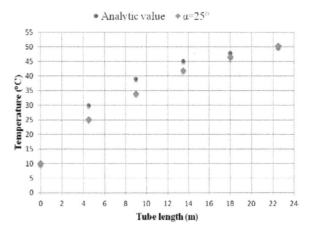

Figure 19. Water temperatures profile through the tube for the inclined diffuser

5. Conclusions

In this work, numerical simulations have been developed to study the turbulent flow inside a heat exchanger using an inclined diffuser. We present all the results from simulation, such as temperature, velocity, total pressure, dynamic pressure, vorticity, turbulent kinetic energy, turbulent dissipation rate and turbulent viscosity. According to the numerical results, fluid flow characteristics decrease in the sides of the first diffuser and in the second diffuser out of the heat exchanger. The evolution of the water temperature profile through the tube results for the inclined diffuser is compared to analytical profile. A good agreement was obtained and confirms the validity of the numerical method. In the future, we propose to study the shape effect on the turbulent flow inside the heat exchanger.

Acknowledgement

This work is done in collaboration with the Tunisian drilling company.

References

[1]   Hatami, M., Ganji, D.D., Gorji-Bandpy, M., "A review of different heat exchangers designs for increasing the diesel exhaust waste heat recovery", Renew. Sust. Energ Rev. 37 168-181, 2014.

[2]   Saidur, R., Rahim, N.A., Ping, H.W., Jahirul, M.I., Mekhilef, S., Masjuki, H.H., "Energy and emission analysis for industrial motors in Malaysia", Energy Policy, 37 (9), 3650-3658, 2009.

[3]   Hasanuzzaman, M., Rahim, N.A., Saidur, R., Kazi, S.N., "Energy savings and emissions reductions for rewinding and replacement of industrial motor", Energy, 36 (1), 233-240, 2011.

[4]   Caputo, A.C., Pelagagge, P.M., Salini, P., "Heat exchanger design based on economic optimization", Applied Thermal Engineering, 28, 1151-1159, 2008.

[5]   Dagdas, A., "Heat exchanger optimization for geothermal district heating systems: a fuel saving approach", Renewable Energy, 32, 1020-1032, 2007.

[6]   Borekal, M.U., Sapkal, V.S., Sapkal, R.S., Study of optimum design-parameters of a condensing heat exchanger for waste heat recovery, Advances in Energy Research 245-249, 2006.

[7]   Navarro, A.H., Gomez, L.C., "A new approach for thermal performance calculation of cross-flow heat exchangers", Int. Journal of Heat and Mass Transfer, 48, 3880-3888, 2005.

[8]   Sun, S.Y., Lu, Y.D., Yan, C.Q., "Optimization in calculation of shell tube heat exchanger", Int. Communications in Heat and Mass Transfer, 20 (5) 675-687, 1993.

[9]   Soylemez, M.S., "On the optimum heat exchanger sizing for heat recovery", Energy Conversion and Management, 41, 1419-1427, 2000.

[10]   Deng YD, Liu X, Chen S, Tong NQ. "Thermal optimization of the heat exchanger in an automotive exhaust-based thermoelectric generator". J Electron Mater, 42-47, 2013.

[11]   Ramesh Kumar C, Ankit Sonthalia Rahul Goel. "Experimental study on waste heat recovery from an internal combustion engine using thermoelectric technology", Thermal Sci, 15 (4): 1011-22, 2011.

[12]   Zhang, Z., Li, Y.Z., "CFD simulation on inlet configuration of plate-fin heat exchanger", Cryogenics, 43, 12, 673-678, 2003.

[13]   Wen, J., Li, Y.Z., Zhou, A., Zhang, K., Wang, J., "Study of flow distribution and its improvement on the header of plate-fin heat exchanger", Cryogenics, 44, 823-831, 2004.

[14]   Wasewar, K. L., Hargunani, S., Atluri, P., Kumar, N., "CFD simulation of flow distribution in the header of plate-fin heat exchanger", Chemical Engineering and Technology, 30, 10, 1340-1346, 2007.

[15]   Gan, G., Riffat, S.B., Shao, L., "CFD modelling of pressure loss across tube bundles of a heat exchanger for closed-wet cooling towers", International Journal Ambient Energy, 21, 77-84, 2000.

[16]   Sheik Ismail, L., Ranganayakulu, C., Shah, R., "Numerical study of flow patterns of compact plate-fin heat exchangers and generation of design data for offset and wavy fins", International Journal of Heat and Mass Transfer, 52, 17, 3972-3983, 2009.

[17]   Driss, Z., Abid, MS., "Numerical investigation of the aerodynamic structure flow around Savonius wind rotor", Science Academy Transactions on Renewable Energy Systems Engineering and Technology, 2, 196-204, 2012.

[18]   Driss, Z., Abid, MS., "Numerical and experimental study of an open circuit tunnel: aerodynamic characteristics", Science Academy Transactions on Renewable Energy Systems Engineering and Technology, 2, 116-123, 2012.

[19]   Driss, Z., Damak, A., Karray, S., Abid, MS., "Experimental study of the internal recovery effect on the performance of a Savonius wind rotor. Research and Reviews", Journal of Engineering and Technology, 1, 1, 15-21, 2012.

[20]   Driss, Z., Bouzgarrou, G., Chtourou, W., Kchaou, H., Abid, MS., "Computational studies of the pitched blade turbines design effect on the stirred tank flow characteristics", European Journal of Mechanics B/Fluids, 29, 236-245, 2010.

[21]   Ammar, M., Chtourou, W., Driss, Z., Abid, M.S., "Numerical investigation of turbulent flow generated in baffled stirred vessels equipped with three different turbines in one and two-stage system", Energy, 36, 5081-5093, 2011.

[22]   Driss, Z., Ammar, M., Chtourou, W., Abid, M.S., "CFD Modelling of Stirred Tanks", Engineering Applications of Computational Fluid Dynamics, 5, 145-258, 2011.

[23]   Driss, Z., Abid, MS., "Use of the Navier-Stokes Equations to Study of the Flow Generated by Turbines Impellers", Navier-Stokes Equations: Properties, Description and Applications, 3, 51-138, 2012.

10

Optimization of Process Parameters of Manual Arc Welding of Mild Steel Using Taguchi Method

A.O. Osayi, E.A.P. Egbe, S.A. Lawal[*]

Department of Mechanical Engineering, School of Engineering and Engineering Technology, Federal University of Technology, PMB 65 Minna, Nigeria
*Corresponding author: lawalbert2003@yahoo.com

Abstract This study was based on design of experiment (DOE) using Taguchi method with four welding parameters namely; welding current, (ii) welding speed, (iii) root gap and (iv) electrode angle considered for experimentation. An orthogonal array of L_9 experimental design was adopted and ultimate tensile strength was investigated for each experimental run. The tensile test was carried out on extracted welded and unwelded specimens using universal testing machine (UTM). Microstructures of the welded specimens were carried out and analyzed. Statistical analysis (ANOVA) and signal to noise ratio were used to study the significant effect of input parameters on ultimate tensile strength and optimized conditions for the process performance respectively. The results showed that experiment number 7 has the highest ultimate tensile strength (UTS) of 487MPa and S/N ratio of 53.74 dB. The S/N ratio of higher value indicates better characteristic of optimum MMAW process performance. The study shows that the optimum condition is $A_3B_1C_3D_2$ at welding current 100A, electrode angle of 70^0, root gap of 3.3 mm and a welding speed of 3.6 mm/s .

Keywords: ANOVA, welding speed, current, electrode

1. Introduction

Welding process is very critical to the development of a nation because it is the hub on which modern industries revolve. Messler [1] stated that no secondary process has been and continues to be more important to the survival, comfort and advancement of mankind than welding. According to him, welding has made it possible to build our world. It is in view of this that many researchers have employed various optimization techniques to improve different welding process parameters on both semi and full automated welding processes. Although, the semi and full automatic welding processes are more productive, experience has shown that due to complexity and economic cost of the equipment and their operations, the choice of manual metal arc welding (MMAW) process is very popular in developing countries. This is so because of its several advantages such as low cost and simple operation. The MMAW process is portable and it can easily be used in places where other welding methods are not possible.

Manual metal arc welding is also known as shielded metal arc welding (SMAW), or stick welding process. It is one of the oldest and most widely used arc welding processes. The process involves the use of arc current to strike an arc between the base material and a consumable electrode rod. The electrode rod is made of a metal that is compatible with the base material being welded and it is covered with a flux. The heat generated melts a portion of the tip of the electrode, its coating and the base metal just below the arc. As the coating on the electrode melts, the flux gives off vapours that serve as a shielding gas and provide layer of slag, both of which protect the weld from atmospheric contamination. The slag formed during welding is chipped off from the weld after cooling. MMAW process can operate with both direct current (DC) or alternating current (AC) power supply depending on coating design. The process is portable, versatile, inexpensive equipment and requires little operator training. Also, the electrode produces and regulates its flux, it has lower sensitivity to wind and draft than gas shielded welding process and the process is applicable in all positions. On the hand, the process is slow and time wasting due to frequent changing of electrode and chipping of slag. Also, it is characterized with excessive spatter, arc stability and rough surface of weld bead and provides limitation deposition rates compared to other arc welding processes.

With the incorporation of automation into the arc welding process, many production companies adopted complete experimental designs and mathematical models to investigate the relevant process parameters to obtain quality weld [2]. Ajay et al., [3] stated that high quality can be achieved by optimizing various quality attributes or by selecting an optimal process environment that is efficient enough to fetch the desire requirements for quality. Taguchi method has been found to be a powerful tool to improve overall process quality by optimizing the welding process parameters in a way that variation is reduced to the barest minimum. Design of experiment (DOE) techniques had been used to carry out such optimization in the last two decades with a view to improving on the mechanical properties of weld materials. Yoon et al., [4] optimized the parameters of welding 7075-T6 aluminum alloy using Taguchi method. Among other investigators who have also worked on the optimization of welding variables using Taguchi method are Kim and Lee [5]. They used the method to suggest optimal combinations for process factors of hybrid welding methods to optimize the welding parameters of resistance spot welding process.

In this study, the application of Taguchi L_9 orthogonal array for the selection of manual metal arc welding process parameters of welded mild steel plates was investigated. The ultimate tensile strength and the microstructures analysis were carried out on each samples and analyzed. Signal-to-noise (S/N) ratio to determine the optimal parameters that affect the response and ANOVA analysis to determine the significant effect of the input variables on the ultimate tensile strength were both investigated.

2. Materials and Methods

2.1 Materials

The base material used for this study is mild steel (AISI C1020) of 100 x 75 x 5 mm plate. Its chemical composition analyzed by the Defence Industry Corporation of Nigeria, Kaduna is approximately 0.23% C,

0.35% Mn, 0.28% Si, 0.02% S, 0.04% P and 99.08% Fe. Mild steel was considered for this study because of its availability in the market and low cost. Mild steel electrodes of 350 mm long and 3.25 mm diameter (E6010 and E6013 steel grade 2, Oelikon) were used for the root running and weld deposits respectively.

2.2. Methods

2.2.1. Welding Process

A mild steel plate was cut into $100 \times 75 \times 5$ mm (length and breadth and thickness) using cut-off machine. This was followed by edge preparation and a single groove butt joint was selected for the joining process in flat position. The surfaces and the prepared edges of the samples were thoroughly cleaned with wire brush to remove any dirt or unwanted inclusion that could affect the weld. In this study, the welding process was a bit different from the normal convectional manual welding process as the electrode (electric arc) was constrained in stationary position while the workpiece moved relative to it. This help to maintain relative stable welding speed and improve the quality of the weld. A 400A capacity manual metal arc welding machine with direct current (DC) straight polarity was used for the welding operations. This is because with DC, it is easier to maintain short arc in the starting stage of welding operation. The E6010 electrodes were first applied for the root running. Thereafter, the grooves were cleaned before the E6013 electrodes were used for the weld deposits in 2 – pass. Fifty four workpieces were used for this experiment that was carried out in the Department of Mechanical Engineering workshop, Federal University of Technology Minna, Nigeria.

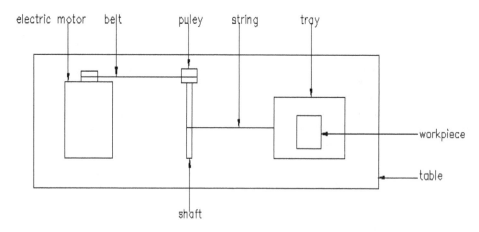

Figure 1. Welding Rig

2.2.2. Welding Rig

The welding rig consists of a speed reduction electric motor (1hp), three different sizes of belts and corresponding pulleys, a shaft, tray, string, six ball bearings and the stand. The main feature of the rig is the variable speed of workpiece relative to a stationary electrode or electric arc. The rotation of the electric motor was converted to linear motion through the belt drive system and guided string as shown in Figure 1.

Input welding parameters selected from the structural codes of American Welding Society [6], and manuals of

metal arc welding system for this study are: welding current, welding speed, root gap and electrode angle. Even though, arc length is one of the critical welding parameters, it cannot be used as welding parameter in manual metal arc welding process. An orthogonal array of L_9 was selected for experimentation as shown in Table 1. Based on the Taguchi orthogonal array designed, nine (9) experimental runs were conducted and each welding process was repeated three (3) times under the same conditions. These welding input parameters and their levels are shown in Table 2.

Table 1. An orthogonal array of L_9 (3^4) matrix

Experimental trial	Process parameters			
	A	B	C	D
1	1	1	1	1
2	1	2	2	2
3	1	3	3	3
4	2	1	2	3
5	2	2	3	1
6	2	3	1	2
7	3	1	3	2
8	3	2	1	3
9	3	3	2	1

Table 2. Process parameters and their levels

Input parameter	Symbol	Level 1	Level 2	Level 3
Welding current (A)	A	80	90	100
Electrode angle ((0^0))	B	70	75	80
Root gap (mm)	C	3.0	3.2	3.3
Welding speed (mm/s)	D	2.6	3.6	4.4

2.2.3. Determination of Ultimate Tensile Strength

The ultimate tensile strength of welded samples was determined using universal tensile machine (UTM) (model: TERCO^{CE} MT 3037) in the Department of Mechanical Engineering laboratory, Federal University of Technology, Minna- Nigeria. The tensile test specimens were extracted from the welded base metal with hacksaw and prepared to standard size according to American Welding Society [7] and universal tensile machine manual. The weld reinforcement and backing strip were removed, flushed with the surface of the specimen by grinding and filing with a smooth file. The stress- strain curve was used to determine the force and the ultimate tensile strength (UTS) was evaluated using equations 1 and 2

$$UTS \; (\sigma) = \frac{F}{S} \qquad\qquad 1$$

where F is the maximum loaded force, S is the cross-sectional area and for rectangular test specimen used in this experiment

$$Cross\,sec tional\,area\,(S) = wt \qquad\qquad 2$$

where w is the width and t is the thickness

2.2.4. Determination of Microstructure of Welded Joint

The micro-examination operation includes: grinding, polishing, etching and viewing. Hand hacksaw was used to cut test specimens from the welded plates as required. Grinding of each specimen was carried out using hand grinding deck of abrasive papers of different grade. Universal rotary wheels polishing machine of emery sheet type was used to carry out the polishing of the surface of each specimen to mirror-like in nature. The polished surfaces were etched with a natal (2% HNO_3 + 98% alcohol) regent and thereafter, the specimens were washed in running water and dried. An Optika (N- 400 POL) metallurgical microscope (bench type) was used to examine the specimens under magnification of x 400. The limitation of Optika (N-400 POL) metallurgical microscope is the none availability of micron marker to identify details in the microstructures.

3. Results and Discussion

The results of ultimate tensile strength as obtained using equations 1 and 2 for the experimental runs are show in Table 3. The values shown are the average of three reading recorded for each experimental trial. The signal- to- noise ratios for each ultimate tensile strength are equally included.

Table 3. Ultimate Tensile Strength (UTS) and S/N ratio values

Experimental trial	UTS (Mpa)				
	L_1	L_2	L_3	Average value(Mpa)	S/N ratio (dB)
1	435.0	396.0	415.0	416.0	52.35
2	420.0	442.0	426.0	429.0	52.65
3	429.0	433.0	438.0	433.0	52.74
4	467.0	440.0	442.0	453.0	53.11
5	471.0	460.0	442.0	458.0	53.20
6	453.0	476.0	455.0	461.0	53.27
7	489.0	487.0	484.0	488.0	53.74
8	471.0	435.0	438.0	448.0	53.13
9	435.0	433.0	449.0	439.0	52.85
Unwelded sample	429.0	411.0	-	420.0	-

3.1. Analysis of Variance (ANOVA)

The effect of each input parameters on the ultimate tensile strength was evaluated using variance analysis (ANOVA). The level of contribution of each of the input parameter to the strength of the welded point is shown in Table 4. In the analysis of variance in this study, pooling method was adopted; this is because pooling is a process of disregarding an individual's parameter's contribution and thereafter adjusting the contributions of other process parameters [8]. Pooling is employed when there is indeterminate situation and the effect of parameter in a process is insignificant. This is done to obtain new non-zero estimates of sum of square and DOF of variance respectively. Pooling process increases the percentage contribution error because the sum of square for the parameter being pooled is usually added to the sum of square error. In this study, the pooling effect is on parameter B (electrode angle) being the least effect on the welding process.

Table 4. Analysis of Variance (ANOVA) for Ultimate Tensile Strength

Process parameter	Symbol	DOF	SS	V	SS'	F	P
Welding current	A	2	0.38	0.19	0.36	19	31.86
Electrode angle	B					pooled	
Root gap	C	2	0.18	0.09	0.16	9	14.16
Welding speed	D	2	0.20	0.10	0.18	10	15.93
Error	E	2	0.35	0.01	0.43		38.05
Total		8	1.13				100

3.2. Signal – to - noise (S/N) Ratio

The choice of the S/N ratio to be used depends on the performance quality characteristics required. In welding process, the higher the strength of the weld, the better and hence, the signal –to –noise (S/N) ratio of higher the better (HB) was used in this study as expressed in equation 3.

$$\frac{S}{N} = -10\log\frac{1}{n}(\sum_{i-1}^{n}\frac{1}{y_i{}^2}) \qquad 3$$

where y = responses for the given factor level combination, n = number of responses in the factor level combination and y_i is the experimental results.

Table 5. Main effect of process parameters

Level	Welding current A	Electrode angle B	Root gap C	Welding speed D
1	52.58	53.07	52.92	52.80
2	53.19	52.99	52.87	53.22
3	53.24	52.95	53.22	52.99
Max-Min	0.66	0.12	0.35	0.42
Rank	1	4	3	2

Table 5 depicts the corresponding values of S/N ratios obtained from the conversion of UTS results. It was observed that experiment number 7 has the highest S/N ratio of 53.71dB, which indicates the best performance characteristic among the 9 runs of experiments conducted.

It was observed that the welding current has significant effect on the welding process while the electrode angle has the least effect. It also indicates that the combination process parameters $A_3B_1C_3D_2$ have the highest or maximum S/N ratios respectively and therefore signifies the optimum condition for the welding process.

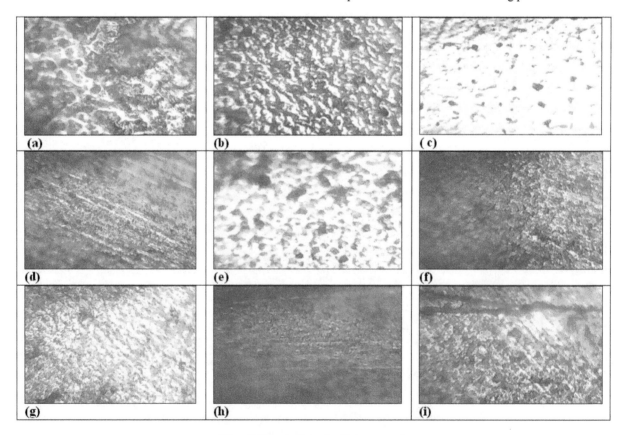

Figure 2. Microstructure of welded component

3.3. Microstructure of Welded Joint

Figure 2 (a- i) shows the microstructure of the welding joint under different welding conditions. Figure 2a shows

the microstructure obtained under the condition of welding current (80 A), electrode angle (70°), root gap (3.0 mm) and welding speed (2.6 mm/s). The grains were large and showing spheroidal globular of the photomicrograph due to low temperature and short time for nucleation. Figure 2b shows microstructure obtained under the welding condition of welding current (80 A), electrode angle (75^0), root gap (3.2 mm) and welding speed (3.6 mm/s). The micrograph shows deformation that produced elongated grains.

While Figure 2c shows the microstructure obtained under welding condition of welding current (80 A), electrode angle (80^0), root gap (3.3 mm) and welding speed (4.4 mm/s). The micrograph shows a mixture of pearlite (dark) and ferrite (light). The grains were widely spaced.

In the same vein, Figure 2d show the microstructure of the welding condition of welding current (90 A), electrode angle (70^0), root gap (3.2 mm) and welding speed (4.4 mm/s). The grains were cohesively arranged and fine. And Figure 2e shows the microstructure for welding condition of welding current (90 A), electrode angle (75^0), root gap (3.3 mm) and welding speed (2.6 mm/s). The micrograph shows partially grain-refined. Figure 2f shows the microstructure obtained under the welding condition of welding current (90 A), electrode angle (80^0), root gap (3.0 mm) and welding speed (3.6 mm/s). The grains were adhesively arranged and fine.

Similarly, Figure 2g shows the microstructure for welding condition of welding current (100 A), electrode angle (70^0), (3.3 mm) and welding speed (3.6 mm/s). The micrograph shows grain- refined, appears tiny and uniform pattern due to fast cooling rate. Figure 2h depicts the microstructure obtained under the welding condition of welding current (100 A), electrode angle (75^0), root gap (3.0 mm) and welding (4.4 mm/s). The grains are fine structurally. And Figure 2i shows the microstructure obtained under welding condition of welding current (100 A), electrode angle (80^0), root gap (3.2 mm) and welding speed (2.6 mm/s). The grains were coarse due to overheating in the weldment.

4. Conclusions

This study employed Taguchi method to optimize process parameters of manual metal arc welding (MMAW) process for mild steel products. From the tensile test carried out on welded samples it was observed that experiment number 7 has the highest ultimate tensile strength (UTS) and S/N ratio of 487 MPa and 53.74 dB respectively. The study indicates that the optimum condition is $A_3B_1C_3D_2$ which coincide with experiment number 7 in the orthogonal array. It was also observed that the failures of all the test specimens extracted from the welded samples did not occur at the weldment point which signifies a quality and strong weld joint. The microstructures show the various effects of the welding parameters on the weld joint.

Moreover, it was noted from results of ANOVA that the welding current, root gap and welding speed are significant parameters in the welding process while the electrode angle has least significant. Computation of the projected optimum performance of the study using S/N ratio (Y_{opt}) is 53.74 dB which is the same value with experiment number 7 and it serve as confirmation test. A confirmation test would have been conducted if the optimum condition was not among the experimental runs. Results from this study, indicated that Taguchi method can actually be used to optimize or improve the quality of MMAW process.

References

[1] Messler, R.W., *Principles of welding processes.* Wiley – VCH Verlag GmbH and Co. KGaA, Weinheim. 2004.

[2] Ill-Soo, K., Joon-Sik, S., Sang-Heon, L., Prasad K.D.V., *Optimal Design of Neural Networks for Control in Robotic arc Welding.* Robotic and Computer-Integrated Manufacturing, 20, 57-63, 2004.

[3] Ajay, Saurav, Swapan, Gautan, *Application of Vikor Based Taguchi Method for Multi-Response Optimization. A Case Study in Submerged Arc Welding (SAW),* Proceedings of International Conference on Mechanical Engineering (ICME), 2009.

[4] Yoon, H., Byeong Hyeon, M., Chil Soon, L., Hyoung, K.D. Kyoun, K.Y. and Jo, P.W., *Strength Charateristics on Resistance Spot Welding of Aluminium Alloy Sheets by Taguchi Method,* International Journal of Modern Physics B, 4, 297-302, 2006.

[5] Kim, H.R., Lee, K.Y., *Application of Taguchi Method to Determine Hybrid Welding Condions of Aliminium Alloy.* Journal of International Research, 68, 296-300, 2009.

[6] American Welding Society, 550 N.W. LeJeune Road, Miami, FL33126, 1997.

[7] American Welding Society, 550 N.W. LeJeune Road, Miami, FL331, 2007.

[8] Roy, R.K., *A Primer on Taguchi Method.* Reinhold International Company Ltd, 11 New Lane, London EC4P4EE, England. 1990.

11

Virtual System to Simulate the Performance of Various Categories of Machine Tools during the Design Stage

Mounir Muhammad Koura[1], Muhammad Lotfy Zamzam[2], Amr Ahmed Sayed Shaaban[3,*]

[1]Prof. Dr. in Faculty of Engineering, Ain Shams University, Egypt
[2]Asst. Prof. in Faculty of Engineering, Ain Shams University, Egypt
[3]PhD. Student in Faculty of Engineering, Ain Shams University, Egypt
*Corresponding author: Amr.Ahmed@eng.asu.edu.eg

Abstract This paper presents a simulation system designed to evaluate the static and dynamic performance of machine tools. The design considerations of the evaluation system are discussed and the system is then employed in order to compare between various categories of milling machine's structure adapted for end milling operation. The machine performance is identified in terms of static loop stiffness in both x and y directions, mode shapes, and frequency response function (FRF) at tool center point (TCP). The advantage of such a reliable model is that it could replace the many experimental tests that must otherwise be carried out each time the parameters affecting the machine tool performance are changed.

Keywords: *machine tools, static performance, natural frequencies, dynamic behavior, finite element method, virtual system*

1. Introduction

The experimental approach to study machining process and machine tool performance is expensive and time consuming, especially when a wide range of parameters are included. Hence, virtual prototyping is used instead. The performance of a machine tool depends on parameters related to cutting process, such as: cutting speed, feed rate, radial and axial depth of cut, and end mill and work piece characteristics, and other parameters related to machine structure, such as: structure category, supporting webs, guide ways stiffness, bolted connections, drive's bearing stiffness and spindle head position.

Among various mathematical models used for simulation, finite element method (FEM) is proved to be useful and widely used. Basic ideas of the finite element method were studied at the beginning of the 1940s. Courant (1943) developed finite element method and he used piecewise polynomial interpolation over triangular sub regions to model torsion problems [1]. Clough (1960) was the first to use the term "finite element" [2]. Zienkiewicz and Cheung (1967) wrote the first book on finite element theory [3]. Also other theory books were written by Cook, et al. (1989) [4], Mohr (1992) [5] and Chandrupatla and Belegundu (2002) [6]. Researchers usually wrote their own FE codes for specific process until the mid-1990s. A foundation and comprehensive information related to the field of virtual machine tool design were presented by Y. Altintas, et al; titled as "A

study on virtual machine tool" [7]. On the other hand, in order to check the process constraints as well as optimal selection of the cutting conditions for high performance milling, E. Budak developed "Analytical models for high performance milling". Milling force, part and tool deflection, form error and stability models have been presented [8]. Furthermore, other study that focuses on applying virtual prototyping to design a machine tool element is entitled "Virtual design of machine tool feed drive system" achieved by R.C. Parpala [9]. "Integrated dynamic modeling, design optimization and analysis on 5-axis ultra-precision micro milling machine" were developed by D. Huo and K. Cheng. In the same context, "Finite element analysis of bolted joints" was developed by I. piscan, N. predincea and N. pop [10]. This paper presents a theoretical model and a simulation analysis of bolted joint deformations. "FEA of high-speed motorized spindle based on ANSYS" was developed by D. Liu, et al [11]. In this paper, the finite element model of the high-speed motorized spindle was derived and presented. Moreover, the paper entitled "Modal analyses of machine tool column using FEM" was concerned in providing designers with useful information about static and dynamic behavior of various categories of machine tool columns [12].

In this paper, a system that employs the concept of virtual prototyping is created to provide designers of machine tools with useful information, and to facilitate improvement decisions in the early design stage. The system is then employed in order to compare between open and closed structures of milling machine tools during end milling operation.

2. Definition of the Evaluation Aspects

In this section, the machine tool performance evaluation aspects employed in this paper are discussed. The following sections present a detailed discussion of the function of each module, the connection between the modules, the data entry needed for each of them, and the generated results.

2.1. Static Performance of Machine Tool

Static analysis calculates the effects of steady loading conditions on the machine, while damping effects are neglected. Based on that evaluation aspect, the structural loop stiffness which characterizes the machine's overall static performance is calculated.

2.2. Dynamic Performance of a Machine Tool

2.2.1. Modal Analysis

The natural frequencies and mode shapes are obtained by solving the eigen-value problem:

$[K][X] = \omega^2[M][X]$, where $[K]$ is the stiffness matrix, $[X]$ is the displacement matrix that contains all degrees of freedom and consequently depends on number of nodes, $[M]$ is the mass matrix, and ω is the angular frequency (rad/s). The problem can be satisfied by either: $[X] = 0$, which is a trivial solution, or; $|[K] - \omega^2[M]| = 0$, where $|\ |$ is the determent of a given matrix. The eigen values ω^2 yield the natural frequencies ω of the system, while the eigen vectors $[X]$ define the mode shapes. The first frequency is usually called the fundamental frequency.

2.2.2. Frequency Response Function

Harmonic analysis is performed to quantitatively determine the steady-state response of the machine towards sinusoidal loads. Although the cutting forces are varying with time over each tooth interval in a complex manner that is not actually sinusoidal, the cutting forces are repeated for each tooth with a certain frequency that depends on cutting speed. As a result, harmonic analyses are helpful to verify whether or not the design will successfully overcome resonance and harmful effects of forced vibrations.

3. Modeling of the Mechanical Structure

3.1. 3D Modeling of Mechanical Structure

The mechanical structure of a machine tool center can be considered to have the major contribution of its rigidity. It mainly consists of the column, bed, table, saddle, slider, and spindle head. When the cutting process is established on hard materials, the dynamic characteristics of the mechanical structure of a machine tool center become crucial. The CAD model is created using any of the commercial CAD software and imported to Ansys® which is the analysis tool package used in this work. Lumped masses are used to simulate the effect of the feed drive and spindle housing. The joints at the guide ways are assumed to be rigid, as well as, the joint between screws and their bearing supports. Bolts are simulated with spring beam elements, and they are preloaded with initial tension of 5 KN for each [8]. A fixed support was assigned to the lower surface of the machine base as a boundary condition. Any topology that has no significant effect was removed for simplicity and to minimize computation time

3.2. FE Model of Mechanical Structure

The FE model of the mechanical substructures are generated from their perspective CAD models using tetrahedron elements. The tetrahedron elements are proved to be more suitable than the bricks elements to simulate machine tool structures. The main characteristic of interest when evaluating a FE model is the mesh size. The fine mesh size leads to converged results, but on the other side, it increases the DOFs which increases the computation cost as well. However, local refinements in mesh size can be specified at critical regions such as TCP and work table using the sphere of influence.

To obtain a mesh independent model with the minimum computation time, constitutive iterations are carried out to properly select mesh characteristics in order to achieve convergence of the results to the exact value within an accepted residual error. The h-method in which the mesh size is refined until convergence is employed in this work. This method is preferred for its ease of execution rather than the P-method in which the degree of polynomial used for the shape function is changed until convergence.

4. Modeling of the Cutting Process

To obtain a realistic integrated simulation system, modeling of the cutting process is integrated with the modeling of machine tool mechanical components discussed in the previous subsections. This integration leads to a realistic representation of the overall performance of the machine tool.

Many works have been carried out on different aspects linked to cutting forces prediction. Recently, researchers used Finite element method to simulate the cutting process and hence, predict the cutting forces and the chip morphology [13]. However, Simulation of cutting process using finite element method is not crucial to this paper. Therefore, analytical method has been used instead. Tlusty and McNeil's cutting force model was developed for conventional end-milling operations [14].

5. Construction of the Evaluation System

Based on the concepts discussed on the previous subsections, the theory based on which the evaluation system is designed can be represented by the chart shown in Figure 1, while the GUI (graphic user interface) of the evaluation system is illustrated in Figure 2. The system's GUI consists of three areas. The system inputs and outputs are defined in Area1, the user-defined design points at which the system runs appear in Area2, and finally, Area3 shows the charts that represent relationships between specified parameters. All input data and design points that are defined in Area1 and Area2 are automatically transferred to carry out the corresponding analyses without any need to log in any of the modules; then the generated results are transferred back to be displayed. The system logic and the interconnections between the modules are represented in Figure 3. The data entry of each module and the results generated from each are represented in Table 1. The data is classified to that

related to the machine tool such as: 3D model file, spindle head position, characteristics of spring elements, heat generation at hot spots and bolts pretension, others related to the cutting process such as: static load at TCP, amplitude and exciting frequency of harmonic loads, T_c and data sheet of time varied loads along one tooth interval.

Table 1. The evaluation sytem entries and obtained results from each module

	System entries	Obtained results
Geometry	Spindle head position	-
Static	Cutting load at TCP and worktable in x and y-directions Bolt's pre-stress Static stiffness and damping coefficient of the spring elements	Directional deformation at TCP relatively to worktable at x and y-directions
Modal	Only the 3D model	Natural frequencies
Harmonic	Range of exciting frequencies of cutting loads Amplitude of harmonic load at TCP and worktable in x and y-directions	Frequency response function (FRF) at TCP relatively to worktable at x and y-directions
Time-varried	Tabular data contains time-varied cutting loads within the time period at which one tooth is engaged	Time-varied values of x and y deformations at TCP and worktable

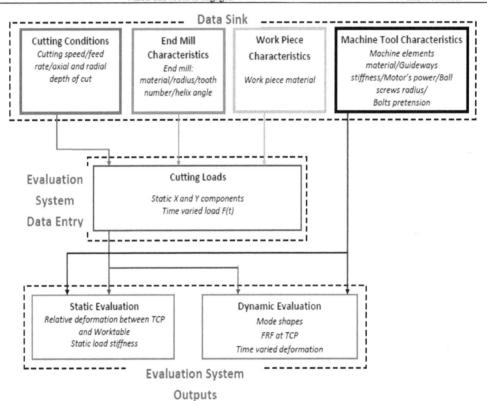

Figure 1. Flow chart of the virtual evaluation system

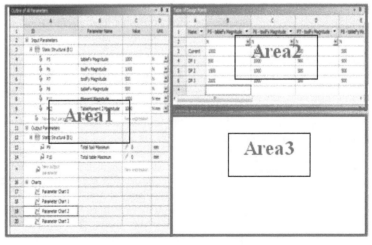

Figure 2. A sample of the input/output panel of the designed evaluation system

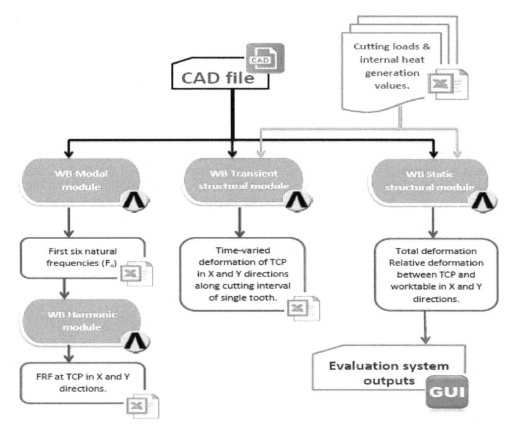

Figure 3. The logic chart of the evaluation system

6. Comparison Investigation on Open and Closed Milling Machine Tool Categories

6.1. Case Definition

The virtual evaluation system is applied on both 3-axis milling machine tools of open and closed categories. The open frame machining center mainly consists of the column, bed, table, saddle, and spindle housing, as well as, three screws to provide the traverse motion in three axes as shown in the CAD model in Figure 4 (a), while the closed frame machining center consists of two columns, bed, table, X-guide, X-slider, Z-slider, and spindle housing as

shown in the CAD model in Figure 4(b). Both two categories are within the same volumetric size 1200 mm* 1790 mm* 1920 mm. The material of the machine structural components was assigned as gray cast iron with modulus of elasticity of 89 GPa; density of 7250 kg/m^3; and Poisson's ratio of 0.25. On the other hand, the cutting process is carried out on work piece material called Inconel718, with cutting conditions [radial depth of cut (a) =4 mm, axial depth of cut (b) =4 mm, feed rate (u) =100 mm/min, feed per tooth (f$_z$) =0.0125 mm/tooth, and cutting speed (s) =2000 rpm], using an end mill of radius (r) =10 mm, number of teeth (n) =4, and helix angle (ß) =pi/6.

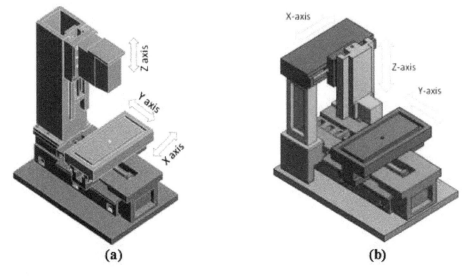

Figure 4. 3D model for (a) open structure, and (b) closed structure

6.2. Preprocessing

6.2.1. Adjusting Spindle Head Position

Before running the system, all data stated in the previous subsection are entered and transferred to each module of the designed system according to its requirements. The spindle head position is adjusted according to the specified depth of cut so the end mill lower surface is 4 mm below the work piece surface.

6.2.2. Cutting Loads Generation

According to the cutting conditions previously stated in the case definition, the cutting load generated using analytical methods is shown in Figure 5. The graph shows the cutting load values with respect to the end mill angle of rotation along one tooth interval.

6.2.3. Preprocessing of Static and Dynamic Analyses

For the static module, the TCP is subjected to static load of 1000 N in X-direction and 500 N in Y-direction. For the harmonic module, sinusoidal loads of 1N amplitude with exciting frequency ranges from 30 to 300 HZ frequency are subjected to the TCP. Finally, cutting load values are transferred in a tabulated form to the transient structure module where they are subjected to the TCP through a specified number of time steps so as to obtain the deformation of the TCP during one tooth interval.

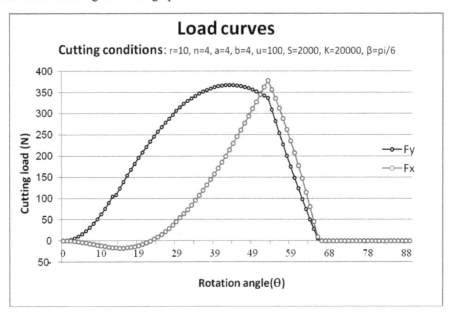

Figure 5. Cutting forces generated along one tooth interval

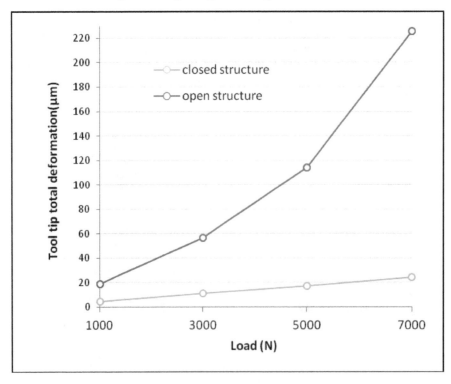

Figure 6. Tooltip deflection at different loads for open and closed structure

6.3. Static Analysis Results

Figure 6 shows the relative total deformation between TCP and worktable at various static loads. It can be noticed that the closed structure gives much better results compared to the open one, and that the larger the load, the greater the deviation between the results of closed and open categories. The supporting columns in the closed category raises the static loop stiffness in YZ plane from 32.5 in the open category to 52.9 N/μm, and in XZ plane from 70.4 in the open category to 294N/μm, as shown in Figure 7. The closed category shows greater rigidity in YZ plane with enhancement of 62.7 % as the supporting columns greatly decreases the tilting of the spindle head that occurs in the open category due to the X component of the cutting load. However, the main advantage of the closed category appears in the increase of rigidity in XZ plane by 317 %, which gives the structure great rigidity against bending [5].

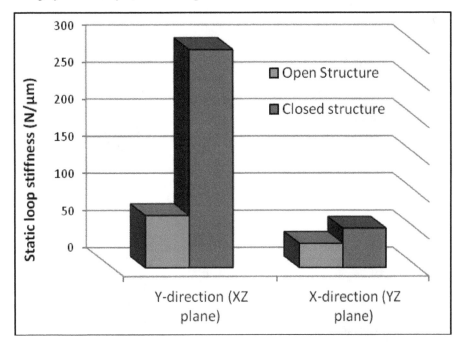

Figure 7. Static loop stiffness in X and Y direction for open and closed categories

6.4. Modal Analysis Results

Table 2 shows the first six natural frequencies for each category, and the position of maximum deformation for each mode shape. The first six mode shapes of each closed and open structure are illustrated in Figure 8 (a), and (b).

The results show that the lowest natural frequency in both open and closed categories occurs at the spindle head, and hence, it can be marked as a weak point from the aspect of dynamic behavior. The first natural frequency in the closed category is slightly lower than the open one. It is also noticed that the range of natural frequencies in the open category is wider than that in the open one, i.e.: at 228 HZ, six vibration modes occurred in the closed category, while five occurred in the open one. This indicates higher sensitivity to vibration in the closed category than in the open one. The explanation of such results may be due to the increase occurred in the mass from 4555 kg in the open case to 5436 kg in the closed one, and also due to the overhang effect of the Z-slider that carries the spindle housing in the closed frame. However, in the open category, the spindle housing show the maximum deformation in the first three mode shapes and the fifth mode shape, while that occurs only on the first and third mode shapes in the closed category. That indicates lower probabilities for resonance in the spindle housing in the closed case. Besides, the table and bed shows low deformation in all mode shapes of the closed category except the fifth one, while in the open category they show high deformation in the last three mode shapes especially the forth one. That indicates higher dynamic rigidity for the bed and table in the closed category.

Table 2. First six natural frequencies for both open and closed categories and the position of maximum deformation at each mode shape

| Mode shape | Open Structure | | Closed Structure | |
	Natural frequencies	Position of max deformation	Natural frequencies	Position of max deformation
1	56	Spindle housing	53.7	Spindle housing
2	63.5	Spindle housing	57.2	Z-slider
3	129.9	Spindle housing	90.5	Spindle housing
4	208.15	Table	96.8	X-Guide way
5	210.47	Spindle housing	199.9	Table
6	246.76	Column	228	Z-slider

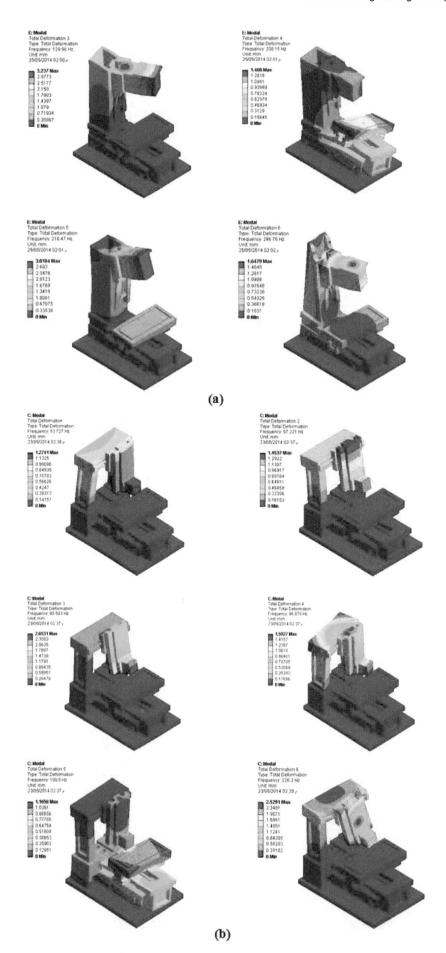

Figure 8. First six mode shapes for the (a) Open category, and (b) Closed category

6.5. Harmonic Analysis Results

As shown in Figure 9, for the open category, the maximum dynamic compliance of about 0.12 μm/N in X-direction and 0.05 μm/N in Y-direction occurred at 60 HZ, which corresponds to a dynamic stiffness in XZ and YZ planes of 8 N/μm, and 20 N/μm respectively. For the closed category, the maximum dynamic compliance of about 0.044 μm/N in X-direction and 0.97 μm/N in Y-direction occurred at 60 HZ and 90 HZ respectively, which corresponds to a dynamic stiffness in XZ and YZ

planes of 22.7 N/μm, and 1.03 N/μm respectively. By comparing the maximum compliance and the compliance all over the frequency range for both categories, it can be concluded that the closed structure has much better dynamic performance in X-direction up to 220HZ especially at 160HZ at which it gives the least compliance, while the open category shows better performance in the Y-direction from 70 HZ to 190 HZ, especially at 180 HZ at which it shows the least compliance.

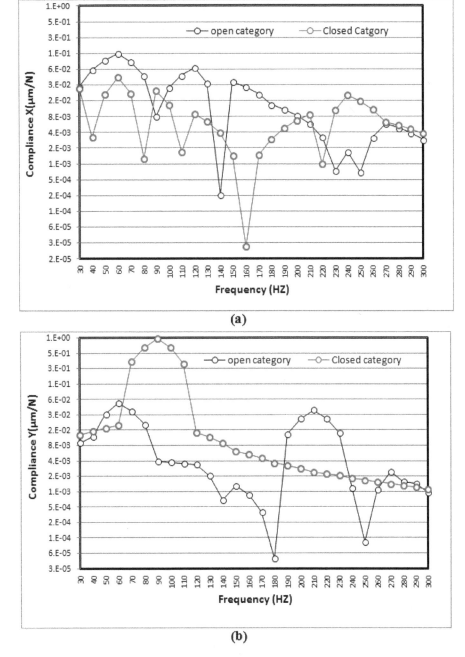

Figure 9. FRF at the TCP at (a) X-direction and (b) Y-direction

7. Conclusion

An integrated modeling and design approach is developed to create a virtual evaluation system capable to

evaluate the static and dynamic performance of machine tools during its design stage without the need of prototyping. The modeling of the machine tool mechanical structures together with the modeling of the cutting process are all integrated together to give a comprehensive evaluation of the static and dynamic performance of the entire machine tool. The system is then applied to a case study at which a comparison investigation based on static and dynamic performance is established by a series of analysis on open and closed milling machine tool categories. The closed category proved to have better dynamic performance than the open one, especially in the x-direction. It also shows better static performance in y-direction. However, better results are expected if the overhang problem is treated.

References

[1]   Courant R. Variational Methods for the Solution of Problems of Equilibrium and Vibrations. Bulletin of American Mathematical Society 1943; 48: 1-23.

[2]   Clough RW. The Finite Element Method in Plane Stress Analysis. Proceedings of American Society of Civil Engineers 1960; 23: 345-37.

[3]   Zienkiewicz OC, Cheung YK. The Finite Element Method in Structural and Continuum Mechanics. London: Mc-Graw Hill; 1967.

[4]   Cook RD, Malkus DS, Plesha ME. Concepts and Applications of Finite Element Analysis. New York, USA: John Wiley & Sons; 1989.

[5]   Mohr GA. Finite Element for Solids, Fluids and Optimization. Oxford University Press, UK; 1992.

[6]   Chandrupatla, Belegundu. Introduction to Finite Element in Engineering. New Jersey, USA: Prentice-Hall; 2002.

[7]   Altintas Y, Brecher C, Weck M, Witt S. Virtual machine tool. CIRP ANN-MANUF TECHN 2005; 54: 115-138.

[8]   Budak E. Analytical models for high performance milling. Part I: Cutting forces, structural deformations and tolerance integrity. INT J MACH TOOL MANU 2006; 46: 1478-1488.

[9]   Parpală R C. Virtual design of a machine tool feed drive system. U.P.B. Scientific Bulletin, Series D 2009; 71: 131-140.

[10]   Piscan I, Predincea N, Nicolae P. Finite element analysis of bolted joint. Proceedings in Manufacturing Systems 2010; 5: 167-172.

[11]   Huo D, Cheng K, Wardle F. Design of a five-axis ultra-precision micro-milling machine—UltraMill. Part 2: Integrated dynamic modelling, design optimisation and analysis. INT J ADV MANUF TECH 2010; 47: 879-890.

[12]   Assefa M. Modal analysis of machine tool column using finite element method. International Journal of Mechanical, Industrial Science and Engineering 2013; 7: 51-60.

[13]   Man X, Ren D, Usui S, Johnson C, and Marusich T D, "Validation of Finite Element Cutting Force Prediction for End Milling," Procedia CIRP, vol. 1, pp. 663-668, 2012.

[14]   Saffar R J, Razfar M, Zarei O, Ghassemieh E. Simulation of three-dimension cutting force and tool deflection in the end milling operation based on finite element method. SIMUL MODEL PRACT TH 2008; 16: 1677-1688.

12

Analytical Modeling of a Piezoelectric Bimorph Beam

A. Lebied[1,*], B. Necib[1], M. SAHLI[2]

[1]Mechanical Engineering Department, Faculty of Technology Sciences, University Constantine 1, Algeria
[2]Departement de Mécanique Appliquée, ENSMM, 24 chemin de l'Epitaphe, 25030 Besançon, France
*Corresponding author: lebied.aziz@yahoo.fr

Abstract Smart structures based on piezoelectric materials are now finding applications in a wide variety of environmental conditions. Such materials are capable of converting mechanical energy into electrical energy. Indeed, when subjected to mechanical stress become electrically charged at their surface and vice versa. In the current paper, the research is focused on a simple analytical model based on Euler–Bernoulli beam theory with the following assumptions was proposed: (a) the piezoelectric layer thickness in comparison to the length of the beam is very thin and (b) the electrical field between the upper surface and lower surface of the piezoelectric layer is uniform. We have applied this model to study its static responses and predict the ambient deformations into usable electrical energy from a cantilever piezoelectric beam. The proposed model was numerically investigated for validity. Analytical data showed that the proposed model simulations are in good agreement with the FE results. The results of the modeling are very promising.

Keywords: *smart structure, piezoelectric material, modeling*

1. Introduction

Piezoelectric materials, such as Lead Zirconate Titanate (PZT), have the ability to convert mechanical forces into an electric field in response to the application of mechanical stresses or vice versa [1,2,3]. This effect is due to the spontaneous separation of charge within certain crystal structures, thereby producing an electric dipole. For a piezoelectric material, it is known that the output voltage of the material is a function of stress. Typically, stress is achieved through the displacement or bending of the piezoelectric beam. These novel properties of the materials have found applications in sensor and actuator technologies, and recently, in the new field of energy harvesting [4,5,6,7].

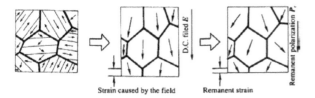

Figure 1. Schematic illustration of the poling process

The origin of the piezoelectric effect is related to an asymmetry in the unit cell and the resultant generation of electric dipoles due to the mechanical distortion. These materials do not possess any piezoelectric properties owing to the random orientations of the ferroelectric domains in the ceramics before poling. During poling, an electric field is applied on the ferroelectric ceramic sample to force the domains to be oriented or poled. After poling, the electric field is removed and a remnant polarization and remnant strain are maintained in the sample, and the sample exhibits piezoelectricity. A simple illustration of the poling process is shown in Figure 1.

In recent years, there have been a considerable number of publications using various models for the electromechanical behavior of piezoelectric energy harvester beams, as can be seen in Anton and Sodano [8]. Eggborn [9] developed the analytical models to predict the energy harvesting from a cantilever beam and a plate using Bernoulli-beam theory and made a comparison with the experimental result. Ajitsaria et al. [10] developed modeling and analysis of a bimorph piezoelectric cantilever beam for voltage generation using on the analytical approach based on Euler–Bernoulli beam theory and Timoshenko beam equations, which is then compared with two previously described models in the literature: the electrical equivalent circuit and energy method. Lead Zirconate titanate (PZT) is the most used piezoelectric material because of its high electromechanical coupling characteristics in single crystals [11,12]. Sodano et al. [13] developed an analytical model of a beam with attached PZT elements to provide an accurate estimate of the power generated by the piezoelectric effect.

Erturk and Inman have recently published a series of papers on energy harvesting using the cantilever model and their provide a broad coverage of several important modeling aspects that were validated with experimental data [14,15,16,17,18].

Works presented in this report concerns the problems the characteristic phenomena of piezoelectricity. In the current paper, the research is focused on the different deformations effect by voltage generation of the piezoelectric beam. The relation between the voltage imposed and the curvature is derived which is used to explain the effect of placement on voltage generation is investigated. The predictive models are validated by being compared to numerical data.

2. Theory and System Modeling

Piezoelectricity involves the interaction between electrical and mechanical behaviors of the material. Meanwhile, the static linear relations between the electrical and mechanical variables can approximate this interaction. In strain-charge form, these relations can be given by the following equations:

$$\varepsilon = s^E \sigma + dE_3 \tag{1}$$

$$D = d\sigma + \varepsilon^\sigma E \tag{2}$$

Where s^E, ε^σ and d are the elastic compliance coefficient, electric field related electrostriction coefficient and linear permittivity respectively and E_3 are the electric field intensity [19].

In equations (1) and (2) above, the electrical quantities (E, D) have vector nature, while the mechanical quantities (σ, s) have tensor nature of six components. In piezoelectric materials, the constants d, e and ε depend on the directions of electric field, displacement, as well as stress and strain. The piezoelectric strain constant (d) can be presented as d_{ij} which can be interpreted as charge created in the ith direction under a stress applied in the jth direction.

The linear constitutive equations for a piezoelectric material [20] have been employed in terms of the piezoelectric coefficient e_{31} and the electric field applied across the thickness of the layer E_3.

3. Cantilever Beam Analysis

The following section describes the development of the PZT models and the analytical estimations of local deformation. The schematic view of the beam undertaken in this study is shown in Figure 2. If the thickness of the both host substrate layers equal to 200μm, respectively hp_1, hp_2 are far greater than the thickness of the electrode patch (500nm), and the PZT beam length (L) equal 10mm is approximately the length of the electrode layer. The PZT beam width (w) equal 1mm is approximately thirty times the width of the electrode layer (w_0=30μm, neglected in the analytical model proposed). A thin electrode layer was attached to the PZT layer using two-part epoxy glue in a typical bimorph configuration (see Figure 3). The PZT PCI-151 was used to manufacture the cantilever beam device. The dimensions and properties of the materials are shown in Table 1.

Based on piezoelectric cantilever, the most important piezoelectric strain coefficients when designing harvesting devices are d_{31} and d_{33} (Roundy et al., 2003). In the 3-1 mode, imposed strain in the 1-direction is perpendicular to the electric field in the piezoelectric material (i.e. in the 3-direction). In the 3-3 mode, the strain and electric field are parallel to each other (in the 3-direction).

The electrical–mechanical coupling of the 3-3 mode (or z-z mode) is higher than the 3-1 mode (or z-x mode) and the 3-3 mode piezoelectric cantilever will allow for a much higher open circuit voltage compared to a similarly sized 3-1 cantilever. Another key advantage of the 3-1 mode of operation over the 3-3 system is that the 3-1 system is much more compliant, hence larger strains can be produced with smaller input forces. The piezoelectric constitutive equation (2) can be rewritten as:

$$D_3 = d_{31}\sigma_1 + \varepsilon_{33}^\sigma E_3. \tag{3}$$

Table 1. The Dimensions and Properties of Electrode and Pzt Beam

	Parameters	Value
PZT Beam	length	10 mm
	width	1 mm
	thickness	200 μmm
	Young modulus	66 GPa
	piezoelectric constant d_{33}	-210.10^{-12} m/V
Electrode	length	10 mm
	width	30 μm
	thickness	500 nm
	Young modulus	200 GPa

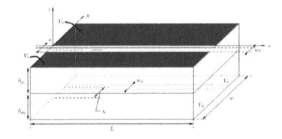

Figure 2. Schema a piezoelectric cantilever beam and electrode patchs

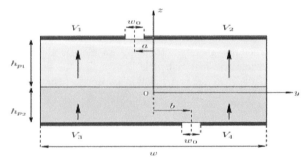

Figure 3. The attachment of electrode on the PZT beam

4. Analytical Results and Discussions

The modeling of a cantilever beam by means of the substitution potential V_1, V_2, V_3, V_4, by the two command variables of the beam axes V_y et V_z of such so that:

$$V_1 = V_z + V_y, \quad V_2 = V_z - V_y$$
$$V_3 = V_z - V_y, \quad V_4 = V_z + V_y.$$

The moment equilibrium equation along y is expressed by:

$$\iint \sigma_1(x, y, z) dA = 0 \tag{4}$$

The relative deformation of the beam due to the elasticity is given by:

$$\varepsilon_1 = \frac{z}{\rho_z} \qquad (5)$$

Furthermore, the slope at an abscissa x of the beam is in a first approximation:

$$\frac{dz}{dx} = \frac{x}{\rho_y} \qquad (6)$$

The resolution of equation (4) by using equation (6) we obtain the following deflection z, δ_z:

$$\delta_z = \frac{3}{8} \frac{d_{31_1} L^2 h_{p_1} \{(w - w_0 + 2a) V_1 + (w - w_0 - 2a) V_2\}}{s_{11}^{p_1} w B_1}$$

$$+ \frac{3}{8} \frac{d_{31_2} L^2 h_{p_2} \{(w - w_0 + 2b) V_3 + (w - w_0 - 2b) V_4\}}{s_{11}^{p_2} w B_1}.$$

With $B_1 = \left(\frac{h_{p_2}^3}{s_{11}^{p_2}} + \frac{h_{p_1}^3}{s_{11}^{p_1}} \right).$

Taking our case the two piezoelectric plates are the same, therefore if the centering error of the electrodes is the same on both faces of the bimorph, or a = b or a=-b.

$$\delta_z = \frac{3}{4} \frac{d_{31} L^2}{w h_p^2} \left[(w - w_0) V_z - 2b V_y \right].$$

Moreover, it is possible to draw on the same graph of the deflection curve of the bimorph beam obtained for different values of a and b errors. The superposition of these curves to quantify the value of the deflection at the end of the cantilever. It was found that this deflection is directly proportional to the value of the electric field V_y by setting the voltage $V_z = 100V$ (cf.Figure 4).

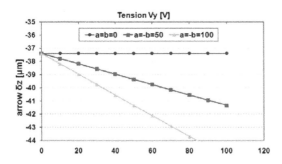

Figure 4. Deflections δ_z according to V_y for several centering errors as a=-b and V_z =100V

In this case we observe that for a = b = 0 there is a stabilization of the deflection for any value of the tension Vy against to Vy=100 V the results obtained without taking account the error values a and b (cf.Figure 5).

However, the values of a and b depend primarily on the technical means of realization implemented for the development of these actuators. From a technological point of view, it is difficult to strictly satisfy the condition a = b = 0, or even a = b.

Let us now try fixing the two tensions and traced the deflections based on centering errors electrodes a and b on the coupling in the case of a =-b.

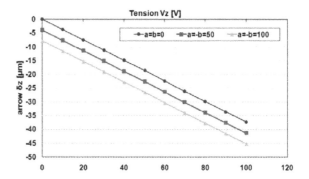

Figure 5. Deflections δ_z according to V_z for several centering errors as a=-b and V_y =100V

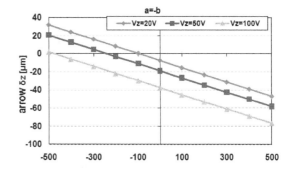

Figure 6. Deflections δz according to V_z for several centering errors as a=-b and Vy =100V

The Figure 6 presents the deflection as a function a=-b, for V_y = 100V and for several values of V_z. In these conditions, we observed not only a decrease of the deflection when a = b increases, but also a coupling V_z on δ_z.

5. FE Modeling Validation

As illustrated in Figure 7, the analytical solution in terms of the evolution of the deflection as a function of voltage conforms to the numerical simulation with the piezoelectric material cited below. The numerical curve being almost equivalent to the analytical.

We also assume that there is no centering error, or a = b = 0 We then obtain for a degree of freedom in z, the following equation:

$$\delta_z = \frac{3 d_{31} L^2 (w - w_0)}{4 w h_p^2} V_z \qquad (5)$$

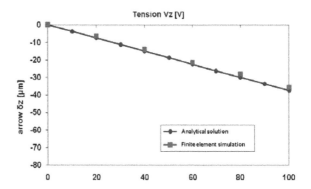

Figure 7. Free deflections δ_z of the piezoelectric beam depending on the control voltage V_z

6. Conclusion

Modeling the clamped-free piezoelectric beam small displacement was developed and presented in this article. The electromechanical coupling has been considered. The analytical results were valid for comparisons with numerical results find in the literature. The analytical results showing a good correlation and agree very well. It was also concluded that the actuator and the sensor will be better placed at underrun because it is the position or the actuator has the greatest impact and where the sensor gives the greatest signal. They are co-located as said Colles one below the other on either side of the beam.

This modeling is again applicable to structures of any shape and may include piezoelectric components to form a control or vibration. In addition, they allow us to know the magnitude of the force imposed equivalent to the applied field for a given displacement. These results are necessary for the design, optimization and development of systems for micromanipulation and assembly of micro objects.

References

[1]   S. Roundy, P.K. Wright, "A piezoelectric vibration based generator for wireless electronics, Smart Mater. Struct., Vol.13, pp.1131-1142, 2004.

[2]   S.R. Anton, H.A. Sodano, "A review of Power harvesting using piezoelectric . materials (2003-2006)," Smart Materials and Structures, Vol.16, pp.R1-R21, 2007.

[3]   E. F. Crawley, J. de Luis, "Use of Piezoelectric Actuators as Elements of Intelligent Structures," Present as Paper 86-0878 at the AIAA/ASME/ASCE/AHS Active Structures, Structural Dynamics and Materials Conference, San Antonio, TX, May 19-21, 1986.

[4]   P. Muralt, "Ferroelectric thin films for micro-sensors and actuators: a review," Journal of Micromechanics and Microengineering, Vol. 10, no. 2, pp. 136-146, 2000.

[5]   C. Niezrecki, D. Brei, S. Balakrishnan, and A. Moskalik, "Piezoelectric actuation: state of the art," Shock and Vibration Digest, Vol. 33, no. 4, pp. 269-280, 2001.

[6]   J. F. Tressler, S. Alkoy, and R. E. Newnham, "Piezoelectric sensors and sensor materials," Journal of Electroceramics, Vol. 2, no. 4, pp. 257-272, 1998.

[7]   J. Ajitsaria, S.Y. Choe, D. Shen, D.J. Kim, "Modeling and analysis of a bimorph piezoelectric cantilever beam for voltage generation," Smart Materials and Structures, Vol.16, pp.447-454, 2007.

[8]   S.R. Anton, H.A. Sodano, "A review of Power harvesting using piezoelectric materials" Smart Materials and Structures, Vol.16, pp. 1-21, 2007.

[9]   T. Eggborn, "Analytical models to predict power harvesting with piezoelectric materials," Dissertação de Mestrado - Virginia Polytechnic Institute and State University, 2003.

[10] J. Ajitsaria, S.Y. Choe, D. Shen, D.J. Kim, "Modeling and analysis of a bimorph piezoelectric cantilever beam for voltage generation," Smart Materials and Structures, Vol.16, pp.447-454, 2007.

[11] H. A. Sodano, G. Park, D. Leo, D.J. Inman, Use of piezoelectric energy harvesting devices for charging batteries. Proceedings of SPIE 5050, 10 – 108, 2003.

[12] H. A. Sodano, J. Lloyd, D.J. Inman, An experimental comparison between several active composite actuators for power generation. Proceedings of SPIE, 370-378? 2004.

[13] H. A. Sodano, G. Park, D.J. Inman, Estimation of electric charge output for piezoelectric energy harvesting. Journal of Strain, 40(2), 49-58? 2004.

[14] A. Erturk, D.J. Inman, A distributed parameter electromechanical model for cantilevered piezoelectric energy harvesters, Journal of Vibration and Acoustics, Vol. 130, 041002, 2008.

[15] A. Erturk, D. J. Innam, Issues in mathematical modeling of piezoelectric energy harvesters, Smart Materials and Structures, Vol.17, 065016 (14pp), 2008.

[16] A. Erturk, D. J. Innam, On mechanical modeling of cantilevered piezoelectric vibration energy harvesters, J. Intell. Mater. Syst. Struct., Vol. 19, pp.1311-1325, 2008.

[17] A. Erturk, J.M. Renno, D. J. Innam, Modeling of piezoelectric energy harvesting from an L-shaped beam-mass structure with an application to UAVs," Vol. 20, pp. 529-544, 2009.

[18] A. Erturk, D. J. Innam, An experimentally validated bimorph cantilever model for piezoelectric energy garvesting from base excitations, Vol. 18, 025009 (18pp), 2009.

[19] A. Nechibvute, A.R. Akande and P.V.C. Luhanga, Modelling of a PZT Beam for Voltage Generation, Pertanika J. Sci. & Technol. 19 (2): 259-271, 2011.

[20] F. Lu, H. P. Lee, S. P. Lim, Modeling and analysis of micro piezoelectric power generators for micro- electromechanical-systems applications, Smart Materials and Structures, Vol.13, pp. 57-63, 2004.

13

Analysis of Stress Mitigation through Defence Hole System in GFRP Composite Bolted Joint

Khudhayer J. Jadee[1,*], A.R. Othman[2]

[1]Technical Engineering College-Baghdad, Middle Technical University, Baghdad, Iraq
[2]School of Mechanical Engineering, Universiti Sains Malaysia, Malaysia
*Corresponding author: khudhayer1970@yahoo.com

Abstract The effect of the defence hole system (DHS) on the stress distribution around the bolt-hole in glass fibre reinforced polymer (GFRP) composite bolted joint has been investigated using a finite element method. The analyses have been carried out on a double-lap composite bolted joint with various geometric parameters for two cases of without and with DHS. The analyses have taken into account a 3D stress plane condition, in which the circumferential and radial stresses at the bearing region, shear-out and net tension regions of the bolt-hole have been determined. Results showed that adding auxiliary hole near the bolt-hole has contributed in reducing the stresses in the vicinity of the bolt-hole.

Keywords: *defence hole system, bolted joint, GFRP, finite element analysis*

1. Introduction

Several methods are available for the use in the assembly between the composite parts, including adhesively bonded joint, mechanical fastening joint, or combination of them. Bolted joint is one of the mechanical fastening methods, which is preferred due to low cost, free from surface treatment [1,2], and the simplicity in repair and maintenance [3]. The stress concentration around the bolt-hole is critical in the design of the bolted joint. These stresses affect the failure load as well as the bearing strength that a structure could sustain, thus the focus should consider on how the stress concentration around the bolt-hole could be reduced to avoid undesirable catastrophic failure.

Various techniques are available from the literatures in reducing the stresses in the vicinity of the hole in the composite plate, such as by increasing the thickness of the plates [4], reinforcement of the hole [5,6], applying trajectorial fiber steering aligned with the stress vectors [7] and optimization the hole shape [8,9,10,11,12]. The latter technique, however could not be applied in the case of bolted joints due to the hole shape being related to the bolt.

One of the methods known as a defence hole system (DHS), has been used to redistribute the stresses around the main hole by introducing an auxiliary holes in the low stress area near the main hole. The method allows the flow of the stress trajectories to be smoothened around the main hole. The previous studies have investigated the use of DHS in the isotropic plates with hole by optimizing the shape, size, and distance of the auxiliary hole to achieve maximum stress reduction, with some results have successfully reported the reduction of 30% [13-23]. However,

only limited studies have considered the DHS for the use in orthotropic plates [24,25,26,27] to achieve reduction of the stress concentration around the main hole. The work has proven that the maximum stress could be well reduced up to 31.7% [27].

Several investigations on the DHS in isotropic and/or orthotropic plates with main central hole have been analysed using finite element method to examine the effect of the configuration in mitigating the stresses around the primary hole [17,18,21,24,25,26,28,29,30]. However, the question remains on how the DHS could improve the stress distribution and reduction, hence increasing the bearing capacity of the composite bolted joints. Therefore, this study presents the numerical investigation on the effect of introducing auxiliary hole near the bolt-hole on the improvement of the stress distribution around the bolt-hole in single-bolt, double-lap composite bolted joint.

2. Finite Element Analysis

The performance of composite bolted joint with single-bolt, double-lab joint as in Figure 1, subjected to tensile loading was analysed using the finite element method (FEM) based on ANSYS code. The analysis was carried out on the two different configurations of without the DHS and with the DHS.

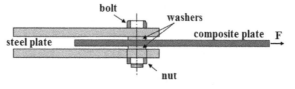

Figure 1. Single-bolt, double-lab joint configuration of composite plate

2.1. Geometry and Material Properties

The configuration of the coupons for the cases of without DHS and with DHS was illustrated in Figure 2. The different in width to bolt-hole diameter ratio (W/D) as well as the edge distance to bolt-hole diameter ratio (E/D) were studied in both cases. The effects of W/D ratios at 2, 3, 4 and 5 and the E/D ratios at 1, 2, 3, 4 and 5 on the stress distribution were examined. For the coupons with the DHS, two additional parameters were further investigated, the first was the defence hole diameter (DHD), while the second parameter involved the defence hole distance from the bolt-hole (DS). Both DHD and DS were selected relative to the bolt-hole diameter (D), in which the values of DHD were 0.625D and 0.75D, while the DS values were varied as 1.5D, 2D, and 2.5D. As a result, 20 models have been evaluated for the laminates without the DHS, and 120 models for the DHS counterparts.

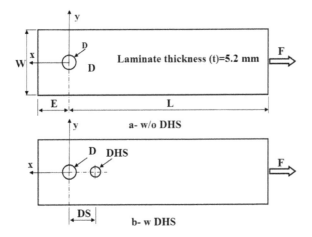

Figure 2. The geometry of composite coupons with and without the DHS

The composite plate was made of plain weave 800g/m² glass fibre reinforced polymer (GFRP) consisted of eight plies oriented in the load direction as [0]₈. In general, the balanced plain weave structure composes of two fibre directions (weft and warp), assuming one direction was aligned parallel to the applied load and the other was perpendicular to it. The mechanical properties of the 800g/m² plain weave GFRP lamina were determined experimentally, and listed in Table 1.

Table 1. Mechanical properties of 800g/m² plain weave GFRP laminate

Parameters	Symbol (Units)	Values	Standard test method
Longitudinal modulus	E_1 (MPa)	17129	ASTM D3039
Transverse modulus	E_2 (MPa)	17129	ASTM D3039
Shear Modulus	G_{12} (MPa)	3693	ASTM D3518
Poisson's ratio	v_{12}(-)	0.24	ASTM D3039
Longitudinal tension strength	X_t (MPa)	222.14	ASTM D3039
Longitudinal compression strength	X_c (MPa)	96.7	ASTM D3410
Transverse tension strength	Y_c (MPa)	222.14	ASTM D3039
Transverse compression strength	Y_c (MPa)	96.7	ASTM D3410
Laminate shear strength	S (MPa)	65.27	T-specimen [31,32,33,34]

SOLID185 elements were utilized to model the laminate. This element is a 3-D layered structural solid element defined by eight nodes with three degrees of freedom at each node, translated in the nodal x, y, and z directions. It is available in two options; a homogeneous structural solid (KEYOPT K3=0) and a layered structural solid (KEYOPT K3=1) which usually used to simulate the layered thick shells or solids. Higher concentration of the elements around the holes has been set to provide the superior accuracy in the high stresses region.

2.2. Boundary Conditions

Due to the symmetrical condition of the model with respect to x-z plane, only half of the model has been constructed for the analysis. The boundary conditions for the laminates for both cases were shown in Figure 3. A uniform distributed tensile load was applied to the nodes at the far end of the composite edge. In order to simulate the condition of rigid bolt, a radial boundary condition was applied at all nodes in the area that in contact with the bolt; these nodes are constrained in the radial direction and free in the circumferential direction [2,31,32,33,34]. Finger tight torque was applied to the bolt; the torque restrained the nodes on the surface under the washers in the z-direction. In the symmetry axis, the displacement of the nodes was constrained in the y-direction.

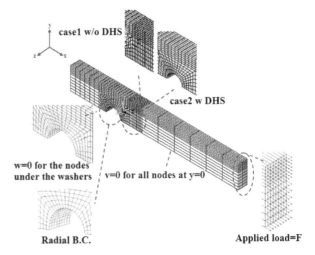

Figure 3. Finite element model of laminate with and without DHS

2.3. Model Verification

In order to verify the herein model, a previous work on the stress distribution around the bolted-hole of the [0/45/90/-45]s carbon fibre epoxy laminate was selected for the verification [35]. The laminate consisted of large values of W/D and E/D (equal to 5), with the geometry of the bolted-hole specimen is illustrated in Figure 4.

In those work, the authors have applied a full 3-D contact problem using ABAQUS program where three-dimensional solid brick elements (C8D3) were used for this purpose, assuming the bolt joint as a pinned joint. The stress distribution around the bolt-hole was examined as a polar stresses (radial and circumferential) normalized by the bearing stress (σ_b), as computed using the following Equation 1:

$$\sigma_b = \frac{F}{D.t} \qquad (1)$$

where, F is the applied load (N), D is the bolt-hole diameter (mm), and t is the laminate thickness (mm).

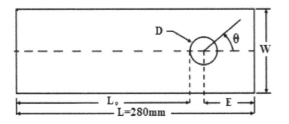

Figure 4. The geometry of the test specimen for pinned-loaded joint

2.4. Stress Analysis

Since the bolt-hole provides the critical region in the joints, the state of the stresses around the bolt-hole requires to be considered extensively. A plane stress condition was assumed for the model, in which only the in-plane stresses were considered in the stress analysis (i.e. τ_{13}, τ_{23}, $\sigma_3 = 0$). A tensile load of 2000N was applied to the model; this value was chosen to be less than the minimum

experimental yield failure load for the weakest geometry (W/D=2 with E/D=1), to ensure that the analysis was completed before the failure initiation. For the laminates without the DHS, the effect of the geometric parameters (W/D and E/D) on the stress distribution around the bolt-hole was also investigated. The effect the DHS on the stress distribution was also analysed extensively.

3. Results and Discussion

3.1. Model Verification Results

Figure 5 shows the comparisons of circumferential and radial stresses distribution around the bolt-hole between the current model and the previous work. The results verified similar tendency of the stress distribution obtained between the comparative results, with some variations were observed in the stress values. The maximum disparity between those two models was computed as 19% for the circumferential stress ($\sigma_{\theta\theta}/\sigma_b$) at ply 1 of $\theta=90°$, while up to 13.7% variation was observed for radial stress (σ_{rr}/σ_b) at ply 4 of $\theta=-45°$. These variations were attributed to the differences of the types of elements used and the differences in the modelling techniques of the pin hole contact simulation.

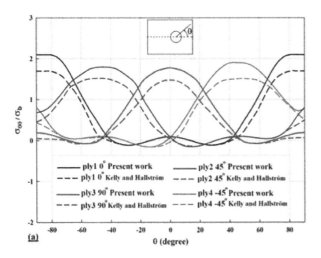

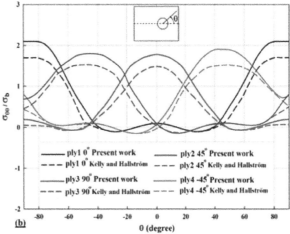

Figure 5. (a) Circumferential and (b) radial stresses distribution around the bolt-hole of the [0/45/90/-45]s laminates at an applied load of F = 1.2 kN

3.2. Laminated Bolted Joint without the DHS

The distribution of stresses around the hole at applied load F=2000N were examined as shown in Figure 6, in which the effects of W/D and E/D were evaluated. It was clear that the maximum value of the normalized circumferential stress ($\sigma_{\theta\theta}/\sigma_b$) was observed at $\theta=90°$ for all W/D and E/D ratios, indicating a region of high stress concentration at the hole boundary in the net-tension plane.

The maximum value of the normalized radial stresses (σ_{rr}/σ_b) for the laminates with short edge distance (E/D=1) was located at $\theta=52.5°$ for all W/D ratios, whilst for the laminates with E/D=2, the (σ_{rr}/σ_b)max was observed at $\theta=22.5°$ for W/D=2 to 3 and $\theta=30°$ for W/D≥4. These results indicated that all laminates experienced a shear-out failure mode around the bolted joint. However, for E/D≥3 at all the tested W/D ratios, the value of the (σ_{rr}/σ_b)max was found at $\theta=15°$, indicating a high stress concentration at

the bearing region. The behavior of the above results was found to completely agree with the results of stress distribution presented in the previous technical paper [36].

To further illustrate the effects of the geometrical dimension on the stress distribution of the bolted joint, the normalized circumferential stresses at $\theta=90°$ for all E/D and W/D ranges at an applied load of 2000N was plotted as in Figure 7. From the analysis, it was apparent that a large decline in the ($\sigma_{\theta\theta}/\sigma_b$) has taken place for the laminates with the edge distance ratio of E/D=1 to 3 (Figure 7a) and width ratio of W/D=2 and 3 (Figure 7b). Within the range of the ratio of 1≤ E/D ≤2, the joint were subjected to high stresses, thus causing shear out failure mode. Beyond the values, a small decline in the circumferential stress was observed and the specimens have experienced less stress concentration, which was expected to reduce the probability of failure in shear-out or net-tension. Except for the joint with W/D=2, all the laminates have suffered from the bearing mode; a desirable failure for the bolted joint configuration.

Similar phenomena were observed when the stress profile was plotted as a function of W/D. For smaller configurations of W/D=2 and 3, the joint with the ratio of 1≤E/D≤2 experienced a shear-out mode, whilst for larger ratio of E/D≥3, the laminates suffered net tension mode

for W/D<3. But beyond these values, the failure has changed to the bearing mode due to the reduce stress values around the main hole. It was also apparent that the ($\sigma_{\theta\theta}/\sigma_b$) has decreased by increasing the width and edge ratios. Interestingly, for those smaller geometries of E/D, the reduction in the circumferential stress was very minimal with the failure remained as the shear-out mode.

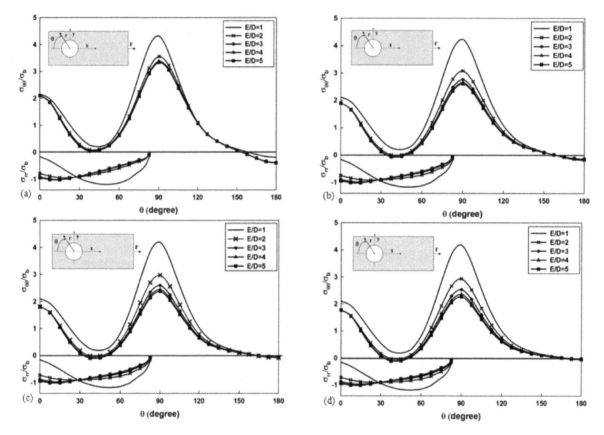

Figure 6. Normalized stresses in laminate at F=2000N for (a) W/D=2; (b) W/D=3; (c) W/D=4 and (d) W/D=5

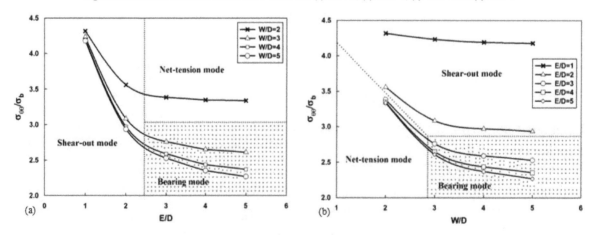

Figure 7. Normalized circumferential stresses (at θ=90°) as a function of (a) E/D and (b) W/D at F=2000N

Figure 8 summarizes the normalized shear stresses (τ_{xy}/σ_b) at θ=90° for all E/D and W/D ratios subjected to loading of F=2000N. For the plot of (τ_{xy}/σ_b) versus E/D, the shear stress was found to drop abruptly for E/D=1 and E/D=2. Higher stress was observed for that range of smaller edge ratios, contributing to the shear out failure mode. As the E/D was increased to 3, a less accentuated inclination was observed, and beyond that the inclination

was barely recognized. The failure has changed as the E/D was increased beyond E/D>2, to the bearing mode for all the joints, except for the laminates with W/D=2 (i.e. net-tension).

It has also been pointed out that the maximum shear stresses was propagated in the laminates with short edge distance (i.e. E/D=1), contributing to the occurrence of the shear-out failure mode. The reduction of about 50% in the normalized shear stresses was computed for the laminates with E/D=2 in comparison to that with E/D=1. This reduction percentage has increased slightly as the edge ratio increased beyond E/D≥3. As a result, the effect of W/D on shear stress was found small or barely noticeable.

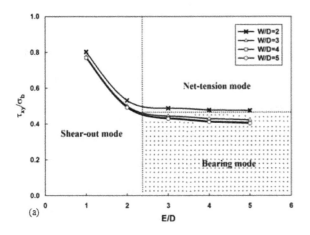
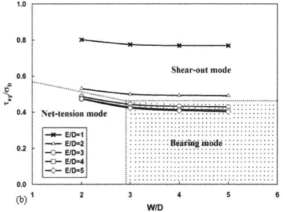

Figure 8. Normalized shear stresses as a function of (a) E/D and (b) W/D subjected to load of F=2000N for shear-out plane (Y=R)

3.3. Laminated Bolted Joint with the DHS

In the defense hole system, an auxiliary hole is introduced in the low stress region near the main hole to redistribute the stresses or to smoothen the stress flow lines around the main hole. The examples of stress distribution (σ_x, σ_y, and σ_{xy}) of laminated composite bolted holes without and with the DHS at 2000N applied load are shown in Figure 9 - Figure 11. It was apparent that the stress contours highlighted the high and low stress regions. The maximum tensile stress ($+\sigma_x$) was observed in the area between the hole boundary and the width edge (net-tension plane) and the maximum compression stress ($-\sigma_x$)

was obtained in the area opposite to the bolt shank (bearing area). Whilst, for the shear stress (σ_{xy}), the maximum concentration was clearly found in the area between the hole boundary and the end edge (shear-out plane). The introduction of the auxiliary hole in the bolted joint configuration has helped to redirect the stress profiles to wider area around the main and defense holes. This then contributed to the reduction in the maximum stress contours especially at the bearing region, hence improving the loading capacity of the joint before failure. As a result, lower stress distribution could be observed around the auxiliary hole as well as in the area behind the bolt shank, but then, it did not posses as a critical region in the joint.

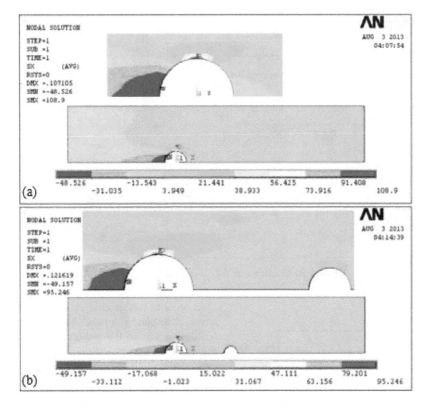

Figure 9. Stress distribution (σ_x) around the bolt hole for laminates with W/D=5 and E/D=5 at F=2000N;(a) Laminate without the DHS (b) Laminate with the DHS

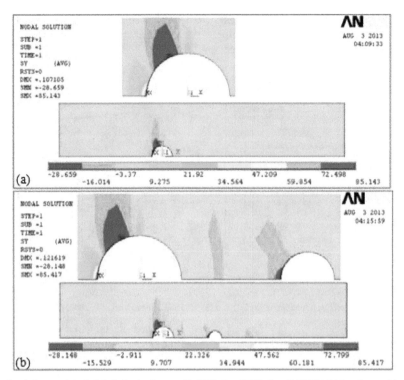

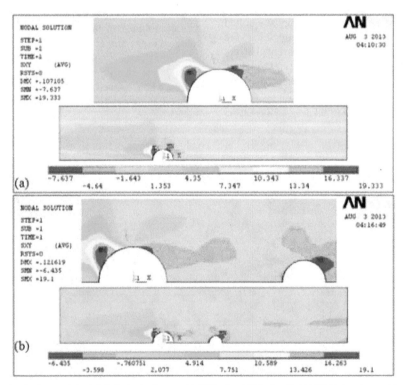

Figure 10. Stress distribution (σ_y) around the bolt hole for laminates with W/D=5 and E/D=5 at F=2000N;(a) Laminate without the DHS (b) Laminate with the DHS

Figure 11. Stress distribution (σ_{xy}) around the bolt hole for laminates with W/D=5 and E/D=5 at F=2000N;(a) Laminate without the DHS (b) Laminate with the DHS

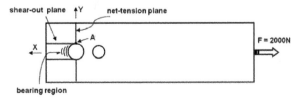

Figure 12. Main failure planes in the composite specimen

In order to further analyze the effect of the DHS on the stress distribution around the bolt-hole, the Cartesian tensile stresses in the x-direction (σ_x) within the critical region of the net-tension plane (point A in Figure 12) was evaluated. For all specimens, a comprehensive comparison of the stress with respect to laminates without the DHS was attained for all the geometric parameters of W/D and E/D, and the DHS parameters of DHD and DS, as shown in Figure 13 and Figure 14.

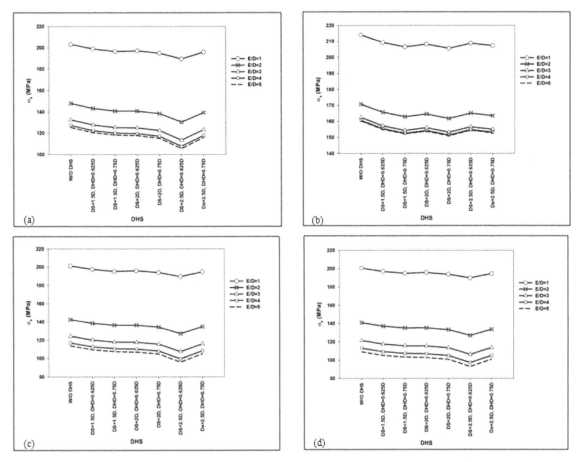

Figure 13. Tensile stresses (σ_x) at 2000N applied load of laminates with (a) W/D=2, (b) W/D=3, (c) W/D=4 and (d) W/D=5

It was found that the general trend of the results revealed that the stresses have decreased with the edge distance ratio (E/D). All the laminates have demonstrated the minimum stresses (σ_x) when the DHS of diameter of DHD=0.625D was introduced at a distance of DS=2.5D, except for those narrow laminates with W/D=2, in which the minimum stresses occurred at DS=2D and DHD=0.75D. For the laminates with the defence hole distances of 1.5D and 2D, more stress reduction could be achieved by increasing the DHD. Moreover, similar behaviour was observed for those of the narrow laminates (i.e. W/D=2) with a defence hole distance of 2.5D. However, for wide laminates (i.e. W/D≥3) with the defence hole distance of 2.5D, the maximum stress reduction was achieved when the DHD decreased from 0.75D to 0.625D.

In contrast, Figure 14 shows that the narrow laminates of W/D=2 could retain high stresses, and the effect of the DHS on the stress reduction was found insignificant. But when the laminate width has increased, these stresses have shown a remarkable decrease. For most of the cases, the maximum tensile stress reduction was observed when the DS=2.5D and DHD=0.625D. The configuration has enabled the stress to be effectively re-distributed at the area behind the bolt shank. In contrast, when the DHD has been increased to 0.75D, the stress value on the net tension region has increased again, diminishing the benefits of the DHS in controlling the maximum bearing capacity of the bolted joint. However, it was noticed that for the narrow laminates (W/D=2), the minimum stresses has occurred at DS=2D and DHD=0.75D.

Table 2. σ_x reduction (%) at 2000N applied load

Configuration	DS	DHD	E/D =1	E/D =2	E/D =3	E/D =4	E/D =5
			\multicolumn{5}{c}{W/D=2}
A	1.5D	0.625D	-2.3	-3.1	-3.3	-3.3	-3.4
B	1.5D	0.75D	-3.7	-4.9	-5.2	-5.3	-5.3
C	2D	0.625D	-2.8	-3.8	-4.1	-4.2	-4.2
D	**2D**	**0.75D**	**-4.1**	-5.6	-6.0	-6.1	**-6.1**
E	2.5D	0.625D	-2.5	-3.5	-3.7	-3.7	-3.7
F	2.5D	0.75D	-3.3	-4.5	-4.8	-4.8	-4.8
			\multicolumn{5}{c}{W/D=3}
A	1.5D	0.625D	-2.2	-3.3	-3.8	-4.0	-4.1
B	1.5D	0.75D	-3.4	-5.0	-5.8	-6.0	-6.1
C	2D	0.625D	-3.1	-5.1	-6.0	-6.4	-6.5
D	2D	0.75D	-4.3	-6.9	-8.2	-8.7	-8.9
E	**2.5D**	**0.625D**	**-7.2**	-13.6	-16.8	-18.1	**-18.7**
F	2.5D	0.75D	-3.8	-6.3	-7.5	-8.0	-8.1
			\multicolumn{5}{c}{W/D=4}
A	1.5D	0.625D	-1.9	-3.0	-3.5	-3.7	-3.8
B	1.5D	0.75D	-3.1	-4.6	-5.3	-5.6	-5.8
C	2D	0.625D	-2.7	-4.6	-5.6	-6.1	-6.4
D	2D	0.75D	-3.8	-6.3	-7.5	-8.1	-8.5
E	**2.5D**	**0.625D**	**-6.2**	-12.2	-15.7	-17.5	**-18.4**
F	2.5D	0.75D	-3.4	-5.9	-7.2	-7.8	-8.2
			\multicolumn{5}{c}{W/D=5}
A	1.5D	0.625D	-1.9	-2.8	-3.3	-3.5	-3.7
B	1.5D	0.75D	-3.0	-4.3	-5.0	-5.3	-5.5
C	2D	0.625D	-2.6	-4.2	-5.2	-5.7	-6.0
D	2D	0.75D	-3.6	-5.8	-7.0	-7.6	-8.0
E	**2.5D**	**0.625D**	**-5.7**	-11.0	-14.3	-16.2	**-17.2**
F	2.5D	0.75D	-3.2	-5.5	-6.8	-7.5	-7.9

Table 2 has further distinguished the percentage in the stress reduction at 2000N applied load for the tested composite bolted joints with the DHS, in comparison with those of without the DHS. For the narrow laminates of W/D=2, the stress reduction ranged from 2.3% (for the laminates with small edge distance of E/D=1) to 6.1% (for the laminates with large edge distance of E/D=5). Within the range of E/D ratios, the benefits of the auxiliary hole was clearly marked at DS=2D and DHD=0.75.

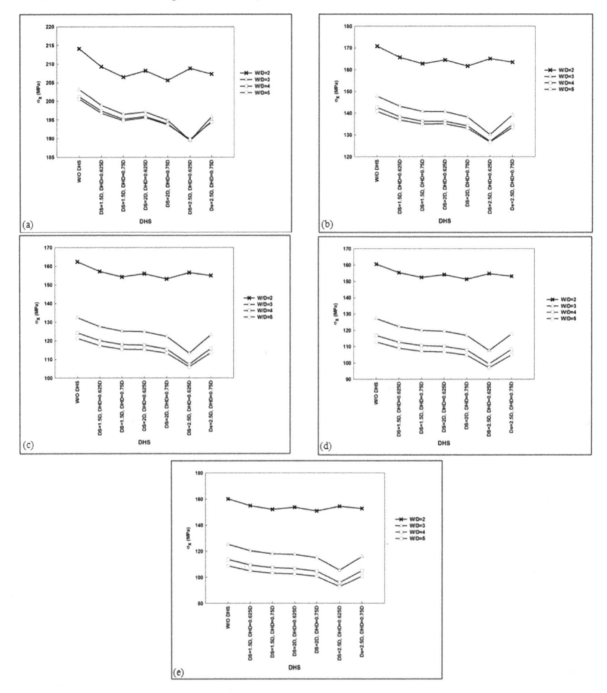

Figure 14. Tensile stresses (σ_x) at 2000N applied load of laminates with (a) E/D=1, (b) E/D=2, (c) E/D=3, (d) E/D=4 and (e) E/D=5

In contrast, for those of wide laminates (i.e. W/D≥3) the stress reduction has ranged from 1.9% (for laminates with small edge distance of E/D=1 at W/D=4 and 5) to remarkably 18.6% (for the laminates with large edge distance of E/D=5 at W/D=4). It was clear that by introducing the DHS in the design of composite bolted joint, the maximum stress around the main hole (especially at the net-tension region) has been successfully reduced with the benefits of the auxiliary hole was more apparent for large geometries of W/D and E/D configurations.

4. Conclusions

Finite element analysis on double-lap single-bolt joint in GFRP composite structure for laminates with and without the DHS was carried out to study the stress distribution around the bolt-hole. The effect of introducing auxiliary hole in the low stress region near the bolt-hole on the stress concentration has been investigated using

ANSYS program. The results have shown that the auxiliary hole contributed in reducing the stress concentration in the vicinity of the bolt-hole; this reduction related to the geometric parameters of the laminate (W/D and E/D) and the size and position of the auxiliary hole (DHD and DS). Optimum stress mitigation was obtained for the wide laminates (W/D >2) with the DHS of auxiliary hole diameter of DHD=0.625D at a distance of DS=2.5D from the bolt-hole, which obtained a maximum stress reduction of 18.6% than those of the counterparts without the DHS.

References

[1]   Choi JH, Ban CS, Kweon JH, Failure load prediction of a mechanically fastened composite joint subjected to a clamping force, *Journal of Composite Materials*, 42(14). 1415-1428. July 2008.

[2]   Ryu CO, Choi JH, Kweon JH, Failure load prediction of composite joints using linear analysis, *Journal of Composite Materials*, 41.865-878. April 2007.

[3]   Goswami S. A finite element investigation on progressive failure analysis of composite bolted joints under thermal environment, *Journal of Reinforced Plastics and Composites*, 24. 161-171. January 2005.

[4]   Tenchev TR, K. Nygard M, Echtermeyer A, Design procedure for reducing the stress concentration around circular holes in laminated composites, *Composites*, 26 (12). 815-828. December 1995.

[5]   Giare GS, Shabahang R, The reduction of stress concentration around the hole in an isotropic plate using composite materials, *Engineering Fracture Mechanics*,32(5). 757-766. 1989.

[6]   Muc A, Ulatowska A, Local fibre reinforcement of holes in composite multilayered plates, *Composite Structures*, 94(4). 1413-1419. March 2012.

[7]   Tosh MW, Kelly DW, On the design, manufacture and testing of trajectorial fibre steering for carbon fibre composite laminates, *Composites Part A: Applied Science and Manufacturing*, 31(10). 1047-1060. October2000.

[8]   Dhir SK, Optimization in a class of hole shapes in plate structures, *Journal of Applied Mechanics*, 48(4). 905-908. January 1981.

[9]   Wu Z. Optimal hole shape for minimum stress concentration using parameterized geometry models, *Structural and Multidisciplinary Optimization*, 37(6). 625-634. February 2009.

[10] Wu H-C, Mu B, On stress concentrations for isotropic/orthotropic plates and cylinders with a circular hole, *Composites Part B: Engineering*, 34(2). 127-134. March 2003.

[11] William LK, Stress concentration around a small circular hole in the HiMAT composite plate. *NASA technical memorandum 86038*, 1-16. December 1985.

[12] Toubal L, Karama M, Lorrain B, Stress concentration in a circular hole in composite plate, *Composite Structures*, 68(1). 31-36. April 2005.

[13] Durelli AJ, Rajaiah K, Optimum hole shapes in finite plates under uniaxial load, *Journal of Applied Mechanics*, 46(3). 691-695. 1979.

[14] Rajaiah K, Naik NK, Hole-shape optimization in a finite plate in the presence of auxiliary holes, *Experimental Mechanics*, 24(2). 157-161. June 1984.

[15] Erickson PE, Riley WF, Minimizing stress concentrations around circular holes in uniaxially loaded plates, *Experimental Mechanics*, 18(3). 97-100. March 1978.

[16] Naik NK, Photoelastic investigation of finite plates with multi-holes, *Mechanics Research Communications*, 15(3). 141-146. May 1988.

[17] Akour SN, Nayfeh JF, Nicholson DW, Defense hole design for a shear dominant loaded plate, *International Journal of Applied Mechanics*, 2(2). 381-398. June 2010.

[18] Sanyal S, Yadav P, Relief holes for stress mitigation in infinite thin plates with single circular hole loaded axially, *ASME 2005 International Mechanical Engineering Congress and Exposition*, ASME . 717-720.

[19] Providakis CP, Sotiropoulos DA, A BEM approach to the stress concentration reduction in visco-plastic plates by multiple holes, *Computers and Structures*, 64(1). 313-317. July 1997.

[20] Nagpal S, Optimization of rectangular plate with central square hole subjected to in-plane static loading for mitigation of SCF, *International Journal of Engineering Research and Technology*, 1(6). 1-8. August 2012.

[21] Nagpal S, S.Sanyal, Jain N, Mitigation curves for determination of relief holes to mitigate stress concentration factor in thin plates loaded axially for different discontinuities, *International Journal of Engineering and Innovative Technology*, 2(3). 1-7. September 2012.

[22] Ulrich TW, Moslehy FA, A boundary element method for stress reduction by optimal auxiliary holes, *Engineering Analysis with Boundary Elements*, 15(3). 219-223. January 1995.

[23] Meguid SA, Shen CL, On the elastic fields of interacting defense and main hole systems, *International Journal of Mechanical Sciences*, 34(1). 17-29. January 1992.

[24] Rhee J, Rowlands RE, Stresses around extremely large or interacting multiple holes in orthotropic composites, *Computers and Structures*,61(5). 935-950. December 1996.

[25] Rhee J, Cho H-K, Marr DJ, Rowlands RE, Local compliance, stress concentrations and strength in orthotropic materials, *The Journal of Strain Analysis for Engineering Design*, 47(2). 113-128. February 2012.

[26] Jain NK, The reduction of stress concentration in a uni-axially loaded infinite width rectangular isotropic/orthotropic plate with central circular hole by coaxial auxiliary holes, *IIUM Engineering Journal*, 12(6). 141-150. 2011.

[27] Akour SN, Al-Husban M, Nayfeh JF, Design and optimization of defense hole system for hybrid loaded laminates, *Technology Engineering and Management in Aviation: Advancements and Discoveries*, IGI Global, Hershey, 151-160. 2012.

[28] Jindal UC, Reduction of stress concentration around a hole in a uniaxially loaded plate, *The Journal of Strain Analysis for Engineering Design*, 18(2). 135-141. April 1983.

[29] Akour SN, Nayfeh JF, Nicholson DW, Design of a defence hole system for a shear-loaded plate, *The Journal of Strain Analysis for Engineering Design*, 38(6). 507-517. January 2003.

[30] Meguid SA, Gong SX, Stress concentration around interacting circular holes: a comparison between theory and experiments, *Engineering Fracture Mechanics*, 44(2). 247-256. January 1993

[31] Okutan B, The effects of geometric parameters on the failure strength for pin-loaded multi-directional fiber-glass reinforced epoxy laminate, *Composites Part B: Engineering*, 33(8). 567-578. November 2002.

[32] Karakuzu R, Çalışkan CR, Aktaş M, İçten BM, Failure behavior of laminated composite plates with two serial pin-loaded holes, *Composite Structures*, 82(2). 225-234. January 2008.

[33] Karakuzu R, Taylak N, İçten BM, Aktaş M, Effects of geometric parameters on failure behavior in laminated composite plates with two parallel pin-loaded holes, *Composite Structures*, 85(1). 1-9. September 2008.

[34] Karakuzu R, Gülem T, I ten BM, Failure analysis of woven laminated glass-vinylester composites with pin-loaded hole, *Composite Structures*, 72(1). 27-32. January 2006.

[35] Kelly G, Hallström S, Bearing strength of carbon fibre/epoxy laminates: effects of bolt-hole clearance, *Composites Part B: Engineering*, 35(4). 331-43. January 2004.

[36] Crews JH, Jr., Hong CS, Raju IS, Stress-concentration factors for finite orthotropic laminates with a pin-loaded hole, *NASA Technical Paper 1862*, 1-40. May1981.

14

Design of a Thermally Homeostatic Building and Modeling of Its Natural Radiant Cooling Using Cooling Tower

Peizheng Ma[1,*], Lin-Shu Wang[1], Nianhua Guo[2]

[1]Department of Mechanical Engineering, Stony Brook University, Stony Brook, United States
[2]Department of Asian and Asian American Studies, Stony Brook University, Stony Brook, United States
*Corresponding author: peizheng.ma@alumni.stonybrook.edu

Abstract Thermal Homeostasis in Buildings (THiB) is a new concept consisting of two steps: thermal autonomy (architectural homeostasis) and thermal homeostasis (mechanical homeostasis). The first step is based on the architectural requirement of a building's envelope and its thermal mass, while the second one is based on the engineering requirement of hydronic equipment. Previous studies of homeostatic building were limited to a TABS-equipped single room in a commercial building. Here we investigate the possibility of thermal homeostasis in a small TABS-equipped building, and focus on the possibility of natural summer cooling in Paso Robles, CA, by using cooling tower alone. By showing the viability of natural cooling in one special case, albeit a case in one of the most favorable locations climatically, a case is made that the use of cooling tower in thermally homeostatic buildings should not be overlooked for general application in wider regions of other climatic zones.

Keywords: *thermally homeostatic building, building design, building energy modeling, TABS, cooling tower, hydronic radiant cooling, small commercial building*

1. Introduction

Thermal Homeostasis in Buildings (THiB) is a new concept developed in two recently published articles [1,2]. Its development consists of two steps: thermal autonomy (architectural homeostasis) first and then thermal homeostasis (mechanical homeostasis). This new approach was called process assumption-based (dynamic) design method [1]. Thermal autonomy [1] is an "architectural" step that determines a building's passive features for keeping indoor operative temperature within a prescribed temperature range without HVAC equipment; a building that meets a constraint of a maximum indoor operative temperature variation under a given ambient temperature amplitude was called thermally autonomous building. The mechanical step engineers a building's active features for keeping indoor operative temperature at desirable temperature level with heat extraction equipment (i.e., heat pump, cooling tower, solar thermal panel, etc.) through thermally activated building systems (TABS). The possibility of achieving thermal homeostasis in an office building was investigated by applying a cooling tower in one summer in seven selected U.S. cities [2]. Instead of sizing equipment as a function of design peak hourly temperature as it is done in heat balance design approach of selecting HVAC equipment, it was shown that the conditions of using cooling tower depend on both

"design-peak" daily-mean temperature and the diurnal temperature amplitude. The study indicated that homeostatic building with natural cooling (by cooling tower alone) is possible in locations of special meso-scale climatic condition, such as Sacramento, CA. In other locations the use of cooling tower alone can only achieve homeostasis partially.

The investigation in the previous articles [1,2] was a modeling study using an RC model that was built in Matlab/Simulink originally developed in Ref. [3]. It was applied to a TABS-equipped room in a large multi-story building located in Zürich, Switzerland, the design of which was already done. [4,5] In this paper, instead of a single room, a small commercial building in Paso Robles, California, is designed in Autodesk Revit, and then an RC model of the building is built to investigate the building thermal behavior. Here is the structure of this paper: the special geographical and climatic conditions of Paso Robles are introduced in Section 2; the detailed design of the building in Autodesk Revit is presented in Section 3; Section 4 models the building system in Simulink; the simulation results are in Section 5; after the discussions in Section 6, main conclusions are summarized at the end of this paper.

2. Special Geographical And Climatic Conditions Of Paso Robles

Large diurnal temperature amplitude is preferred for buildings to achieve thermal homeostasis and Paso Robles is an ideal location for thermally homeostatic buildings.

Paso Robles is located at 35°37'36"N and 120°41'24"W, approximately halfway between Los Angeles and San Francisco [6,7]. It is at the southern end of the Salinas River Valley [8], which is "approximately 75 km in length and 20-30 km in width, oriented in a NW-SE direction. The valley opens into Monterey Bay on the Pacific coast in the northwest and in the southeast it gradually merges into the coastal mountains. It is bounded by the Gabilan mountain range on the east and Sierra de Salinas Mountain on the west (Figure 1). Elevations of the surrounding mountains are typically near 900-1000 m above mean sea level. Fremont Peak (960.61 m) is NE of the city of Salinas within a distance of 15 km. Mt. Toro (1056.36 m) is WSW of Salinas within a distance of 10 km. North Chalone Peak (1001.21 m) is SSE of Salinas at a distance of ~40 km." [8]

"Typical daytime up-valley and nighttime down-valley winds prevail in the Salinas Valley. However, since it is a coastal valley this diurnal pattern is strongly influenced by the coastal winds and the land and sea breeze systems. During nighttime in the northern part of the valley, the land breeze regime results in the winds descending down the mountain slopes toward the Pacific Ocean. In the middle and southern part of the valley, the down-valley wind blows towards Monterey Bay (Figure 1). However, nighttime meteorological data for the study period shows significant periods of up-valley flows." [8]

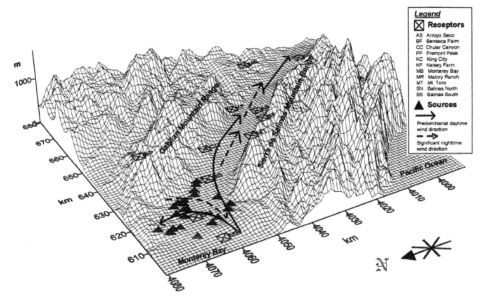

Figure 1. 3D view showing the sources, receptors and the main wind patterns in the Salinas river valley, CA [8]

Due to its special geographical condition, the Paso Robles area consists of two different climate types and classifications [6]: KCC type *BSk* and KCC type *Csb*. The types are based on the Köppen climate classification (KCC) system. The type *BSk* is a semi-arid, dry, steppe-type climate, and the type *Csb* is the typical, coastal Californian & "Mediterranean" type. The primary climate of the area is defined by long, hot, dry summers and brief, cool, sometimes rainy winters.

The long-lasting, mild autumns and occasional early springs give Paso Robles a unique climate. Summers are usually very hot, with daily temperatures frequently exceeding 100°F (38°C) and even exceeding 110°F (43°C) occasionally. The diurnal temperature swing in summers of Paso Robles is unusually very large: as much as 50°F (28°C). Winters in Paso Robles are often very cool and moist. The lowest temperature can reach to 25°F (-4°C).

According to a minimum 30-year weather record of Paso Robles, [9] the hottest month is July and the coldest month is December. In July, the average high and low temperatures are 93°F (34°C) and 54°F (12°C), respectively; the monthly mean temperature is 74°F (23°C); and the record high temperature is 115°F (46°C) occurred in 1961. In December, the average high and low temperatures are 59°F (15°C) and 34°F (1°C), respectively; the monthly mean temperature is 47°F (8°C); and the record low temperature is 8°F (-13°C) occurred in 1990.

Extending the record period, according to Ref. [10], the all-time record high temperature is 117°F (47°C) on August 13, 1933, and the record low temperature is 0°F (-18°C) on January 6, 1913. On average, there are 81.0 days with high temperatures of 90°F (32°C) or higher and 64.0 days with low temperatures of 32°F (0°C) or lower. At the Paso Robles FAA Airport [11], the record high temperature is 115°F (46°C) on June 15, 1961 and July 20, 1960, and the record low temperature is 8°F (-13°C) on December 22, 1990. There is an average of 86.7 days with highs of 90°F (32°C) or higher and an average of 53.6 days with lows of 32°F (0°C) or lower.

In summer time, Paso Robles has large diurnal temperature amplitudes because of the sea breeze from the Monterey Bay, which is critical for thermally homeostatic buildings. The real-time hour-by-hour dry-bulb temperatures of Paso Robles were requested by email from the website of the U.S. Department of Energy (DOE) [12]. In the four summer months from June to September in 2007, the amplitude distribution of the dry-bulb temperatures in Paso Robles is shown in Figure 2. As seen from the figure, nearly two thirds of the amplitudes are equal or greater than 20°C. This is the main reason that Paso Robles is selected as the building location.

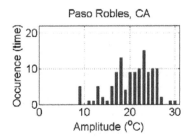

Figure 2. Amplitude distribution of dry-bulb ambient temperatures in Paso Robles (summer, 2007)

3. Design Of A Small Commercial Building In Autodesk Revit

The designed building is a stand-alone, one-story, south-facing, small commercial building located in Paso Robles, which is in Climate Zone 3 [13]. The building has a large lobby, a waiting room, a reception room, three offices and two restrooms. The total cooled and heated floor area is 2310 ft^2 (214.6 m^2). Figure 3 shows its main dimensions.

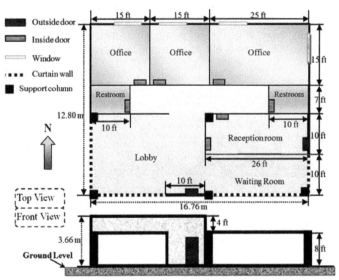

Figure 3. Schematic of the small commercial building

The small commercial building can be divided into two zones: the Front Zone (consisting of the lobby, the waiting room, the reception room and the two restrooms) and the Office Zone (consisting of the three offices). In the exterior walls of the front zone, large curtain walls are installed, which means that this zone is almost transparent to the outdoor environment.

The small commercial building is designed in the Autodesk Revit 2013. The configurations of the building are as follows.

3.1. Exterior Walls

The exterior walls are selected as "Basic Wall: CW 102-85-100p." The U-Factor of this kind of wall is 0.3463 W/m^2-K, which meets the Climate Zone 3's requirement (maximum 0.701 W/m^2-K for mass walls) in the ASHRAE Standard 90.1-2010 [13]. However, in this investigation, the U-Factors of the building envelope are expected to be as close as the values recommended by the Standard 90.1. Therefore, the insulation thickness will be modified and the differences of the building U-Factors and the recommended values will be smaller than 5%. After change the air layer thickness from 0.050 m to 0.014 m, the U-Factor of the wall becomes 0.6907 W/m^2-K (corresponding R-value is 1.4477 m^2-K/W). The thickness of the concrete in the walls is 0.100 m, which is exactly the recommended thickness for exterior walls in Ref. [1]. From exterior side to interior side, the exterior walls consist: 0.102m common brick, 0.014m air, 0.035m cavity fill, vapor retarder membrane layer, 0.100m concrete masonry units, and 0.012m gypsum wall board. The total thickness of the walls is 0.263m.

3.2. Floor

Below the ground level, the floor is selected as "Floor: Concrete Domestic 425mm 2," whose U-Factor is 0.6084 W/m^2-K that does not meet the Zone 3's requirement (maximum 0.606 W/m^2-K for mass floors). After adding a 0.005 m thick carpet, the U-Factor changed to 0.5790 W/m^2-K (corresponding R-value is 1.7270 m^2-K/W). The new floor type is renamed as "Floor: Concrete Domestic 430mm 2." It consists: 0.005m carpet, 0.050m sand/cement screed concrete, 0.175m cast-in-situ concrete, damp-proofing membrane layer, 0.050m rigid insulation, and 0.150m site-hardcore. For thermal protection purpose, the exterior walls are extended to the bottom of the floor.

3.3. Foundation walls

Below the exterior walls, there are 3 ft (0.914 m) depth foundation walls, which are selected as "Basic Wall: Foundation - 300mm Concrete." As its name implies, the walls are constructed by 0.300m-thick cast-in-situ concrete. The thermal resistance of the concrete walls is 0.2868 m^2-K/W.

3.4. Roofs

There are two roofs in the building: the upper level roof, which covers the lobby area, and the lower level roof, which covers other areas of the building. The two roofs have a 2 ft (0.610 m) extension from the outside surface of the exterior walls. The roofs have the same roof type—"Basic Roof: Warm Roof - Concrete." The U-Factor of this roof type is 0.5861 W/m²-K, which does not meet the Zone 3's requirement (maximum 0.273 W/m²-K for insulation entirely above deck roofs). The concrete thickness is 0.225m but is expected to be 0.250m for a TABS-equipped building.

Two modifications are made: the cast-in-situ concrete layer is changed from 0.175m to 0.200m; and the rigid insulation layer is thickened from 0.050m to 0.118m.

Now the U-Factor of the new roof type becomes 0.2723 W/m²-K (corresponding R-value is 3.6731 m²-K/W). The new roof type is renamed as "Basic Roof: Warm Roof – Concrete 250mm." It consists: 0.038m tile roofing, 0.118m rigid insulation, 0.020m asphalt-bitumen, roofing felt membrane layer, 0.050m sand/cement screed concrete, and 0.200m cast-in-situ concrete.

3.5. Interior walls

The type of the interior walls is selected as "Basic Wall: Interior – Blockwork 190," which is made of 0.190m concrete masonry units with 0.012m gypsum wall board on both sides.

3.6. Doors

There are two exterior doors (south and east) and seven interior doors in the building. All the doors are selected as "M_Single-Flush 0915 × 2134 mm," whose U-Factor is 3.7021 W/m²-K (Zone 3's requirement is maximum U-3.975 for swinging opaque doors).

3.7. Windows

In the office zone, there are four exterior windows (three in the north wall and one in the east wall). All the windows are the type "M_Fixed 2134 × 1524 mm 2," which is modified from the basic "M_Fixed" window type. According to the Standard 90.1 [13], in Zone 3 for nonmetal framing vertical glazing that is 0%-40% of wall area, the assembly maximum U-Factor is 3.69 W/m²-K and the assembly maximum SHGC (solar heat gain coefficient) is 0.25. The following glazed panel meets these two requirements: "Double glazing - 1/4 in thick - gray/low-E (e = 0.05) glass," whose U-Factor is 1.9873 W/m²-K that is much smaller than the recommended value and the SHGC is 0.24. The visual light transmittance of this glazed panel is 0.35.

3.8. Curtain Walls

In Revit, no detailed thermal properties of the curtain wall type are presented. According to Ref. [14], "A standard clear insulated double glazing unit has a U-Factor of 2.76 W/m²-K at center-of-glass. When the edge-of-glass and frame are taken into account, the overall U-Factor will become even higher." Comparing to other building components, curtain wall has a higher U-Factor, which may "lead to a number of potential problems, such as high-energy consumption, thermal discomfort to occupant in the perimeter zones, and condensation risk." [14] However, "the typically large continuous span of glazing in curtain walls can provide occupants with pleasant view, contact with outdoors and natural lighting." [14] Many architects prefer large glazing in their designs.

The Zone 3's requirements of metal framing curtain wall are: maximum U-Factor is 3.41 W/m²-K and maximum SHGC is 0.25, which are almost the same as that of nonmetal framing vertical glazing (windows). Therefore, the glazed panels of the curtain walls in the designed building are selected the same as the glazed panels of the windows.

Under the lower level roof, the length of the east-facing curtain wall is 2.743 m and the length of the south-facing curtain wall is 7.010 m; under the upper level roof, the south-facing curtain wall is 5.486 m and the west-facing curtain wall is 5.486 m. Notice that between the upper and lower roofs, there is standard exterior wall, rather than curtain wall. Therefore, including the windows and the curtain walls, the total WWR (window to wall ratio) is 35.2% (25.7% east, 59.0% south, 34.6% west and 18.9% north) and people outside of the building can see most of the front zone through the curtain walls. In Paso Robles, the hottest month is July, and in this month the ambient mean temperature is 23°C with mean peak-to-peak amplitude of 22°C. According to Ref. [15], when the ambient temperature is 22°C and the exterior wall U-Factor is 0.691 W/m²-K, the recommended maximum WWR is about 33% for thermally autonomous buildings. Our design is very close to the recommended value.

3.9. Roof Support Column

Because the curtain walls cannot support the roofs, five roof support columns are added between the foundation and the roofs: four columns supporting the upper level roof are at the corners of the roof and one column supporting the lower level roof is at the southeast corner of the building. The columns are selected as "M_Rectangular Column 610 × 610 mm" and the material is sand/cement screed concrete. These support columns are assumed to be well-insulated to avoid thermal bridge and are not considered in the building models in Section 4 and 5.

The southeast view of the designed small commercial building is shown in Figure 4. Some thermal and physical properties of the materials used in the building are summarized in Table 1.

Table 1. Thermal properties of materials in the building

Category	Material	Conductivity	Capacity	Density
		k (W/m K)	C (kJ/kgK)	ρ (kg/m³)
Brick	Common brick	0.540	0.840	1550
Concrete	Concrete masonry units	1.300	0.840	1800
	Sand/cement screed concrete	1.046	0.657	2300
	Cast-in-situ concrete	1.046	0.657	2300
Insulation	Rigid insulation	0.035	1.470	23
Membrane	Vapour retarder	0.167	1.674	1500
	Damp-proofing	1.150	0.840	2330
	Roofing felt	0.500	1.000	1700
Curtain wall		0.391	—	—
Misc.	Gypsum wall board	0.650	0.840	1100
	Carpet	0.060	1.360	190
	Tile roofing	0.840	0.800	1900
	Asphalt-bitumen	1.150	0.840	2330
	Cavity fill	0.058	0.840	350
	Air	0.025	0.001	1.2
	Site-hardcore	No thermal properties are presented.		

Figure 4. Southeast view of the small commercial building

4. Modeling of the Building in Simulink And Cooling Tower Cooling

4.1. One-zone Model

The small commercial building is modeled by the RC (resistor-capacitor) method used in Refs. [1,2,3,15] in Matlab and Matlab/Simulink, as shown in Figure 5. In this one-zone model, the building envelope (including roofs, exterior walls, windows, curtain walls and doors) is connected to the outdoor air and indoor air with surface thermal resistors; the floor is connected to the indoor air and the earth; inside of the building, there are internal walls and other interior thermal mass (assumed as wood with dimensions of 0.1m × floor area); the indoor air is considered as a small capacitor and its temperature is assumed to be uniform; the internal heat gain is put into the indoor air directly; the solar energy gain is calculated according to the solar geometry on July 15 in Paso Robles, and its distribution is 80% on the floor surface and 20% on the upward surface of other interior thermal mass; the ambient mean dry-bulb temperature is 23°C with the peak-to-peak amplitude of 22°C; the simulation time step is 60 seconds.

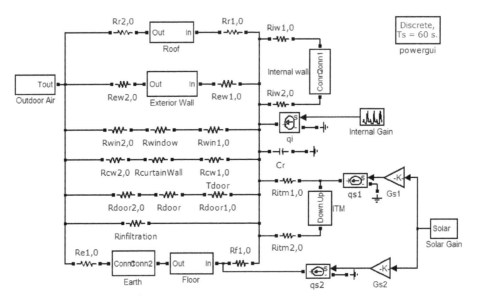

Figure 5. Model of the small partially homeostatic building in Simulink

In order to maintain the operative temperature level, as suggested in Ref. [2], a hydronic system—a cooling tower (CT) combined with thermally active building slabs (TABS)—can be used for summer cooling of a partially homeostatic building in locations with large diurnal temperature variation. The TABS was reviewed in detail in Ref. [16]. The one-zone model in Figure 5 is modified by adding a wet cooling tower and TABS, as shown in Figure 6.

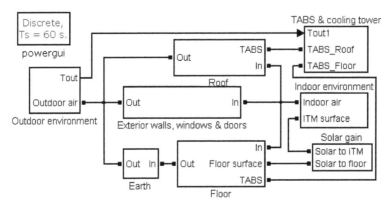

Figure 6. Model of the small homeostatic building in Simulink: one zone

Notice that the detailed components are hidden in the corresponding subsystems in order to make the figure more clearly. The hydronic system works in the nighttime from 8:00PM to 4:00AM to cool down the water in the system by the cold ambient air in the cooling tower. The cold water from the cooling tower is split into two branches: one goes to the pipes embedded in the roof concrete slabs, and the other goes to the floor. Thus the coldness is stored into the large amounts of building thermal mass in the nighttime. Then the water in the two branches is mixed together and goes back to the cooling tower. When the hydronic system does not need to work, the water pump is off and the water in the system is still. In the daytime, the stored coldness is released from the building thermal mass and the indoor environment is thus maintained in the comfortable zone.

Detailed modeling of the hydronic system can be found in Ref. [2]. Here only the two most important parameters of cooling tower, which will be used and discussed later, are defined:

Effectiveness ε (or thermal efficiency), which is between 0 and 1:

$$\varepsilon = \frac{T_{win} - T_{wout}}{T_{win} - T_{wbin}} = \frac{1 - \exp\left[-NTU\left(1 - \dot{C}_w/\dot{C}_a\right)\right]}{1 - \left(\dot{C}_w/\dot{C}_a\right)\exp\left[-NTU\left(1 - \dot{C}_w/\dot{C}_a\right)\right]} \quad (1)$$

where T_{win} is the cooling tower inlet water temperature, T_{wout} is the outlet water temperature, T_{wbin} is the wet-bulb temperature of the inlet air, $NTU = UA_e/\dot{C}_w$ is the Number of Transfer Units, U is the cooling tower overall heat transfer coefficient, and $A_e = A\overline{c}_{pe}/c_p$ is the equivalent heat transfer surface area, A is the heat transfer surface area, $\overline{c}_{pe}$ is the mean specific heat of the moist air treated as an equivalent ideal gas, c_p is the specific heat of moist air, $\dot{C}_w = \dot{m}_w c_{pw}$ and $\dot{C}_a = \dot{m}_a \overline{c}_{pe}$, $\dot{m}_w$ is the

mass flow rate of water, $\dot{m}_a$ is the mass flow rate of air, c_{pw} is the specific heat of water, and $\overline{c}_{pe}$ is the mean specific heat of the moist air treated as an equivalent ideal gas.

Approach is an important indicator of cooling tower performance and defined as "the difference between the cooling tower outlet cold-water temperature and ambient wet bulb temperature" [17]. Lower approach means better cooling tower performance. "As a general rule, the closer the approach to the wet bulb, the more expensive the cooling tower due to increased size. Usually a 2.8°C approach to the design wet bulb is the coldest water temperature that cooling tower manufacturers will guarantee." [17] In this paper, the minimum approach of the wet cooling tower will be kept at 2.8°C.

4.2. Two-zone Model

It is better to model the building into two zones: the Front Zone (the lobby, waiting room, reception room and restrooms) and the Office Zone (the three offices). The zones are separated by the internal walls and doors. As the envelopes of these two zones are quite different, the thermal behavior of the zones should be different. For the hydronic system, the only difference in the two-zone model is that the cold water from the cooling tower is divided into four branches to go to the roof and the floor of each zone. The two-zone model is shown in Figure 7.

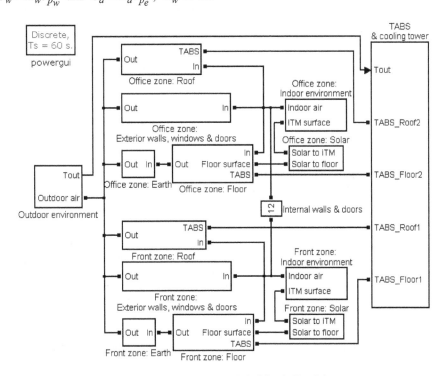

Figure 7. Model of the small homeostatic building in Simulink: two zones

From the energy gain point of view, buildings can be divided into two types [18]: internally load dominated buildings and envelope (externally) load dominated buildings. Commercial and office buildings usually belongs to the first type, which have a large amount of internal heat gains (produced by people, lights and equipment in buildings) and solar energy gains (through windows and curtain walls). In this type of building,

cooling is much more important than heating. In the previous papers [1,2,3,15], all the windows and curtain walls are assumed to have good shading devices and only 8% of solar energy goes into the building. The solar energy gain and the internal heat gain are shown in Figure 8 (a). With good shading devices, the designed building belongs to a moderately internally-load-dominated building type. However, the assumption of the building with good shading may not be practical when occupants of the building choose to take advantage of the large fenestration for aesthetics and natural lighting. If the shading devices are removed and all the available solar energy enters the building, the building becomes a strongly internally-load-dominated one, as shown in Figure 8 (b).

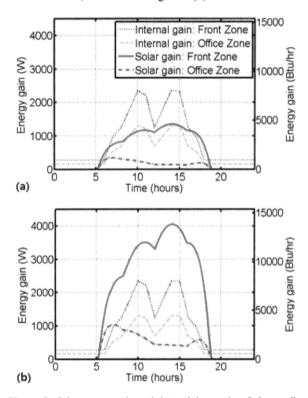

Figure 8. Solar energy gain and internal heat gain of the small commercial building: (a) with shading and (b) without shading

5. Simulation Results

5.1. One-zone Cases

Using the one-zone model, simulation results show that with good shading the mean value of the indoor operative temperature is 28.59°C (without cooling process) and the variation is 1.84°C, which is smaller than the 2°C constraint [1]. Therefore, the designed building is a good thermally autonomous building as the operative temperature range is well maintained. Actually this 2°C per day constraint is pretty strict, as mentioned in Ref. [19] "people find temperature drifts within the comfort zone acceptable up to a rate of 4 K/h (7.2°F/h)." If there is no shading device for the building, the operative temperature variation increases to 2.31°C, while the mean value increases to 32.09°C, because of the bigger influence of the solar energy gain.

Because the diurnal ambient temperature variation is big (which means low nighttime temperature) and the mean value is not high (23°C), a very small cooling tower with effectiveness of 0.031 can cool the indoor operative temperature to the most comfortable summer temperature with the mean value of 25.25°C. Comparing with the previous case without cooling (the hydronic system), the operative temperature variation only increases from 1.84°C to 1.87°C. If there is no shading device, the cooling tower effectiveness should be increased to 0.072

in order to reach the 25.25°C mean operative temperature, while the variation becomes 2.46°C due to the larger solar gain variation.

The simulation results using the one-zone model are summarized in Table 2. Several conclusions are as follows: due to the large fenestration area in the building envelope, shading effects are big on the operative temperature (both variation and mean value) and the cooling tower effectiveness; the hydronic system can keep the operative temperature level in the comfortable range almost without enlarging the operative temperature range.

Table 2. Summarization of the one-zone cases

T_{out} (°C)	Shade/Cool	ΔT_{op} (°C)	Mean T_{op} (°C)	Required CT ε
23.00 ± 11.00	Yes / No	1.84	28.59	NA (not available)
	No / No	2.31	32.09	NA
	Yes / Yes	1.87	25.25	0.031
	No / Yes	2.46	25.25	0.072

5.2. Two-zone Cases

Table 3 summarizes the simulation results of the two-zone model. It can be seen that the indoor operative temperature variation in the Front zone in each case is much bigger than that in the Office zone, and the corresponding mean operative temperature in the Front zone is a little higher than that in the Office zone. The results are expected, because the fenestration area in the Front zone is much larger than that in the Office zone and thus the effect of the solar energy gain should be bigger in the Front zone.

Table 3. Summarization of the two-zone cases

T_{out} (°C)	Shade/Cool	Front zone (°C)		Office zone (°C)		
		ΔT_{op}	Mean T_{op}	ΔT_{op}	Mean T_{op}	CT ε
23.00 ± 11.00	Yes / No	2.06	28.20	1.41	27.71	NA
	No / No	2.72	31.75	1.59	30.06	NA
	Yes / Yes	2.12	25.25	1.47	24.92	0.031
	No / Yes	2.90	25.24	1.74	24.14	0.074

5.3. Two Worst Cases

The hour-by-hour real-time weather data of Paso Robles were requested by email from the website of the U.S. Department of Energy (DOE). [12] The data were collected by the National Weather Service (NWS) from weather stations and stored in a database at the National Renewable Energy Laboratory (NREL). [20] The dry-bulb temperatures in the four summer months from June to September in 2007 are presented in Figure 9.

Analyzing the ambient temperature data, it can be found that in 2007, the mean ambient temperature is from 12.75°C to 29.70°C and the peak-to-peak diurnal amplitude is from 8.90°C to 30.00°C. Therefore, for controlling the operative temperature range, the worst case is 29.70 ± 15.00°C (big amplitude), and for maintaining the operative temperature level, the worst case is 29.70 ± 4.45°C (high mean temperature and small amplitude). Although these worst values may not occur simultaneously, the first worst case may be taken for the simulation of the system without hydronic system (the first two cases in Table 4), and the second worst case may be taken for the simulation of the system with hydronic system (the last two cases in Table 4). The first case is used to test the thermal autonomy of the building and the second one is to test its thermal homeostasis. In Worst case 2, the minimum cooling tower approach is 2.8°C following Ref. [2]. As shown in Table 4, except the operative temperature range in the Office zone, others cannot be well maintained under these worst conditions. Therefore, several possible methods that may improve the system performance will be investigated in the following two sub-sections.

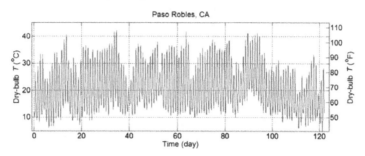

Figure 9. Dry bulb temperatures of Paso Robles in the summer of 2007

Table 4. Summarization of the two-zone cases: worst cases

T_{out} (°C)	Shade/Cool	Front zone (°C)		Office zone (°C)		
		ΔT_{op}	Mean T_{op}	ΔT_{op}	Mean T_{op}	CT ε
Worst case 1: 29.70 ± 15.00	Yes / No	2.47	34.90	1.67	34.41	NA
	No / No	3.12	38.45	1.85	36.76	NA
Worst case 2: 29.70 ± 4.45	Yes / Yes	1.67	26.43	1.25	25.98	0.319
	No / Yes	2.46	27.93	1.48	26.76	0.359

5.4. Heat Distribution Between Zones with Hydronic System in Daytime

It is expected that the TABS system can distribute heat effectively between zones and thus reduce the difference of the operative temperatures. Therefore, instead of shutting down the hydronic system in the daytime, the water in the four branches of the TABS system is circulated in order to distribute heat between the two zones. The results under the worst condition for controlling of the operative temperature level are given in Table 5. However, comparing with the corresponding results in Table 4, there is almost no change of both the operative temperature level and range, although the corresponding cooling tower effective can be a little higher. The results show that the heat cannot be distributed effectively because of the building thermal inertia due to the large amount of the thermal mass in the floors and roofs, as well as the thermal resistance due to the water pipes, the concrete, the carpet on the floor, and the convection on the surfaces.

Table 5. Summarization of the two-zone cases with daytime hydronic circulation

T_{out} (°C)	Shade/Cool	Front zone (°C)		Office zone (°C)		
		ΔT_{op}	Mean T_{op}	ΔT_{op}	Mean T_{op}	CT ε
Worst case 2: 29.70 ± 4.45	Yes / Yes	1.67	26.41	1.25	26.04	0.321
	No / Yes	2.45	27.84	1.50	26.95	0.365

5.5. Heat Distribution Between Zones with Internal Ventilation

It may be better to distribute heat through internal ventilation between the two zones, such as open all the internal doors or add a fan to circulate the interior air forcedly. Manipulating the indoor air temperatures of the two zones to be equal in the model, the simulation results are shown in Table 6. Comparing with the corresponding results in Table 5, there is some improvement, especially the operative temperature range: the range difference of the two zones is reduced from 0.42°C to 0.11°C in the case with shading and from 0.95°C to 0.22°C in the case without shading. With internal ventilation, the corresponding cooling tower effective can be even higher while the minimum cooling tower approach is kept at 2.8°C. However, the mean operative temperatures in the zones are still higher than 25.25°C even when the cooling tower works fully.

Table 6. Summarization of the two-zone cases with daytime hydronic circulation and internal ventilation

T_{out} (°C)	Shade/Cool	Front zone (°C)		Office zone (°C)		
		ΔT_{op}	Mean T_{op}	ΔT_{op}	Mean T_{op}	CT ε
Worst case 2: 29.70 ± 4.45	Yes / Yes	1.56	26.30	1.45	26.22	0.323
	No / Yes	2.18	27.56	1.96	27.43	0.370

5.6. Effects of Ambient Temperature Variation on the System

With both daytime hydronic circulation and internal ventilation, the simulation results are summarized in Table 7. In these cases, there is no shading device; the mean operative temperature in the Front zone is maintained at the optimal value of 25.25°C; the minimum cooling tower approach is kept at 2.8°C; the peak-to-peak diurnal amplitudes of the ambient temperature are from 2°C to 30°C with a 4°C step (the first row in Table 7); the maximum mean values of the ambient temperature that the cooling tower can maintain the optimal operative temperature are calculated (the second row); the operative temperature variations in the two zones are given in the following two rows; the mean operative temperatures in the Office zone and the cooling tower effectiveness are recorded in the last two rows for reference purpose only. From the table, with the increase of the ambient temperature amplitude, the maximum mean ambient temperatures increase steadily. This means that the cooling tower can manage the indoor thermal homeostasis under higher ambient temperature if the amplitude is larger.

The architect J.M. Fitch [21] said that "the central paradox [challenge] of architecture [is] how to provide a stable, predetermined internal environment in an external environment that is in constant flux across time and space…" We argued [2] that the "'external environment that is in constant flux across time and space' is both a challenge and an opportunity": a challenge for controlling the operative temperature range (the operative temperature ranges of the two zones increase steadily) and an opportunity for maintaining the operative temperature level (the cooling tower works well under higher ambient temperature). Notice that there is almost no change of the cooling tower effectiveness. That is to say, the cooling tower size does not need to be increased with higher mean ambient temperature if the amplitude is larger.

Table 7. Summarization of the two-zone cases in the design day

ΔT_{out} (°C)	2	6	10	14	18	22	26	30
Max mean T_{out} (°C)	25.99	26.79	27.60	28.40	29.21	30.01	30.81	31.63
Front ΔT_{op} (°C)	1.87	2.05	2.24	2.42	2.61	2.80	2.99	3.19
Office ΔT_{op} (°C)	1.69	1.85	2.01	2.17	2.34	2.50	2.67	2.83
Office mean T_{op} (°C)	25.15	25.13	25.12	25.10	25.09	25.08	25.07	25.06
CT effectiveness	0.370	0.370	0.371	0.371	0.372	0.372	0.372	0.373

5.7. Summarization of the Building Cooling

In Figure 10, the red line is the maximum mean ambient temperatures from Table 7 against the ambient temperature amplitudes; the blue crosses are the daily mean ambient temperatures vs. temperature amplitudes from the real-time weather data of Paso Robles in the four summer months in 2007. Clearly, there is only one day above the red line, which means that only this day the cooling tower alone cannot maintain the optimal mean operative temperature. In Ref. [2], the most favorable location, Sacramento, CA, has four days that the cooling tower alone cannot meet the cooling requirement. The result in Paso Robles is even better.

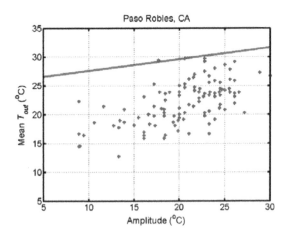

Figure 10. Daily distribution of dry-bulb temperatures vs. temperature amplitudes of Paso Robles in the summer of 2007

6. Discussions

This paper investigated the possibility of whether cooling tower alone can maintain the summer thermal homeostasis in a small commercial building located in Paso Robles, which is one of the most favorable locations for thermally homeostatic buildings. In reality, there are many other factors should be considered, including but not limited to:

(1) The investigation focuses on a certain type of buildings: small commercial building with large fenestration area, heavy thermal mass and hydronic system; the thermal behavior of other kinds of buildings may be different.

(2) The design of the building in fact is not good since there are large glazing area in the south and west walls. In building designs, this should be avoided from thermal comfort and building cooling energy consumption point of view. Here we intentionally designed the building this way. If the building with this "bad" design can be managed, it can be managed more easily with good designs.

(3) The on-off control of the cooling tower is too simplistic for the cooling operation in the whole summer. In fact, as the design day is chosen as the hottest day in the summer, the indoor temperature will be too low in other days if the cooling tower is always on in the nighttime. Control strategy should be carefully designed for maintaining the comfortable level in the whole summer.

(4) As cooling tower of effectiveness 0.370 can maintain 25.25°C even on the hottest design day, it is possible to maintain the indoor comfort with a smaller cooling tower and a thermal energy storage tank for storing the extra coolness in other days.

(5) The continual operation of such a smaller cooling tower over the whole summer has been studied in another paper [22] for Paso Robles as well as the three cities of Sacramento, CA, Albuquerque, NM, and Atlanta, GA.

(6) Paso Robles is one of the most favorable locations for thermally homeostatic buildings; in other non-favorable locations, such as Atlanta GA [2], composite heat extraction systems (CHESS) will be investigated for achieving full thermal homeostasis in the whole year [23].

(7) The proposed building system is more flexible for cooling than conventional buildings: instead of maintaining heat balance on hourly basis, our approach maintains a building's heat balance by recharging its thermal condition on daily basis. Therefore, the cooling operation using heat pump can take place at any favorable time interval during one-day period. Thus building cooling can be powered at times that the whole sale power cost is almost ridiculously low.

(8) The internal heat gain in the building is fixed with a certain pattern and the selected values are moderate; different pattern or values may affect the indoor temperature variation.

7. Conclusion

The development of a thermally homeostatic building consists of two steps: thermal autonomy first, which is based on the architectural requirement of a building's envelope and its thermal mass, and then thermal homeostasis, which is based on the engineering requirement of hydronic equipment. This paper focuses on the possibility of natural summer cooling in Paso Robles, CA, by using cooling tower alone. By showing the viability of natural cooling in one special case, albeit a case in one of the most favorable locations climatically, a case is made that the use of cooling tower in thermally homeostatic buildings should not be overlooked for general application in wider regions of other climatic zones. This paper is limited to answering the question of possibility, rather than providing any details of continual operation of a cooling tower. Some of the operation details are provided in future paper. In the general applications to other climatic zones, composite heat extraction system (CHES) is proposed for achieving full thermal homeostasis. In other words, building thermal mass and the generalized concept of heat extraction are keys to homeostasis in buildings.

References

[1] Wang, L.-S., Ma, P., Hu, E., Giza-Sisson, D., Mueller, G., Guo, N., "A study of building envelope and thermal mass requirements for achieving thermal autonomy in an office building," *Energy and Buildings* 78. 79-88. Aug. 2014.

[2] Ma, P., Wang, L.-S., Guo, N., "Modeling of hydronic radiant cooling of a thermally homeostatic building using a parametric cooling tower," *Applied Energy* 127. 172-181. Aug. 2014.

[3] Ma, P., Wang, L.-S., Guo, N., "Modeling of TABS-based thermally manageable buildings in Simulink," *Applied Energy* 104. 791-800. Apr. 2013.

[4] Gwerder, M., Lehmann, B., Tötli, J., Dorer, V., Renggli, F., "Control of thermally activated building systems TABS," *Applied Energy* 85 (7). 565-581. Jul. 2008.

[5] Lehmann B., Dorer V., Gwerder M., Renggli F., Tötli J., "Thermally activated building systems (TABS): Energy efficiency as a function of control strategy, hydronic circuit topology and (cold) generation system," *Applied Energy* 88 (1). 180-191. Jan.2011.

[6] Wikipedia.org, "Paso Robles, California," [Online]. Available: http://en.wikipedia.org/wiki/Paso_Robles,_California. [Accessed July 08, 2015].

[7] PRcity.com, "Paso Robles," [Online]. Available: http://www.prcity.com. [Accessed July 08, 2015]

[8] Honaganahalli P.S., Seiber J.N., "Measured and predicted air-shed concentrations of methyl bromide in an agricultural valley and applications to exposure assessment," *Atmospheric Environment* 34 (21). 3511-3523. 2000.

[9] Weather.com, "Paso Robles, CA (93446) Weather," [Online]. Available: http://www.weather.com/weather/monthly/l/93446:4:US. [Accessed July 08, 2015]

[10] Wrcc.dri.edu, "Paso Robles, California (046730)," [Online]. Available: http://www.wrcc.dri.edu/cgi-bin/cliMAIN.pl?ca6730. [Accessed July 08, 2015]

[11] Wrcc.dri.edu, "Paso Robles Muni AP, California (046742)," [Online]. Available: http://www.wrcc.dri.edu/cgi-bin/cliMAIN.pl?ca6742. [Accessed July 08, 2015]

[12] U.S. Department of Energy, "Real-Time Weather Data," [Online]. Available: http://apps1.eere.energy.gov/buildings/energyplus/weatherdata_download.cfm. [Accessed July 08, 2015]

[13] ASHRAE Inc., *ANSI/ASHRAE/IES Standard 90.1-2010: Energy Standard for Buildings Except Low-Rise Residential Buildings*, S-I ed. 2010.

[14] Ge H., *Study on overall thermal performance of metal curtain walls*, Ph.D. thesis. Concordia University. Oct.2002.

[15] Ma, P., Wang, L.-S., Guo N., "Maximum window-to-wall ratio of a thermally autonomous building as a function of envelope *U*-value and ambient temperature amplitude," *Applied Energy* 146. 84-91. May.2015.

[16] Ma, P., Wang, L.-S., Guo N., "Energy storage and heat extraction – From thermally activated building systems (TABS) to thermally homeostatic buildings," *Renewable and Sustainable Energy Reviews* 45. 677-685. May.2015.

[17] United Nations Environment Programme, *Electrical Energy Equipment: Cooling Towers*, 2006.

[18] L.-S. Wang, *Radiant Free Heat: Radiant conditioning of buildings with natural cooling and heating*, Stony Brook University Research Report. 2011.

[19] Olesen, B.W., "Thermo active building systems - Using building mass to heat and cool," *ASHRAE Journal* 54 (2). 44-52. Feb.2012.

[20] Long N., *Real-time weather data access guide*, National Renewable Energy Laboratory Contract No.: DE-AC36-99-GO10337. Mar.2006.

[21] Lechner N., *Heating, Cooling & Lighting: Design methods for architects*, 2nd ed., John Wiley & Sons: New York, U.S.A., 2001.

[22] Ma P., Wang L.-S., Guo N., "PWM control of a cooling tower in a thermally homeostatic building," *American Journal of Mechanical Engineering* [to be submitted for publication].

[23] Wang L.-S., Ma P., "The Homeostasis Solution – Mechanical homeostasis in architecturally homeostatic buildings," *Applied Energy* [submitted in June 2015, APEN-D-15-03388].

15

Evaluation of Emitted Particulate Matters Emissions in Multi-cylinder Diesel Engine Fuelled with Biodiesel

Miqdam Tariq Chaichan[*]

Mechanical Engineering Department, University of Technology, Baghdad, Iraq
*Corresponding author: miqdamchaichan@gmail.com

Abstract As an alternative fuel, biodiesel is receiving rising attention for diesel engines. The corn oil biodiesel was prepared from Iraqi produced corn oil through transesterification process, using methanol and sodium hydroxide in the present investigation. Neat corn oil biodiesel, as well as, the blends of varying proportions of it and diesel was used to run a 4-cylinders direct injection CI engine. The effects of some engine variables like load, speed, and injection timing on emitted particulate matters (PM) were studied. The aim was to evaluate the emitted particulate matters from a diesel engine when it fuelled with blends diesel and biodiesel. As the Iraqi conventional diesel has high sulfur content that release high rates of PM, it was taking as the baseline fuel in the tests. The results showed a significant reduction in PM when biodiesel was used. The maximum reductions in PM concentrations observed were 34.96 % in the case of biodiesel operation compared to diesel at full engine loads and constant speed. An increment in the PM concentrations as the timing retarded from the optimum injection timing. Biodiesel has a significant impact on smoke at idle mode with reductions of 8.6%, 18%, and 39.75% for B20, B50, and B100 respectively compared to diesel. The study concludes the possibility of more reduction of PM concentrations in the case of reducing sulfur content in Iraqi diesel fuel significantly.

Keywords: *biodiesel, particulate matters, transesterification process*

1. Introduction

Diesel engines usually emit higher particulate matter (PM) compared to the spark ignition engines. The strict emission regulations in the world have placed design limitations on diesel engines. The worldwide growing trend towards cleaner burning fuel pushed towards many alternative diesel fuels that show better exhaust emissions than traditional diesel ([1,2]).

Biomass energy is considered as one of the renewable energies among these alternatives. Biomass energy includes liquid biofuels derived from vegetable oils with low environmental pollution impact, to replace petroleum-based fuels. Some of the well known liquid biofuels are ethanol for gasoline engines and biodiesel for compression ignition engines or diesel engines [3].

Vegetable oils were used as fuel for diesel engines to some extent since the invention of the compression ignition engine by Rudolf Diesel in the late 1800`s. During the early stages of the diesel engine, strong interest was shown in the use of vegetable oils as fuel, but this interest declined in the late 1950`s after the supply of petroleum products become abundant [4]. During the early 1970`s, oil shock, however, caused a renewed interest in vegetable oil fuels. This interest evolved after it became apparent that the world's petroleum reserves were dwindling. At present, to replace a part of petroleum-based diesel usage, the use of vegetable oil product biodiesel has been starting in some countries [5]. Biodiesel properties are comparable with those of conventional diesel fuels. Therefore, it added to diesel engines as neat or blended, without any engine modifications. However, the variations in fuels chemical nature result in differences in their fundamental properties affecting engine performance, combustion process, and pollutant emissions [6].

Biodiesel exhibits several merits when compared to that of the existing petroleum fuels. Many researchers have shown that using biodiesel as fuel reduced particulate matter, unburned hydrocarbons, carbon monoxide, and sulfur levels significantly in the exhaust gas. However, biodiesel employment increases the oxides of nitrogen levels as reported by [7]. Besides, biodiesel produces less detrimental to human health pollutants as it does not contain carcinogens materials such as polyaromatic hydrocarbons and nitrous poly-aromatic hydrocarbons [8].

The exhaust of diesel engines contains solid carbon soot particles or particulate matter (PM); these observed as smoke emission. Particulate matter (PM) is a primarily emitted emission by diesel engine compared with a gasoline engine. There is a clear consensus that biodiesel fuels result in significant reductions in PM [9,10,11]. The methyl esters current capability to reduce regulated or non-regulated emissions emitted from the engines have not fully cleared. In the same time, the impact of vegetable oils on the particulate formation and production have not well assessed yet. Some researchers claimed that a reduction in PM achieved when vegetable oils used [12]. While, other researchers demonstrated PM concentrations increased when biofuels employed [13]. However, most authors report decreases in aromatic and polyaromatic hydrocarbon emissions [14,15,16].

The majority of researchers worked in this field agreed that the reduction in PM emissions referred to the notable decrease in the insoluble fraction concentration, as a result of the increase in oxygen inside biodiesel structure and aromatic hydrocarbons absence in biodiesel fuels [17]. With the same air at the admission, the oxygen content of the ester molecule allows for a complete combustion. Even with fuel, intensive diffusion flames exist in zones of the combustion chamber, in addition to the support of the already formed soot oxidation. The aromatic hydrocarbons reduction in biodiesel involves a lack of the soot precursor species concentration in the combustion chamber [18,19].

Particulate matters are often fractioned in term of sulfate, soluble organic fraction (SOF) or volatile organic fraction (VOF) and carbon or soot [20]. In the diesel combustion process, some of fuel droplets may never vaporize, and thus never burn. However, the fuel does not remain unchanged; the high temperatures in the combustion chamber cause it to decompose. In the next steps, these droplets may be completely or partly burned in the turbulent flame. If any droplets not burned completely, they emit as a droplet of heavy liquid or carbon particles [21]. The conversion of fuel to PM is mostly occurring in the last part of the injected fuel in a cycle, or when the engine is running at high speed, and Iraq is a country distinguishes by large agricultural areas. Some of these fields are used, and the others are not. Corn is an Iraqi crop while corn oil produced in Iraq in large quantities. From this point of departure, the Mechanical Engineering Department- University of Technology, consider researchers about using corn oil and other Iraqi crops as alternative biodiesel fuels in future seriously.

The evaluation of the particulate matters concentrations emitted from a diesel engine fuelled with blends diesel fuel and biodiesel produced from Iraqi corn oil is the primary objective of this study. The Iraqi conventional diesel fuel was taking as the baseline fuel in the tests.

The method used to measure PM emissions for type approval tests is the filter samples gravimetric analysis.

These filters fixed in a full exhaust flow dilution tunnel. Another notable characteristic of the particle is the particle size distribution or particle number since the particle diameter has an influence on the human health, especially nanoparticles [23,24]. Many studies conducted on the particle size distribution of diesel engine fueling conventional diesel [25,26]. Generally speaking, particle size distribution of diesel engine includes nucleation mode and accumulation mode, the division diameter of modes is 50 nm. Nucleation mode particles related to the soluble organic fraction and sulfate. The accumulation mode particles linked to soot [27,28].

2. Experimental Work

2.1. Materials and Transesterification Process

It is necessary to differentiate between fats and oils or the origin of biodiesel. The difference between a fat and oil is their physical state. The term fat usually defines the solid state, and oil in the liquid state. These terms are changeable depending on the temperature to which the compound exposed. When the state is unimportant, the term fat is usually used [3].

The transesterification process employed to convert the corn oil into its methyl ester. The transesterification is a chemical reaction used in the production of biodiesel. Fatty acid in vegetable oil reacts with an alcohol in a presence of a catalyst to form fatty acid alkyl ester. Simple alcohols like methanol or ethanol used in transesterification process and it is usually carried out in the complete absence of water with a basic catalyst (NaOH, KOH) [21]. In this work, methanol was used. Methanol takes high yield reaction quite easy. An excess of methanol needed to speed up the response, with the occurrence of a ready separation of methyl alcohol glycerol. In the transesterification process, the alcohol combines with the acidic triglyceride molecule to form glycerol and ester, which removed by density separation. The transesterification process decreases the oil viscosity, making its characteristics similar to diesel fuel [2,29]. The method of producing biodiesel illustrated in the block diagram of Figure 1.

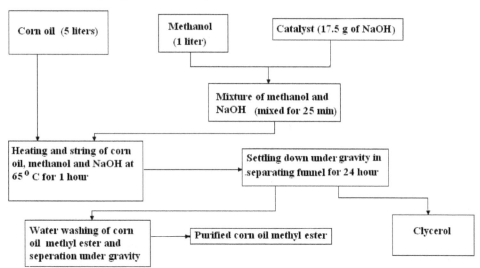

Figure 1. Block diagram for producing corn biodiesel by transesterification process

Several factors were found to affect the yield and quality of the ester significantly. The water content of all materials, including the catalyst and triglyceride, and the acid value of the triglyceride were required to be very low. Acid contents above 0.5% and water content of 0.3% were reported to cause a significant decrease in the ester yield. Other important factors for transesterification are reaction time and temperature. The time required to complete the reaction estimated about one hour. The reaction temperature depends on the used alcohol type and it temperature recommends to be below the boiling point of the used alcohols used with a few degrees (for examples of reaction temperatures for methanol was 60°C) [30].

2.2. Fuel-related Properties of Vegetable Oils

The related fuel properties of vegetable oil vary depending on the fuel fatty acid. The fuel-related tested properties of used biodiesel listed in Table 1. The viscosity of vegetable oils varies in the range of 14.38 to 65 mm²/s at 27°C. The high viscosity of these biodiesel oils is due to the considerable molecular mass and chemical structure. Vegetable oils molecular weights range between of 550 to 900, which are triple or more times higher than diesel. The vegetable oils flash point is considerably high (236°C). The volumetric heating values

of these blends are in the range of 39 to 41.6 MJ/kg that is low compared to diesel (about 44.2 MJ/ kg). The presence of chemically bound oxygen in vegetable oils lowers their heating values. The biodiesel blends cetane numbers are in the range of 38 to 42.9. The cloud and pour points of vegetable oils are higher than that of diesel fuels [31,32].

2.3. Experimental Setup

A four stroke, direct injection, naturally aspirated four-cylinder four-stroke diesel engine employed for the present study. The engine specifications listed in Table 2. The engine coupled to a hydraulic dynamometer, and the engine speed measured by a tacho-generator connected to the dynamometer.

The load and speed of the engine were controlled by adjusting the dynamometer resistance and the injection rate of the fuel pump. The determination of the engine fuel consumption was conducted by the measurement of the fuel level decrement in a scale container for a given period. The volumetric flow rate of the intake air was measured using orifice plate. The exhaust gas temperature measured using a thermocouple connected to the exhaust pipe just downstream of the exhaust manifold. The cooling water temperatures at the inlet and outlet of the engine were measured using calibrated thermocouples.

Table 1. Tested fuels specifications

Fuel type	Calorific value (kJ/kG)	Density (g/dm³)	Viscosity (mm²/s at 27°C)	Cetane No.	Flame point (°C)	Cloud point (°C)	Pour point (°C)
Diesel fuel	44227	810	4.23	49	59	-13.8	-29
Biodiesel (B100)	39873	906	65	38.6	239	-3.7	-12.4
B50	40368	877	44.7	40.6	179	-10.2	-17.833
B20	41654	829	14.38	42.9	112	-11.78	-24.68

The low volume air sampler (type Sniffer L-30) employed to collect emitted PMs. Whatmann-glass micro-filters used to gather PMs. The filters weighted before and after the sampling operation that extend for one hour. The PMs concentrations determined by the equation:

$$PM \ in \left(\mu g/m^3 \right) = \frac{w_2 - w_1}{Vt} \times 10^6$$

Where: PM = particulate matters concentration in ($\mu g/m^3$).
w_1 = filter weight before sampling operation in (g).
w_2 = filter weight after sampling operation in (g).
Vt = drawn air total volume (m³)
Vt found by the equation:

$$Vt = Q_t \cdot t$$

Where: Q_t = elementary and final air flow rate through the device (m³/sec).
t = sampling time in (min).
The filters preserved in individual plastic bags temporarily at the end of collecting samples operation until analyzing and studying the results using a light microscope.

2.4. Experimental procedure

The amount of PM generated was measured using different engine variables strategies, as well as, the fuel impact on these strategies. Experiments were conducted on the engine using diesel fuel to provide baseline data. The engine was warmed up for half an hour. Engine

cooling water temperature maintained at 70°C. Then, 20%, 50% and 100% blends of corn biodiesel were also tested. Physical characteristics of the tested fuels listed in Table 1. The first set of experiments conducted at the engine speed of 1500 rpm. The experiments performed at the designed injection timing of 38° BTDC for no-load, 50% load, and full load for all tested fuels. The injection timing varied from 20 to 45 ° BTDC, each step 5 degrees at constant 1500 rpm engine speed.

Table 2. Tested engine specifications

Engine type	4cyl., 4-stroke in line
Engine model	TD 313 Diesel engine rig
Combustion type	DI, water cooled, naturally aspirated
Displacement	3.666 L
Valve per cylinder	two
Bore	100 mm
Stroke	110 mm
Compression ratio	17
Fuel injection pump	Unit pump, 26 mm diameter plunger
Fuel injection nozzle	Hole nozzle 10 nozzle holes Nozzle hole dia. (0.48mm) Spray angle= 160° Nozzle opening pressure=40 MPa

In the second set of experiments, the engine speed varied between 1250 and 2500 rpm with intervals of 250 rpm while the engine operated at full load. The fuel delivery angle of the fuel injection system was kept constant at 38° BTDC with variable speed tests. Before each blend test, the fuel tank and fuel lines drained. Then, the engine left to operate at least 15 minutes to stable on the new blend. For each speed, the engine run for about 5 minutes until the steady-state conditions achieved, and then the data were collected in the sixth minute. The tests repeated three times to confirm repeatability, and the average of the results of the three trials counted in the study.

3. Results and Discussions

Figure 2 shows the PM concentrations for the tests fuels at a variable engine loads. The smoke contains solid carbon soot particles that generated when the fuel has no enough oxygen to react with all the carbon or in the fuel rich zone of combustion chamber during combustion process as Ref. [7] declared. From the experimental results, the smoke emission from corn biodiesel fuel and the diesel fuel have a few differences in 0% to 25% load level. However, at 50% to 100% load level the smoke emission from all biodiesel blends are lower than that of the diesel fuel. The smoke emission in case of various blends of biodiesel smoke emission is less as compared to diesel as Figure 2 declares. At full load, the maximum reductions in PM concentrations observed were 34.96 %, in the case of biodiesel operation when compared to diesel. There is an obvious reduction in PM emission for all biodiesel blends at all loads. This reduction referred to the soot free biodiesel fuel, and the complete combustion of it.

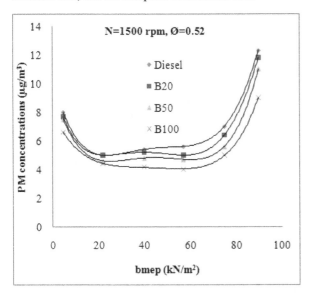

Figure 2. The effect of bmep on PM concentrations at 1500 rpm and constant equivalence ratio

Figure 3 represents the effect of equivalence ratio variation on PM concentrations at 1500 rpm engine speed and medium load. Increasing oxygen content in the fuel-air mixture reduced PM concentrations. Therefore, the PM concentrations are very low at ultra-lean equivalence ratios and increased highly by mixture enrichment with fuel. However, there are still recorded concentrations that mean using biodiesel reduces emitted PM but doesn't annihilate it completely.

Ref. [21] explained the reduction of PM with biodiesel fuelling is mainly caused by reduced soot formation and enhanced soot oxidation. Biodiesel has a nil content of sulfur and aromatic hydrocarbons that are considered soot

precursors. The lower final boiling point of biodiesel, despite its higher initial boiling point and average distillation temperature (Table 1), provides a lower probability of PM formation from the inability to vaporize heavy hydrocarbon fractions. The oxygen bonded in the ester molecules structure allows for a complete combustion and promotes the oxidation of the already formed PM particles.

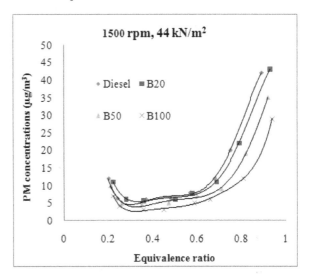

Figure 3. The effect of equivalence ratio variation on PM concentrations at constant load and speed

The injection timing is an important engine variable that has an impact on PM production. The primary cause to characterize is not only the fuel type but also the combustion mode used. Figure 4 illustrates the evaluation of the PM concentrations generated utilizing different injection strategies and the fuel impact on these strategies.

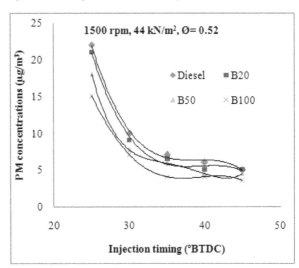

Figure 4. Injection timing variation effect on PM concentrations at constant load and speed

As Figure 4 reveals, comparing the effect of advancing and retarding the timing on PM concentrations shows an increment in these concentrations as the timing retarded from the optimum injection timing. This trend is independent of the used fuel. However, for comparison the emitted PM by B20, B50 and B100 were less than that produced by diesel fuel by 7%, 18.8%, and 33%

respectively on average. The injection timing retardation means reducing available combustion time, and hence causing a higher concentration of PM. It is suitable to advance the combustion start since it enlarges the residence time of PM particles in a high-temperature environment and gives the oxygen presence to promote further oxidation.

Figure 5, Figure 6 and Figure 7 showed the tests when engine speed changed between 1250 and 2500 rpm at intervals of 250 rpm while the engine operated at no load, medium, and full loads. The engine injection timing kept constant at 38° BTDC. Constant equivalence ratio (Ø=0.52) was used, around this Ø the maximum brake power obtained.

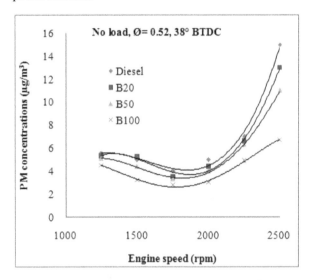

Figure 5. Engine speed effect at no load for Ø=0.52 and 38° BTDC

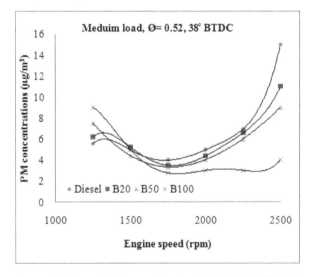

Figure 6. Engine speed effect at medium load for Ø=0.52 and 38° BTDC

The use of biodiesel reduced PM concentrations significantly. Biodiesel declared the greatest effect on smoke at no load mode where B20 reduced smoke by 8.6% and B50 reduced smoke by 18% while at B100 blend the reduction was 39.75%. In this study, biodiesel addition reduced particulate emission in all stage as seen in Figures.

According to the figures, the PM concentrations decreased with biodiesel blends fuelling at practically all loads and engine speeds. At full load, this indicator of PM

emissions was 12.7%, 25% and 52.38% lower for B20, B50 and B100 respectively than for diesel fuel, depending on engine speed. However at partial loads, this difference was attenuated. The impact of the operation mode of the engine on PM concentrations seemed to be fuel sensitive. For biodiesel blends, smoke compartment trend was similar at all loads and engine speeds; it was always less than diesel fuel. Higher loads require more fuel consumptions and higher engine speeds drive to shorter residence times of fuel-air mixture in the combustion chamber leading to higher PM concentrations.

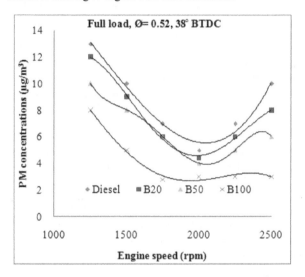

Figure 7. Engine speed effect at medium load for Ø=0.52 and 38° BTDC

Many works of literature attested that PM emissions increase or decrease according to the sulfur concentration. Sulfur in the fuel caused the formation of sulfates that are absorbed on soot particles and increase the PM emitted from diesel engines. Besides, fuel oxidation complete due to the oxygen content increase in the fuel. Oxygen increases even in locally rich zones resulting in a significant decrement in PM as References [33] and [34] revealed. Many researchers record higher reduction in PM concentrations with other types of biodiesel compared with this study. The reason for this difference is the amount of sulfur content in the tested fuel. The Iraqi conventional diesel fuel contains about 10000 to 15000 ppm sulfur [35] in comparison with the free sulfur diesel fuel employed in the other articles.

4. Conclusions

Biodiesel oil manufactured in a laboratory using Iraqi produced corn oil. Biodiesel production conducted using transesterification process. Three blends of biodiesel and diesel were prepared, B20 (contains 20% biodiesel and 80% diesel), B50 (contains 50% biodiesel and 50% diesel and B100 contains 100% biodiesel. The three blends used in operating 4-cylinder direct injection diesel engine and the emitted PM concentrations of these operation modes compared to diesel fuel operation. The tests conducted on several engine variables. The results show:

1. There is a significant reducing in the PM concentrations for all biodiesel blends at the part and full loads. PM concentration reduced with an increase in the blending of biodiesel.

2. PM concentrations reduced by increasing biodiesel percentage in fuel for all range of equivalence ratios.

3. The injection timing retardation increased PM concentrations while advancing it reduced these concentrations. Injection timing has a significant influence on emitted PM.

4. The increasing speed increases PM concentrations due to a reduction in available reaction time for the fuel's molecules with higher weight.

5. Increasing load increased the fuel consumption and hence the emitted PM.

6. At all tested engine variables, using neat biodiesel and its blends reduced PM concentrations for the safe range.

7. It is possible to reduce PM concentration and reach the percentages reported by other researchers if sulfur contents in Iraqi diesel fuel could be removed, or at least reduced.

References

[1] Chaichan, M.T., Ahmed, S.T., "Evaluation of performance and emissions characteristics for compression ignition engine operated with disposal yellow grease," *International Journal of Engineering and Science*, 2 (2). 111-122. 2013.

[2] Suryawanshi, J.G., "Palm oil methyl ester: A new fuel for CI engines," *Science and Technology*, 5 (7). 36-40. 2009.

[3] Myo, T., "The effect of fatty acid composition on the combustion characteristics of biodiesel," Ph D thesis, the Kagoshima University, 2008.

[4] Knothe, G., "*Cetane numbers- The History of Vegetable Oil-based Diesel Fuels.*" The biodiesel handbook, American Oil Chemist's Society Press, 2005.

[5] Subbaiah, G.V., Gopal, K.R., Prasad, B.D., "Study of performance and emission charactristics of a direct injection diesel engine using rice bran oil ethanol and petrol blends," *Journal of Engineering and Applied Sciences*, 5 (6). 95-103. 2010.

[6] Hussain, S.A., Subbaiah, G.V., Pandurangadu, V., "Performance and emission characteristics of a supercharged direct injection diesel engine using rice bran oil," I- manager's Journal on Future Engineering and Technology, 4 l (4). 48-53. 2009.

[7] Heywood, J.B., *Internal Combustion Engine Fundamentals*, 20th Ed. Singapore: Tata McGraw-Hill Publishers. 2002.

[8] Pramanik, K., "Properties and use of Jatropha cruces oil and diesel fuel blends in compression ignition engine," *Fuel and Energy*, 44 (3). 13-16. 2003.

[9] Chaichan, M.T., "Performance and emission study of diesel engine using sunflowers oil-based biodiesel fuels," *International Journal of Scientific and Engineering Research*, 6 (4). 260-269. 2015.

[10] Lapuerta, M., Armas, O., Fernández, J.R., "Effect of biodiesel fuels on diesel engine emissions," *Progress in Energy and Combustion Science*, 34. 198-223. 2008.

[11] Alam, A.M., Song, J., Acharya, R., Boehman, A.L., Miller, K., "Combustion and emissions performance of low sulfur, ultra low sulfur and biodiesel blends in a DI diesel engine," *SAE paper 2004-01-3024*, 2004.

[12] Yoshimoto, Y., Onodera, M., Tamaki, H., "Reduction of NOx, smoke and bsfc in a diesel engine fuelled by biodiesel emulsion with used frying oil," *SAE Paper1999-01-3598*, 1999.

[13] Labecki, L., Ganippa, L.C., "Soot reduction from the combustion of 30% rapeseed oil blend in a HSDI diesel engine," Towards Clean Diesel Engines, TCDE2009, 2009.

[14] Graboski, M.S., McCormick, R.L., "Combustion of fat and vegetable oil derived fuels in diesel engines," *Progress in Energy and Combustion Science*, 24. 125-164. 1998.

[15] Agarwal, A.K., "Biofuels (alcohols and biodiesel) applications as fuels for internal combustion engines," *Progress in Energy and Combustion Science*, 33. 233-271. 2007.

[16] Yamane, K., Ueta, A., Shimamoto, Y., "Influence of physical and chemical properties of biodiesel fuels on injection, combustion and exhaust emission characteristics in a direct injection compression ignition engine," *Int. J. Engine Res.*, 4. 249-261. 2004.

[17] Lapuerta, M., Armas, O., Ballesteros, R., Fernández, J.R., "Diesel emissions from biofuels derived from Spanish potential vegetable oils." *Fuel*, 84. 773-780. 2005.

[18] Boehman, A.L., Song, J., Alam, M., "Impact of biodiesel blending on diesel soot and the Regeneration of particulate filters," *Energy & Fuels*, 19. 1857-1864. 2006.

[19] Schmidt, K., Gerpen, J.V., "The Effect of biodiesel fuel composition on combustion and emissions." *SAE Paper no. 961086*, 1996.

[20] Schmidt, K., Gerpen, J.V., "The Effect of biodiesel fuel composition on combustion and emissions." *SAE Paper no. 961086*, 1996.

[21] Agudelo, J., Benjumea, P., Villegas, A.P., "Evaluation of nitrogen oxide emissions and smoke opacity in a HSDI diesel engine fuelled with palm oil biodiesel," *Rev. Fac. Ing. Univ. Antioquia*, 51. 62-71. 2010.

[22] Kadarohman, A., Khoerunisa, F., Astuti, R.M., "A potential study on clove oil, eugenol and eugenyl acetate as diesel fuel bio-additives and their performance on one cylinder engine," *Transport*, 25 (1). 66-76. 2010.

[23] Donaldson, K., Mills, N., MacNee, W., et al., "Role of inflammation in cardiopulmonary health effects of PM," *Tox Appl Ph*, 207 (2). 483-488. 2005,

[24] Wise, H., Balharry, D., Reynolds, L.J., et al., "Conventional and toxic-genomic assessment of the acute pulmonary damage induced by the instillation of Cardiff PM10 into the rat lung." *Sci Total E*, 360 (1-3). 60-67. 2006.

[25] Abdul-Khalek, I.S., Kittelson, D.B., Graskow, B.R., et al., "Diesel exhaust particle size: measurement issues and trends." *SAE Paper 980525*. 1998.

[26] Liu, S.X., Gao, J.D., Jing, X.J., "The study on development of particulate measurement and evaluation method of vehicle engines." *Small Int Combust Eng Motor*, 34 (2). 43-46. 2005.

[27] Kittelson, D.B., "Engines and Nano-particles: a review," *J Aeros Sci*, 29 (5-6). 575-588. 1998.

[28] Lu, X.M., Ge, Y.S., Han, X.K., "Experimental study on particle size distributions of an engine fueled with blends of biodiesel." *Chinese J Environ Sci*, 28 (4). 701-705. 2007.

[29] Yaliwal, V.S., Banapurmath, N.R., Tewari, P.G., Kundagol, S.I., Daboji, S.R. and Galveen, S.C., "Production of renewable liquid fuels for diesel engine applications – a review," *Journal of Selected Areas in Renewable and Sustainable Energy (JRSE)*, January Edition. 2011.

[30] Knothe, G., "Dependence of Biodiesel Fuel Properties on the Structure of Fatty Acid Alkyl Ester," *Fuel Processing Technology*, 86. 1059-1070. 2005.

[31] Cardonea, M., Mazzoncinib, M., Meninib, S., Roccoc, V., Senatorea, A., Seggianid, M., Vitolod, S., "Brassica carinata as an alternative oil crop for the production of biodiesel in Italy: agronomic evaluation, fuel production by transesterification and characterization," *Int. J. Biomass and Bio-energy*, 25. 623 – 636. 2003.

[32] Naik, M., Meher, L.C., Naik, S.N., Das, L.M., "Production of biodiesel from high free fatty acid Karanja (Pongamia Pinnata) oil," *Biomass and Bio-energy*, 32. 354-357. 2008.

[33] Pedkey, L.R., Hobbs, C.H., "Fuel quality impact on heavy duty diesel emissions- A literature review." *SAE Paper No. 982649*. 1998.

[34] Akasaka, Y., Suzuki, T., Sakurai, Y., "Exhaust emissions of a DI diesel engine fueled with blends of bio-diesel and low sulfur diesel fuel." *SAE Paper No. 972998*. 1997.

[35] United Nation Environment Program (UNEP), "Opening the door to cleaner vehicles in developing and transition countries: The role of lower sulfur fuels." Report of the sulfur working group of the partnership of clean fuels and vehicles (PCFV), Nairobi, Kenya, 2007.

16

Characteristic Entities in PhotoStress Method

Peter Frankovský[1,*], František Trebuňa[2], Ján Kostka[2], František Šimčák[2], Oskar Ostertag[2], Władysław Papacz[3]

[1]Department of Mechatronics, Faculty of Mechanical Engineering, Technical University of Košice, Košice, Slovakia
[2]Department of Applied Mechanics and Mechatronics, Faculty of Mechanical Engineering, Technical University of Košice, Košice, Slovakia
[3]Instytut Budowy i Eksploatacji Maszyn, Uniwersytet Zielonogórski, Zielona Góra, Poland
*Corresponding author: peter.frankovsky@tuke.sk

Abstract The presented paper describes characteristic entities in PhotoStress method - isoclinic fringes, singular points, isostatic lines, and isochromatic fringes. These entities are used in PhotoStress method to visualise and quantify deformation and stress fields of various photoelastically coated structural elements.

Keywords: *isoclinic fringes, singular points, isostatic lines, isochromatic fringes*

1. Introduction

Reflection photoelasticity or PhotoStress method is an experimental method of mechanics which allows quantitative as well as qualitative analysis of directions and magnitudes of principal strains or principal normal stresses on a variety of structural elements.

Being universal, this method can be used in a variety of engineering applications such as structural analysis of engines, analysis of architectural structure solutions, development of prosthetic implants, development of structural components of planes, machines etc.

A photoelastic coating is used for the evaluation of deformation. This coating is applied with a special two-component adhesive on the analysed surface of the object under examination. When the object, which is subjected to loads, is illuminated with light from the reflection polariscope, it is possible to view isoclinic or isochromatic fringes, based on the polariscope settings.

For the purposes of a quantitative evaluation it is necessary to be familiar with parameters of isoclinic and isochromatic fringes in points of measurements. This paper presents the procedure of obtaining individual photoelastic entities, as well as a description of these entities.

2. Isoclinic Fringes

Isoclinic fringes (Figure 1) are defined as geometrical points in which directions of principal normal stresses are parallel to intersected polarisation planes of the polariser and the analyser. They occur at plane-polarised white light. Isoclinic fringes of a particular angle parameter α are related to a chosen direction of intersected polarisation planes of the polariser and the analyser, which is given by angle α or $\alpha + \pi/2$ [1,2].

The creation of isoclines with regards to angle parameter α is a periodical phenomenon with period $\pi/2$. With synchronous rotation of the polariser and the analyser, isoclinic fringes continuously change from the isoclines of angle parameter $\alpha = 0°$ up to angle parameter $\alpha = 90°$ in line with rotation angle of the polariser and the analyser.

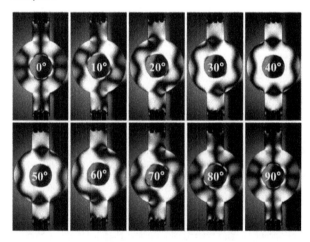

Figure 1. Isoclinic fringes

The creation of isoclines with regards to angle parameter α is a periodical phenomenon with period $\pi/2$. With synchronous rotation of the polariser and the analyser, isoclinic fringes continuously change from the isoclines of angle parameter $\alpha = 0°$ up to angle parameter $\alpha = 90°$ in line with rotation angle of the polariser and the analyser. They are distributed along whole analysed surface of the object under examination simultaneously according to the changes in directions of principal normal stresses. The isoclinic fringe runs through every point of the surface under consideration with a specific angle parameter α, as there is only one direction of principal stresses, i.e. direction α, or $\alpha + \pi / 2$. However, there are points, fringes and surfaces which remain dark even during simultaneous rotation of the polariser and the analyser. Both principal normal stresses in such points, fringes and surfaces have identical magnitude in every direction. Through these points, fringes or surfaces are running isoclines of all angle parameters. These are called singular points, fringes or surfaces [3].

Angle parameter α changes from point to point along any curve S which runs through the point P (Figure 2).

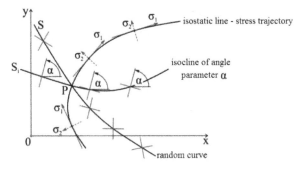

Figure 2. General example of an isocline and stress trajectory (isostatic line)

However, there is the S_i curve which runs through the point P. Along this curve the angle parameter α remains constant, while

$$\frac{\partial \alpha}{\partial S_i} = 0 \qquad (1)$$

Provided that stress σ_x is known, σ_y and τ_{xy}, the equation of such isocline can be expressed mathematically as a relation

$$tg\, 2\alpha = \frac{2\, tg\alpha}{1 - tg^2\alpha} = \frac{2\tau_{xy}}{\sigma_x - \sigma_y} = \text{konst.} \qquad (2)$$

where σ_x, σ_y represent normal stresses, τ_{xy} is shear stress and α represents isoclinic angle parameter. One specific isocline is related to every relevant constant. Considering the equation (1), its can be concluded that

- the isocline with angle parameter $\alpha = 0°$ connects points in which shear stress $\tau_{xy} = 0$ and normal stress $\sigma_x \neq \sigma_y$. On one side, the shear stress $\tau_{xy} > 0$ and on the other side the shear stress $\tau_{xy} < 0$.

- angle parameter α changes on the examined surface continuously and, as a result, other isoclines following each other do not intersect. Only one isocline can run through the analysed point, except for singular points [1,5].

- the edge of the plane object subject to examination, which is not shear-loaded in tangential direction, has only principal normal stress σ_n other than zero $\sigma_n \neq 0$ [1,5].

3. Singular Points

Points, in some cases lines and surfaces, which remain dark permanently even when rotating intersected polarisation filters, are called singular points, lines and surfaces. Shear stresses τ_{xy}, τ_{yx} and the difference of principal normal stresses $\sigma_1 - \sigma_2$ in singular points equal zero. In singular points the object is loaded with hydrostatic tension ($\sigma_1 > 0$) or pressure ($\sigma_1 < 0$), or is in a stressless state ($\sigma_1 = 0$). Any direction in these points can be considered the direction of principal normal stresses. Isoclines of all angle parameters α intersect in singular points. As a result, isostatic lines in the area of singular point are largely curved what makes the drawing more difficult. The position of singular points can be identified on the basis of the picture of isochromatics, which appear as dark points, fringes or surfaces [3,4].

According to rotation direction, singular points can be divided into two groups:

- positive singular points
- negative singular points.

When rotating the analyser clockwise (to the right) and isoclinic fringes are moving equally around the singular point, the singular point is positive (Figure 3).

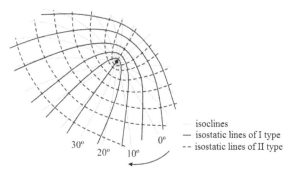

isoclines
— isostatic lines of I type
-- isostatic lines of II type

Figure 3. Positive singular point

Isoclinic fringes are moving in the opposite direction around negative singular point, i.e. anti-clockwise (Figure 4). Singular points which occur on the object's circumference are always negative.

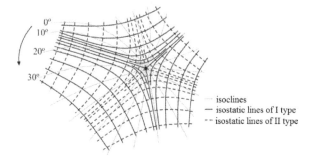

isoclines
— isostatic lines of I type
-- isostatic lines of II type

Figure 4. Negative singular point

Singular points can as well be classified as points of first order, points of second order and points of a higher order. Such classification respects the number of dark lines which run through given singular point. Singular points of second and a higher order are unstable with respect to the nature of mathematical expressions of stress components. It is manifested through decomposition to singular points of lower orders (e.g. point of second order breaks down to two points of first order) even after a small change in the shape of the object subject to examination. Figure 5 depicts singular points of different orders which occur in a network of isostatic curves [3,4].

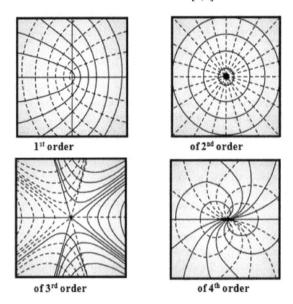

Figure 5. Singular points

4. Isostatic Lines

Isostatic lines are lines which are characteristic for inner stress states. They provide informative overview of the behaviour of the object subject to examination. There are zero shear stresses along isostatic lines. Their tangents give the direction of principal normal stresses in every point of the analysed object [1].

Isoclinic lines do not offer direct representation of the direction of principal stresses. On the other hand, they provide us with data which are necessary to gain a set of isostatic lines (stress trajectories) of I and II type. Stress trajectories can be obtained via graphical construction, while the base for such construction of isostatic lines of I and II type is the set of isoclinic lines. Graphical construction of isostatic lines of I and II type is described in detail in given literature [4,5].

Isostatic lines can be expressed mathematically with equation

$$\frac{dy}{dx} = tg\,\alpha \qquad (3)$$

where angle α is defined by the expression which derives from the Mohr's circle (Figure 6) [3,4].

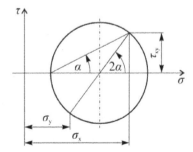

Figure 6. Mohr's circle of stresses

Considering the Mohr's circle (Figure 6):

$$tg\,2\alpha = \frac{2\tau_{xy}}{\sigma_x - \sigma_y} \qquad (4)$$

Adjusting the relation (2) and after substitution for

$$tg\,\alpha = \frac{-1 + \sqrt{1 + tg^2\,2\alpha}}{tg\,2\alpha} \qquad (5)$$

a differential equation of isostatic lines of I type emerges in the following form:

$$\frac{dy}{dx} = \frac{-(\sigma_x - \sigma_y) + \sqrt{(\sigma_x - \sigma_y)^2 + 4\tau_{xy}^2}}{2\tau_{xy}} \qquad (6)$$

Considering isostatic lines of II type which in every point are perpendicular to isostatic lines of I type:

$$\frac{d\overline{y}}{d\overline{x}} = -\frac{1}{\dfrac{dy}{dx}} = \frac{-2\tau_{xy}}{-(\sigma_x - \sigma_y) + \sqrt{(\sigma_x - \sigma_y)^2 + 4\tau_{xy}^2}}. \qquad (7)$$

Given differential equations (6) and (7) define both types of isostatic lines.

Figure 7 shows the set of isostatic lines of I and II type in a half plane loaded by a single force [3].

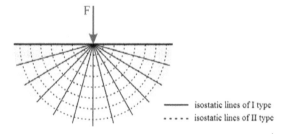

    —— isostatic lines of I type
    - - - - isostatic lines of II type

Figure 7. Set of isostatic lines in a half plane loaded by a single force

5. Isochromatic Fringes

Isochromatic fringes are known as connecting lines of points along which the difference of principal normal stresses $\sigma_1 - \sigma_2$ is constant. They appear on an illuminated photoelastic coating as lines of single (iso) colour (chromos). They can be observed under circular polarised light [3,6]. Figure 8 shows isochromatic fringes during gradual diametrical tensional loading of the photoelastically coated sample [2].

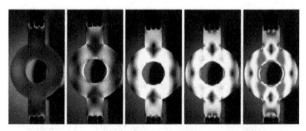

Figure 8. Isochromatic fringes

Linear dependency between temporary birefringence and the difference of principal strains or principal normal stresses enables us to determine the plane stress state on the surface of a photoelastically coated object. There is a constant phase shift of light waves along individual isochromatic fringes, i.e. isochromatics are geometrical locations of constant birefringence points [7,8].

When white light is used, photoelastic pattern appears as a series of gradual isochromatic areas of different colours, while every area represents a different birefringence level related to that particular deformation in the object subject to examination. The colour of every area (isochromatic fringe) uniquely identifies the birefringence or the fringe order (deformation level) throughout the whole area. The characteristics of colourful isochromatic fringes are listed in Table 1 [1,5].

The orders of isochromatic fringes in the photoelastic coating are proportional to the difference between principal strains in the coating and on the surface of the object subject to examination. This linear relation can be expressed in terms of the following relation:

$$\varepsilon_1 - \varepsilon_2 = f \cdot (N - N_0) = \gamma_{max} \qquad (8)$$

where $\varepsilon_1, \varepsilon_2$ represent principal strains in the given photoelastic coating,

f - fringe constant of optical sensitivity of the photoelastic coating,

N_0 - fringe order identified in the measurement point during zero loading of the object under analysis (initial compensator value),

N - fringe order identified in the measurement point during loading of the object under analysis (final compensator value),

γ_{max} - maximum shear deformation.

Table 1. Characteristics of colourful isochromatic fringes

Colour		Approximate relative delay (nm)	Fringe order N (-)
	black	0	0
	grey	160	0,28
	white	260	0,45
	pale yellow	345	0,60
	orange	460	0,80
	dull red	520	0,90
	transitional purple	575	1,00
	dark blue	620	1,08
	dark green	700	1,22
	yellow	800	1,39
	orange	835	1,63
	pink-red	1050	L,82
	transitional red-green	1150	2,00
	green	1350	2,35
	yellow	1440	2,50
	red	1520	2,65
	transitional red-green	1730	3,00
	green	1800	3,1

Engineering praxis more often uses stress than strain, hence equation (8) can be expressed via extended Hooke's law for biaxial stress state:

$$\sigma_1 = \frac{E}{1 - \mu^2}(\varepsilon_1 + \mu\varepsilon_2) \qquad (9)$$

$$\sigma_2 = \frac{E}{1 - \mu^2}(\varepsilon_2 + \mu\varepsilon_1) \qquad (10)$$

Subtracting the relation (9) from the relation (10) gives expression for the difference of principal normal stresses in the following form:

$$\sigma_1 - \sigma_2 = \frac{E}{1 + \mu}(\varepsilon_1 - \varepsilon_2) \qquad (11)$$

Implementing the relation (8) into the relation (11) gives the expression

$$\sigma_1 - \sigma_2 = \frac{E}{1 + \mu} f \cdot (N - N_0) \qquad (12)$$

Hooke's law for plane (biaxial) stress state can be used to apply experimentally gained photoelastic data to calculate the difference of principal normal stresses $\sigma_1 - \sigma_2$ and maximum shear stress τ_{max}. Here applies the following

$$\sigma_1 - \sigma_2 = \frac{E}{1 + \mu}(\varepsilon_1 - \varepsilon_2) = 2\tau_{max} \qquad (13)$$

where σ_1, σ_2 principal normal stresses on the surface of the object under examination,

E - modulus of elasticity of the material of which the analysed object is made,

μ - Poisson's ratio of the material of which the analysed object is made.

Maximum shear stress τ_{max} on the photoelastic coating in any point can be expressed via relation

$$\tau_{max} = \frac{1}{2}\left(\frac{E}{1 + \mu}\right) \cdot f \cdot (N - N_0) \qquad (14)$$

Relations (8) to (14) are principal relations which are used in the analysis of inner forces of the coated object under examination. They provide us only with the difference of principal strains or principal normal stresses and not with their individual magnitudes. In order to determine magnitudes of individual components and signs in case of a biaxial stress state, it is necessary to use numerical or experimental separation methods. In case of a plane (uniaxial) stress state ($\sigma_1 = 0$ or $\sigma_2 = 0$) there is only one non-zero principal normal stress in the plane of the photoelastically coated object subjected to the analysis. This stress can be identified directly from the equation (12) [5]. For instance, the following relation applies for $\sigma_2 = 0$

$$\sigma_1 = \frac{E}{1 + \mu} f \cdot (N - N_0) \qquad (15)$$

6. Conclusion

All photoelastic entities discussed above (isoclinic fringes, singular points, isostatic lines, isochromatic fringes) are necessary for the analysis of deformation and stress fields of various structural elements by means of PhotoStress method.

Parameters of isoclinic as well as isochromatic fringes are still determined manually in the majority of cases, even though we face advancing software support tools. However, PhotoStress software is has currently been developed at the workplace of the authors, which is based on classic methods in combination with digital records and modern hardware. The software will enable to quantify directions and magnitudes of principal strains and principal normal stresses on the basis of records of isoclinic and isochromatic fringes. Furthermore, it will be able to draw isostatic lines of I and II type automatically within particular field of the structural element under consideration.

Acknowledgement

This research was supported by grant project VEGA 1/0393/14, VEGA 1/0937/12 and KEGA 004TUKE-4/2013.

References

[1] Trebuňa, F., Jadlovský, J., Frankovský, P., Pástor, M.: *Automatizácia v metóde Photostress* [Automation in PhotoStress method]. 1. issue. Košice: TU - 2012. 285 p.

[2] Huňady, R., Trebuňa, F., Frankovský, P., "Uplatnenie metódy PhotoStress[Application of PhotoStress method]," In: *Strojárstvo*. Vol. 15, No. 12, 2011, p. 10-12.

[3] Milbauer, M., Perla, M.: *Fotoelasticimetrie a příklady jejího použití* [Photoelasticity and Examples of its Use]. ČSAV, 1961.

[4] Milbauer, M., Perla, M.: *Fotoelasticimetrické přístroje a měřicí metody* [Photoelasticity Devices and Measurement Methods]. ČSAV, 1959.

[5] Trebuňa, F.: *Princípy, postupy, prístroje v metóde photostress* [Principles, Processes, Devices in PhotoStress Method]. Typopress, Košice, 2006.

[6] Frocht, M. M.: *Photoelasticity*, John Wiley and Sons, New York, 1941.

[7] Kobayashi, A. S.: *Handbook on experimental mechanics*. New York: VCH, 1993.

[8] Zandman, F., Redner, S. T., Riegner, E. I.: *"Reinforcing effect of birefringent coatings,"* Experimental Mechanics, (2), pp. 55-64, 1962.

17

Proposed Model of Hand for Designing Ergonomic Vibration Isolation Systems for Hand-held Impact Tools

Michał Śledziński[*]

Chair of Machine Design Fundamentals, Poznan University of Technology, Poznań, Poland
*Corresponding author: michal.sledzinski@put.poznan.pl

Abstract This article presents the methodology of ergonomic design and testing of hand-held impact tools with vibration isolation systems. An operator's hand model is proposed comprising uniform bars positioned at angles adequate to the working position of the upper extremity during soil compacting operation. The model is addressing both the subject operator's anthropometric traits and the tool's interaction with the system. Model testing data of the hand-arm system are in good agreement with the results of experimental testing, in which the level of accelerations measured both at the tool handle and at the most exposed areas of the hand, that is at the wrist and at the elbow, were determined using miniature triaxial vibration sensors. The presented hand-arm system model makes it possible to analyze interaction between the hand and the tool and transmission of vibration from the tool to the human body. The proposed methodology may be used in creating ergonomic working environments and in designing safe impact tools.

Keywords: *hand model, simulation, impact tools*

1. Introduction

The subject of the analysis are pneumatically driven hand-held impact tools and their vibration isolation systems. The first tools of this type were built by Jonathan J. Couch and Joseph W. Fowle as early as between 1848 and 1851. While improving working performance these tools had also adverse effects on human health due to exposure of operators to harmful vibration. The first reports on disease symptoms in pneumatic drill operators were provided by Loriga in 1910 [9]. In 1955 E. Andreeva-Galanina introduced the term "vibration disease. Vibration transmitted to human body may assume the form of the so-called local vibration (hand–arm vibration) characteristic of vibrations having an effect directly on human hands.

Pneumatic impact tools are not only a source of vibration harmful to humans and to the environment but also generate high intensity noise often exceeding the 100-130dB range. The sources of noise include flow of compressed air through nozzles and operating chambers and the hitting the material by the tool foot. The tool designers are required to eliminate or considerably minimize the associated health and safety hazard factors.

2. Methodology of Ergonomic Design of Hand-Held Impact Tools

The purpose the ergonomic design of impact tools is to ensure optimal working environment for the operators. As far as hand-held impact tools are concerned of particular importance are the individual traits of the operator, properties of the tool and of the process, and the material parameters of the working environment such as:

–   the tool (design, working condition),
–   technological process (type of operation, medium undergoing compaction),
–   vibration (intensity, frequency, direction),
–   exposure (duration of impact, protective measures), and
–   individual traits of the operator (skill to damp vibrations, the way of tool handling).

If the allowable limits of the above-mentioned factors are exceeded, the operator may be exposed to potential vibration disease hazard.

The recent technical and technological progress has lead to a shift from the producer's market to the end-user's market, and currently we can speak about Human-Centered Design. This ergonomic approach to design puts man at the centre and sees him as the subject of activity to which main design objectives are subordinated [4,8,18].

In the design of hand-held impact tools however engineers face a difficult problem. On the one hand their aim is to minimize harmful effect of vibration on the operator but, on the other hand, they need to ensure the tool's high vibration intensity necessary for proper performance in the technological process.

In human-centered design of impact tools one should aim at identifying the device's features important for the user and environment (operating comfort, immediate and long-term safety, e.g., effect of vibration and noise). In the light of the requirements imposed on the modern ergonomic design a methodology of design [21] and testing [22] of impact tools and their damping systems was developed. It is presented in the form of diagrams in Figure 1 and Figure 2.

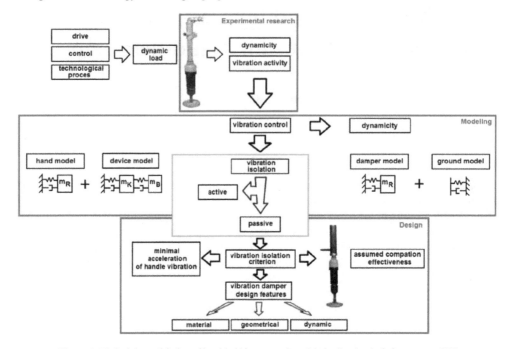

Figure 1. Methodology of design of hand-held impact tools and their vibration isolation systems [21]

The algorithm of methodology of design and testing of impact tools (Figure 1) was developed based on own analyses and research. The status of the tool's dynamic loads is determined at the preliminary stage. The determined level of vibration activity is then compared with the maximum allowable values set out in standards, machine safety regulations and EU directives. When the tool vibrations level exceeds allowable values the solution searching stage starts and a decision on possible interference into the source of vibration or on use of vibration isolation is taken. If a damping system is the solution to be adopted, the type of vibration isolation is specified, meaning it is indicated whether passive or active isolation is to be applied.

The next design stage involves mathematical modeling, comprising all the components of the man – impact tool – ground system. Vibration isolation criteria, i.e. achieving minimum acceleration of vibration of the handle without reducing the effectiveness of the compaction process, are established. Then the design features are selected and a vibration damper prototype is built.

The vibration isolation system prototype needs to be verified experimentally from the anthropo-technical perspective (Figure 2). Work at this stage is particularly oriented to protecting humans by reducing transmission of vibration from the tool to the operator's hand-arm system.

Tool testing, in compliance with the standard recommendations, was conducted by taking measurements of acceleration values on the vibration isolator's handle and at the same time the testing scope was expanded by the measurements of the element's displacement. However, the scope of testing recommended in standard normative procedures was evaluated as incomplete and insufficient. For this reason measurement of accelerations occurring on the directly exposed parts of the hand were taken simultaneously using miniature triaxial acceleration sensors ADXL325 [22].

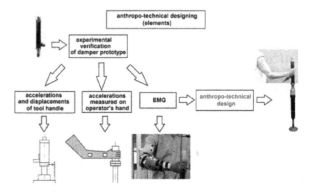

Figure 2. Methodology of testing of impact tool damping system prototypes (from anthropo-technical perspective) [22]

An important part of the ergonomic designing cycle is modeling of the analyzed system. Modeling makes it possible to analyze various structural concepts as early as at the designing stage and to study influence of individual material, geometric and dynamic tool features on its behavior [7,20].

Modeling of biomechanical systems is an important part of the ergonomic engineering process.

3. Biomechanical System Modeling

In the designing process it is necessary to describe biodynamic traits of the human body in mathematical terms.

The conventional approach to hand-arm system modeling with the objective to analyze the system's biodynamics usually boils down to the use of lumped-mass mechanical models with elements of spring-damper type [16,19], or distributed-mass models [25]. However, biomechanical model designing should address also broadly understood ergonomic aspects: working environment, tool's characteristics and its vibration activity, vibration parameters, and also the operator's anthropometric data and his technique of tool handling.

When analyzing the effect of vibration on the human body, the body is treated as a vibrating, elastic system with a considerable ability to damp vibration. In the dynamic analysis of the behavior of the body of a person operating a tool it is necessary to take into account the working position [1] as it is related with the change in elastic properties and the position of the centre of gravity of the body or its part.

The biomechanical parameters used in model testing are given in numerous studies [3,10,17]. However, their values tend to differ significantly and the conditions in which they were determined are usually not specified. This makes an analysis and comparison difficult and the adoption of these parameters in new research impossible. Therefore, one should attempt to determine anthropometric parameters and parameters describing the behavior of the hand-arm biomechanical system that are adequate for the specific tool, technological process and operator and his specific tool handling technique.

Modeling of the hand-tool handle system for tool movement in horizontal plane was the subject of studies [3,10,14,15,26].

Considering the impact tool for soil compaction working in the vertical plane, being the object of the analysis a mathematical model of the hand-arm system with distributed parameters, adequate for the working movement performed, was developed.

One of the first models of a pneumatic drill operator's upper extremity was the model proposed by Kuchna (1953). In this model the hand was modeled with big approximation with a linear damping element. Dieckmann's model (1958) represented the position of the hand characteristic of working with a demolition hammer. The parameters of this model were selected based on experimental studies. Next attempts to develop models were undertaken by Miwa (1964), based on the system's impedance study, Vasilev (1972) - for different angles between the hand and the arm and Reynolds (1975), who developed a three degree-of-freedom discrete system describing the hand's dynamic characteristics. From more recent models, one may mention Fritz model (1991), and Daikoku and Ishikawa model (1990) based on the studies of hand impedance noted for different ways of tool handling. Dong et al. (2005) presented a two-dimensional MES model of fingers of the hand used in the analysis of transmission of vibrations to the hand-arm system. Kazi (2008) developed a two and tree-mass model of the hand for researching biodynamic reactions in a vibrating environment. Joshi and Murray (2010) proposed a model with single degree of freedom in rotation used in biodynamic analysis of connecting elements with bolted joints. Dong and Welcome et al. (2013) dealt with a four-degree-of-freedom lumped-parameter model used in the analysis of vibration damping by personal protection equipment used by workers operating impact tools.

4. Proposed Hand Model

Following numerous attempts to build a model representing as close as possible the actual behavior of the hand-arm system in relation to a given tool, the working conditions and the particular operator a biomechanical model shown in Figure 3 and Figure 4 was developed.

This model has the form of a system of uniform bars with distributed masses representing angular position of the bent upper extremity during operation of an impact tool. In this model the individual bars represent, respectively, the operator's arm, forearm and hand. The articulations connecting the bars correspond with shoulder and elbow joints and represent the wrist and the hand fingers-handle coupling. The action of the upper part of the operator's body (segment I, Figure 4) on the hand-arm system and of the operator's hand on the tool (segment V, Figure 4) is described with viscoelastic connections (Kelvin-Voigt model).

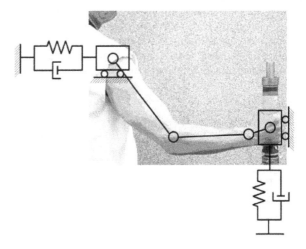

Figure 3. Operator's hand-arm model –three-segment bar system in the form of a triple pendulum

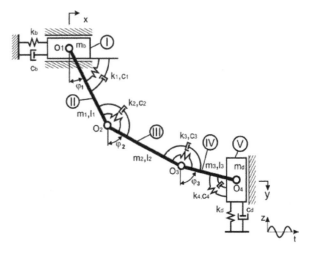

Figure 4. The model's diagram with description of parameters: l – length, m – mass, k – stiffness coefficient, c – damping coefficient; symbols of parameters: b – shoulder, 1 – arm, 2 – forearm, 3 – hand, 4 – hand fingers; symbols of systems: I – shoulder, II – arm, III – forearm, IV – hand, V – fingers of the hand, φ – initial angle

The model was described with equations of motion using the Lagrange method [22].

5. Simulation Analyses and Experimental Verification of the Model

The developed model was used to carry out simulation analyses of the operator's hand action on the tool and to study the degree of vibration reduction by the vibration isolation system. Following numerical solving of differential equations describing the model with the Runge-Kutta fourth order method simulation analyses were made. After the preliminary identification of the model the following parameter values determined for the tool operator participating in the tests were adopted for the analysis: mb = 4 kg – mass of the shoulder, kb = 180 N/m – stiffness of the spring for the shoulder, cb = 8 Ns/m – viscous damping of the shoulder, m1 = 2.45 kg – shoulder mass, k1 = 45 Nm – stiffness of the arm spring, c1 = 5.5 Nms – viscous damping, l1 = 0.32 m – length of bar representing the arm, m2 = 1,3 kg – forearm mass, k2 = 25 Nm – stiffness of the spring, c2 = 4 Nms – viscous damping, l2 = 0.26 m – forearm length, m3 = 0.45 kg – mass of the hand, k3 = 10 Nm – stiffness of the spring, c3 = 2 Nms – viscous damping, l3 = 0.1 m – hand length.

Figure 5 to Figure 9 show selected diagrams of changes in displacements and accelerations of the modeled hand segments.

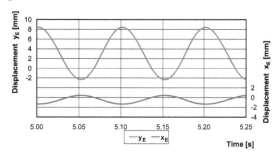

Figure 5. Changes in linear displacements of the elbow of the hand model in horizontal direction x_E and in vertical direction y_E

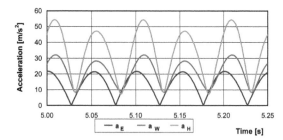

Figure 6. Changes in resultant accelerations of the hand a_H, wrist a_W and elbow of the model of the hand a_E

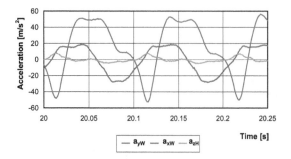

Figure 7. Accelerations from experimental measurements on the wrist: in horizontal direction a_{xW}, vertical direction a_{yW} and in the direction transverse to the hand plane a_{zH}

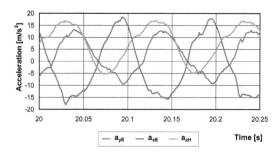

Figure 8. Accelerations from experimental measurements on the elbow: in the horizontal direction a_{xE}, in the vertical direction a_{yE} and in the direction transverse to the plane of the hand a_{zH}

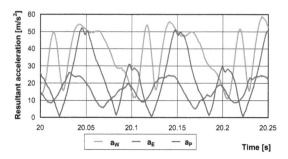

Figure 9. Changes in resultant accelerations on the operator's wrist a_W, elbow a_E and piston of the inductor a_P during experimental testing

Analysis of the model and experimental testing results demonstrated a fairly good agreement between the values of the measured working parameters, which may be a confirmation of the usefulness of the proposed model for the analysis of the hand-arm system behavior and for selection of adequate parameters of the system damping the transmission of vibration to the human body. The noted differences result from unavoidable artifacts related with dynamic measurements (movement of the extremity) and with the fastening of the acceleration sensors on the skin of the hand.

6. Summing up

The developed methodology of design and testing of vibration isolation systems for impact tools was verified by developing and building of vibration isolators allowing for a considerable reduction of vibration transmitted to the human hand-arm system [21,22]. Effective vibration control, and hence minimized exposure of operators to harmful effects while working with impact tools without compromising the effectiveness of soil compaction working process, can confirm correctness of the proposed methodology of designing, testing and modeling of vibration isolation systems. The presented methodological approach to designing of an ergonomic working environment with manually operated impact tools and in a way covering all the elements of the man-tool anthropo-technical system, as well as the proposed hand model adequate for individual traits of a given man, the tool under study and the working process may be used with success in testing and designing new and verification of existing hand-held impact tools. The presented results may contribute to improving the occupational health and safety of operators, better comfort of work, higher accuracy and effectiveness of the technological process, and significantly reduce the impact of vibration on human health, and, as a result, reduce the incidence of occupational diseases, including in particular vibration disease.

References

[1] Adewusi, S., Rakheja, S., Marcotte, P., *Biomechanical models of the hand-arm to simulate distributed biodynamic response for different postures*, International Journal of Industrial Ergonomics, Vol. 42. 2012.

[2] Chaffin, D.B., Anderson, G.B.J., Martin, B.J., *Occupational Biomechanics*, Wiley-Interscience, New Jersey, 2006.

[3] Cherian, T.H., *Control of hand transmitted vibration through development and analysis of a human hand-arm-isolator model*, A thesis, Concordia University, Montreal, Quebec, 1994.

[4] Dietrych, J., *Projektowanie i konstruowanie*, WNT, Warszawa, 1974.

[5] Dong, R.G., Wu J.Z., Welcome D.E., *Recent advances in biodynamics of human hand-arm system*, Industrial Health, 43.

[6] Dong, R.G., Welcome, D.E., McDowell, T.W., Wu, J.Z., *Modeling of the biodynamic responses distributed at the fingers and palm of the hand in three orthogonal directions*, Journal of Sound and Vibration, 332 (4), 2013.

[7] Foster, M., *Sztuka modelowania układów dynamicznych*, Warszawa, WNT, 1996.

[8] Gause, D.C., Minch, E., *Procesy projektowe z perspektywy przestrzeni stanów*, Projektowanie i systemy: zagadnienia metodologii. T. 12, Ossolineum, Warszawa, 1990.

[9] Grzegorczyk, L., Walaszek, M., *Drgania i ich oddziaływanie na organizm ludzki*, PZWL, Warszawa, 1971.

[10] Gurram, R., *A study of vibration response characteristics of the human hand-arm system*, A thesis, Concordia University, Montreal, Quebec, 1993.

[11] Joshi, A., Leu, M.C., Murray, S., *Modeling of hand-arm system for studying the dynamic effect of impact generated by fastening of shear-type fasteners*, Proceedings of 4th Annual ISC Research Symposium ISRS 2010, April 21, 2010, Rolla, Missouri.

[12] Jurczak, M., *Wibracje*, PWN, Warszawa, 1974.

[13] Kaz,i S., As'arry, A., Zain, M.Z. et al., *Biodynamic response of human hand-arm models under vibration*, 10th WSEAS Int. Conf. on Automatic Control, Modelling and Simulation ACMOS'08, Istanbul, 2008.

[14] Kinne, J., Latzel, K., Melzig-Thiel, R., Schatte, M., *Schwingungstechnische Modellierung der beiden Hand-Arm-System von Bedienpersonen für die Anwendung bei der Prüfung von Handmaschinen*, Forschung Projekt Fb 1955, Dortmund, Berlin, Dresden, 2011.

[15] Marcotte, P., Adewusi, S., Boutin, J., Nelisse, H., *Modelling the contributions of handle dynamics on the biodynamic response*, 11th International Conference On Hand-Arm Vibration, Bolonia, 2007.

[16] Mishoe, J.W., *Suggs, C.W., Vibrational responses of human hand*, Journal of Sound and Vibration, 53, 4, 1977.

[17] Meltzer, G., Melzig-Thiel, R., Schatte, M., *Ein mathematisches Schwingungsmodell für das menschliche Hand-Arm-System*, Maschinenbautechnik, 29, 1980.

[18] Pahl, G., Beitz, P., *Nauka konstruowania*, WNT, Warszawa, 1984.

[19] Reynolds, D.D., Falkenberg, R.J., *A study of hand vibration on chipping and grinding operators*, Part II: Four- degree-of freedom lumped parameter model of the vibration response of the human hand, Journal of Sound and Vibration, 95, 1984.

[20] Szücs, E., *Modelowanie matematyczne w fizyce i technice*, Warszawa, WNT, 1997.

[21] Śledziński, M., *Kształtowanie cech konstrukcyjnych tłumika drgań ubijaka pneumatycznego*, Rozprawa doktorska, Politechnika Poznańska, Poznań, 2006.

[22] Śledziński, M., Badania układów „Człowiek-narzędzie udarowe-podłoże", Rozprawy nr 509, Wydawnictwo Politechniki Poznańskiej, Poznań, 2013.

[23] Tytyk, E., *Projektowanie ergonomiczne*, Wydawnictwo naukowe PWN, Warszawa – Poznań, 2001.

[24] Tytyk, E., Butlewski, M., *Ergonomia w technice*, Wydawnictwo Politechniki Poznańskiej, Poznań, 2011.

[25] Wood, L.A., Suggs, C.W., Abrams, C.F., *A distributed parameter dynamic model of the human hand-arm system*, Journal of Sound and Vibration, 57, 2, 1978.

[26] Ziemiański, D., *Identyfikacja i modelowanie wybranych zachowań dynamicznych układu człowiek–otoczenie*, Rozprawa doktorska, Politechnika Krakowska, Kraków, 2011.

18

Prediction of Surface Roughness and Feed Force in Milling for Some Materials at High Speeds

Omar Monir Koura[1,*], Tamer Hassan Sayed[2]

[1]Mechanical Department, Faculty of Engineering, Modern University for Technology & Information, Egypt
[2]Design & Prod. Eng. Department, Faculty of Engineering, Ain Shams University, Egypt
*Corresponding author: koura_omar@yahoo.com

Abstract Machining at relatively high speed perform differently than when traditionally cutting speeds are used. High speed machining affects to a great extent the quality of the manufactured products. The effects differ from material to another. The aim of the present paper is to compare the quality of the machined parts at different cutting speed ranges. The study covers several engineering materials. Neural Network techniques was applied in the prediction of both the resulted surface roughness and the developed feed forces.

Keywords: *high speed cutting, cutting conditions, roughness and tool wear, artificial neural network*

1. Introduction

Milling is currently the most effective and productive manufacturing method for roughing and semi-finishing large surfaces of metallic parts. High speed milling is sometime necessary to mill materials with special characteristics. Milling performance, accuracy and surface texture are tied up with the operating cutting conditions.

Reference [1], built a model to predict surface roughness of milling surface based on cutting speed, feed and depth of cut of end milling operations. The model is based on Genetic expression programming (GEP) which is a solution method that makes a global function search for the problem, developed as a resultant genetic algorithm (GA) and genetic programming (GP) algorithms. The tests were carried out on Aluminum 6061 T8 using a 10 mm diameter HSS end mill. The range of speed used was 23 to 47 m/min, range of feed 135 to 650 mm/min and range of depth of cut 0.25 to 1.27 mm. The model gave the relation between cutting parameters and surface roughness with accuracy of about 91%.

Reference [2], proposed a method for determination of the best cutting parameters leading to minimum surface roughness in end milling mold surfaces of an ortez part used in biomedical applications by coupling neural network and genetic algorithm. A series of cutting experiments for mold surfaces in one component of ortez part are conducted to obtain surface roughness values. The tests were carried out on Aluminum 7075 T6 using a 10 mm diameter Sandvik end mill. The range of speed used was 100 to 300 m/min, range of feed 0.32 to 0.52 mm/rev, range of axial depth of cut 0.30 to 0.7 mm and range of radial depth of cut 1 to 2 mm. A feed forward neural

network model is developed exploiting experimental measurements from the surfaces in the mold cavity. Genetic algorithm coupled with neural network is employed to find optimum cutting parameters leading to minimum surface roughness without any constraint.

Reference [3], developed a mathematical model for determining the optimal machining conditions, so as to obtain a surface with specified properties, taking account of the technological constraints on the following parameters: the residual stress; the roughness and micro-hardness (cold working) of the machined surface; the structural–phase composition of the surface layer (the temperature), the tool life; and the standard machine-tool data. It is found that the surface roughness declines with increase in cutting speed and decrease in the feed and depth. Also, the cutting speed and depth have the greatest influence on the surface micro-hardness. Increasing the cutting speed made the surface properties become more uniform in high-speed end milling.

Reference [4], presented an artificial neural network (ANN) model for predicting the surface roughness performance measure in the machining process. Matlab ANN toolbox was used for the modelling purpose. The tests were carried out on Titanium Alloy (Ti-6A1-4V) using un-coated, TiAIN coated and SNTR tools. The range of speed used was 124 to 167 m/min and range of feed 0.025 to 0.083 mm/tooth. The study concluded that the model for surface roughness in the milling process could be improved by modifying the number of layers and nodes in the hidden layers of the ANN network structure, particularly for predicting the value of the surface roughness performance measure. As a result of the prediction, the recommended combination of cutting conditions to obtain the best surface roughness value is a high speed with a low feed rate and radial rake angle.

Reference [5], developed a statistical model for surface roughness estimation in a high-speed flat end milling process under wet cutting conditions, using machining variables such as spindle speed, feed rate, and depth of cut. First- and second order models were developed using experimental results, and assessed by means of various statistical tests.

Reference [6], proposed a method of determining the optimal cutting conditions in the high-speed milling of titanium alloys. The proposed method based on the thermo-physical data regarding cutting and tool wear. This research identified the optimal cutting speed as a function of the mechanical properties of the machined material and the tool material, the mill diameter, the number of cutting teeth, the cutting depth, and the specified tool life.

2. Experimental Work

Two sets of experiments are planned. One is for the study of the effect of cutting conditions on the surface roughness (Ra) in high speed end milling and the second is for the study of the cutting conditions on the feed force. Four types of materials are used, namely Aluminum alloy, Brass, Phosphorus bronze and Steel. Also tests for checking the rate of wear is given in Table 1. The experimental conditions are given in Table 2. It contains the cutting conditions and the various parameters to be measured in each case.

Table 1. Experimental set up for tool wear

W.P. Mat.	Speed (m/min)	Depth (mm)	Tool Wear Time (sec)	Wear (mm)
Phosphorus Bronze	1257	0.5	0	0
			5	0.28
			11	0.34
			17	0.35
			23	0.38
Steel	943	1	1	0.01
			5	0.1
	1257		1	0.008
			5	0.09

3. Results and Discussion

3.1. Effect of Speed on Tool Wear

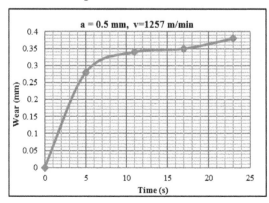

Figure 1. Tool wear when cutting phosphorus bronze

Figure 1 shows the rate of tool wear when cutting phosphorus bronze at speed of 1257 m/min. The wear increased sharply during the first 5 seconds then increases steadily at a low rate. When cutting steal with speeds 943 m/min or 1257 m/min, the tool wear remains, almost, the same at 0.01 mm after 1 sec. even with different depth of

cuts. It reached around 0.1 mm after 5 sec. The better results for steel may be due to the less friction on the rake face of the cutter with the faster flow of the chip.

Table 2. Experimental set up for tool wear

W.P. Mat.	Speed (m/min)	Feed rate (mm/min)	Depth (mm)	Feed Force (N)	Ra (µm)
Aluminum Alloy	628		0.5		1.2
	785	365			0.34
	943				1.07
	1100				1.3
	1257				0.67
	628				0.9
	785	550			0.67
	943				0.75
	1100				1.17
	1257				0.75
	943	145	1	6	
		220		8	
		365		14	
		550		16	
	1257	145		5	
		220		8	
		365		10	
		550		11	
Phosphorus Bronze	943	145		6	
	1100			6.4	
	1257			8	
	943	220		12	
	1100			6.4	
	1257			8	
	628		0.5		0.41
	785				0.46
	943	365			0.28
	1100				0.25
	1257				0.28
	628				0.44
	785				0.39
	943	550			0.6
	1100				0.39
	1257				0.56
	943	75	1		3.39
		145			2.91
		220			1.9
		365			2.25
		550			0.45
	1257	145			0.87
		220			0.28
		550			1.1
Brass	628		0.5		1.1
	785				1.94
	943	365			0.82
	1100				0.86
	1257				0.45
	628				0.4
	785				0.8
	943	550			0.72
	1100				0.73
	1257				1.34
	943	145	1		2.21
	943	220			0.61
	1257	220			2.4
	943	550			1.71
	1257				1.66
Steel	943	50	0.5		0.52
		75			0.5
		145			0.72
		365			0.6
		550			0.45
	1257	50			0.35
		75			0.56
		145			0.87
		365			0.99
		550			1.1
	943	145	1	28	
	1100			15	
	1257			87	
	943	220		29	
	1100			16	
	1257			82	

3.2. Effect of Speed and Feed Rate on Feed Force

Figure 2 & Figure 3 show the effect of speed on the feed force when cutting phosphorus bronze and steel. No effect was noticed when varying the feed from 145 to 220 mm/min. A minimum feed force was obtained at cutting speed 1100 m/min. The increase in the order of magnitude when cutting steel is mainly due to the higher resistivity to cut and the increase in the depth of cut.

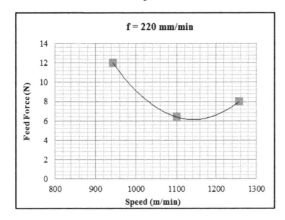

Figure 2. Phosphorus bronze (a = 0.5mm)

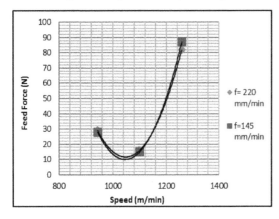

Figure 3. Steel (a = 1 mm)

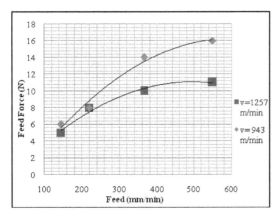

Figure 4. Aluminum Alloy a=1mm

Figure 4 shows the effect of feed on the feed force when cutting aluminum alloy. Feed force increases with increasing the feed rate. At the same time little increase resulted as speed changed from 943m/min to 1257 m/min. But comparing materials such as phosphorus bronze and aluminum alloy with steel a greater increase in the feed

force resulted. Further increase resulted when cutting speed decreased from 1257 m/min to 1257 m/min. This later decrease may be to the lesser friction between the material and the cutter.

3.3. Effect of Speed on Roughness

Figure 5 shows the change of roughness with the increase of cutting speed at two different feed rates when cutting Aluminum alloy. Roughness varied between 0.35 to 1.3 μm when cutting at feed rate 365 mm/min. the variation was limited between 0.65 to 1.15 μm when cutting at feed rate 550 mm/min. So, it may be stated that increasing the feed rate improves the resulted surface roughness.

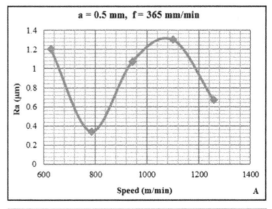

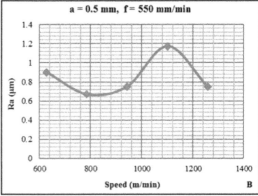

Figure 5. A & B Effect of speed on surface roughness for Aluminum Alloy

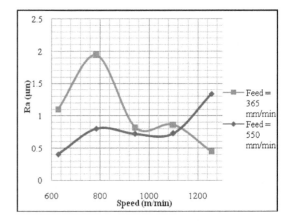

Figure 6. Effect of speed on surface roughness for Brass, a=0.5 mm

Figure 6 & Figure 7 show the same parameters but when cutting brass and phosphorus bronze. Same results may be noticed. But cutting steel, Figure 8, showed a reverse conclusion, that for harder material surface roughness increases with increasing both feed and speed.

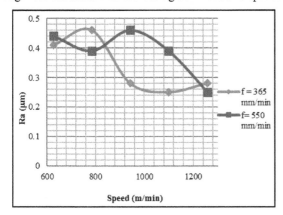

Figure 7. Effect of speed on surface roughness for Phosphorus bronze, a= 0.5 mm

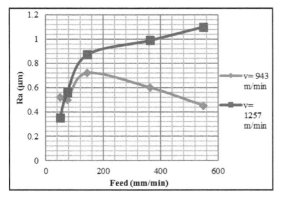

Figure 8. Effect of feed rate on surface roughness for steel, a=0.5mm

4. Artificial Neural Networks (ANNs)

4.1. ANN1 for Prediction of Surface Roughness

In correlating all the variables, neural networks were utilized. The networks were trained to reach the required goal for modeling and predicting the surface roughness (R_a) by adjusting the values of the connections (weights) between elements. The best ANN structure found was that shown in Figure 9. It consisted of 3 layers. The first layer is the input layer which has 4 neurons for the 4 inputs of the network, the second layer is the hidden layer and it consists of 8 neurons, and the third layer is the output layer with 1 neuron for the predicted surface roughness.

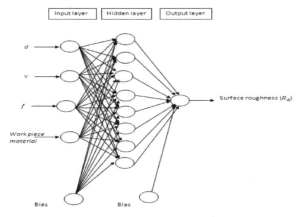

Figure 9. Neural network structure – ANN1

MATLAB 2011b is used to develop the ANN. The processing function for the hidden layer is logsig, and for the output layer is purelin. Feed-forward back propagation ANN used, Leven berg-Marquardt back propagation (TRAINLM) algorithm is used for network training and mean square error (MSE) is used as performance function.

The results show that the correlation coefficient between the measured values and predicted values is 0.94867, as shown in Figure 10.

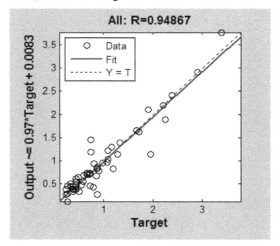

Figure 10. ANN1 regression

The correlated equation for the roughness (R_a) is:

$$R_a = 1.567H + 1.818 \qquad (1)$$

Where:

$$H = w_{33}h_1 + w_{34}h_2 + w_{35}h_3 + w_{36}h_4 + w_{37}h_5 + w_{38}h_6 + w_{39}h_7 + w_{40}h_8 + b_9$$

The values $h_1, h_2, \ldots\ldots h_8$ are given by:

$$h_1 = \frac{1}{1 + e^{-\left(w_1 y_1 + w_2 y_2 + w_3 y_3 + w_4 y_4 + b_1\right)}}$$

$$h_2 = \frac{1}{1 + e^{-\left(w_5 y_1 + w_6 y_2 + w_7 y_3 + w_8 y_4 + b_2\right)}}$$

$$h_3 = \frac{1}{1 + e^{-\left(w_9 y_1 + w_{10} y_2 + w_{11} y_3 + w_{12} y_4 + b_3\right)}}$$

$$h_4 = \frac{1}{1 + e^{-\left(w_{13} y_1 + w_{14} y_2 + w_{15} y_3 + w_{16} y_4 + b_4\right)}}$$

$$h_5 = \frac{1}{1 + e^{-\left(w_{17} y_1 + w_{18} y_2 + w_{19} y_3 + w_{20} y_4 + b_5\right)}}$$

$$h_6 = \frac{1}{1 + e^{-\left(w_{21} y_1 + w_{22} y_2 + w_{23} y_3 + w_{24} y_4 + b_6\right)}}$$

$$h_7 = \frac{1}{1 + e^{-\left(w_{25} y_1 + w_{26} y_2 + w_{27} y_3 + w_{28} y_4 + b_7\right)}}$$

$$h_8 = \frac{1}{1 + e^{-\left(w_{29} y_1 + w_{30} y_2 + w_{31} y_3 + w_{32} y_4 + b_8\right)}}$$

The values of y_1, y_2, y_3 & y_4 are given by:

$$y_1 = 0.666 \; work \; piece \; material - 1.666,$$

Where: Aluminum alloy=1, Brass=2, Phosphorus bronze=3, steel=4

$$y_2 = (2/629)v - (1885/629), y_3 = 0.004f - 1.2,$$
$$y_4 = 4d - 3.$$

Table 3. Weights and bias for Roughness module

w_1= 18.3522	w_2= -3.7422	w_3= 7.9096	w_4= -4.4639
w_5= 0.8670	w_6= 2.4159	w_7= -6.7033	w_8= -0.9764
w_9= 2.8788	w_{10}= -1.5942	w_{11}= -0.1705	w_{12}= -6.3613
w_{13}= -2.1902	w_{14}= 4.0429	w_{15}= -15.1151	w_{16}= -1.1441
w_{17}= 13.0374	w_{18}= -7.2017	w_{19}= -2.4471	w_{20}= -4.4974
w_{21}= 5.2364	w_{22}= 5.1525	w_{23}= -0.0075	w_{24}= 9.4335
w_{25}= 7.3848	w_{26}= 8.3276	w_{27}= 2.5354	w_{28}= 3.7426
w_{29}= -5.0983	w_{30}= 5.5675	w_{31}= 0.7422	w_{32}= -5.8394
w_{33}= -0.6252	w_{34}= -2.9517	w_{35}= 1.4120	w_{36}= 2.5224
w_{37}= -1.0590	w_{38}= 0.4313	w_{39}= 0.1011	w_{40}= -1.9059
b_1= -7.1969	b_2= -4.635	b_3= -3.8938	b_4= -8.2197
b_5= 8.7731	b_6= 0.5382	b_7= 3.1988	b_8= 6.7520
b_9= 1.2656			

The GUI used to create the user interface for ANN1 is shown in Figure 11.

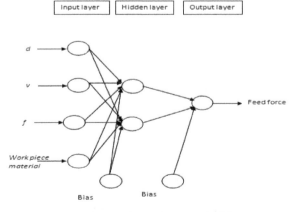

Figure 11. User interface for creating ANN1

4.2. ANN2 for Prediction of Feed Force

Same number of hidden layer as in network ANN1, but the number of neurons was found to 2 neurons to get higher correlation. The structure is shown in Figure 12.

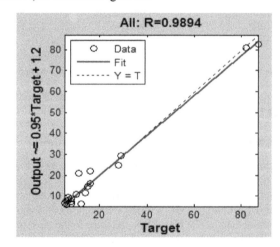

Figure 12. Neural network structure (ANN2)

The correlated equation for the roughness (R_a) is:

$$F = 41H + 46 \qquad (2)$$

Where: $H = w_9h_1 + w_{10}h_2 + b_3$.
The values h_1 & h_2 are given by:

$$h_1 = \frac{1}{1 + e^{-(w_1y_1 + w_2y_2 + w_3y_3 + w_4y_4 + b_1)}}$$

$$h_2 = \frac{1}{1 + e^{-(w_5y_1 + w_6y_2 + w_7y_3 + w_8y_4 + b_2)}}.$$

The values of y_1, y_2, y_3 & y_4 are given by:

$$y_1 = work\ piece\ material - 2,$$

Where: Aluminum alloy=1, Phosphorus bronze=2, steel=3

$$y_2 = 0.006v - 7.006, y_3 = 0.004f - 1.716, y_4 = 4d - 3.$$

Table 4. ANN2 weights and bias for feed force

w_1= 6.6076	w_2= -5.2636	w_3= 1.3689	w_4= 3.1703
w_5= 2.8033	w_6= -0.7136	w_7= 1.2180	w_8= 1.4536
w_9= -3.0902	w_{10}= 3.8920		
b_1= -3.9859	b_2= -1.3478	b_3= -0.9999	

The results show that the correlation coefficient between the measured values and predicted values is 0.9894, as shown in Figure 13.

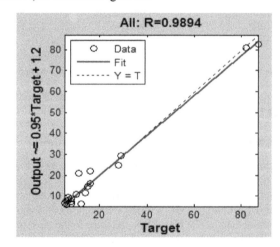

Figure 13. ANN2 regression

The user interface for the feed force model is shown in Figure 14.

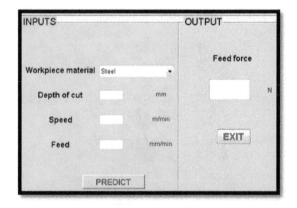

Figure 14. User interface for Feed force

5. Conclusions

1. For materials (aluminum alloy, brass and phosphorus bronze) increasing the cutting speed improves the quality of the surface roughness and v.v. in case of steel.

2. For material such as phosphorus bronze the tool wears much faster than in case of steel.

3. Within the present range there was no change in the feed force as feed rate increased, but it has an optimum when cutting speed was at 1100 m/min.

4. Artificial neural network developed for the prediction of surface roughness has proved efficient. The regression equation showed correlation between the cutting conditions and predicted surface roughness of 94.867%. Equation (1) representing this relation.

5. Artificial neural network developed for the prediction of feed force has good correlation between the cutting conditions and predicted feed force of 98.947%. Equation (2) representing this relation.

References

[1]　Oguz Colak, Cahit Kurbanoglu and M. Cengiz Kayacan, Milling surface roughness prediction using evolutionary programming methods, Materials and Design 28 (2007) 657-666.

[2]　Hasan Oktem, Tuncay Erzurumlu and Fehmi Erzincanli, Prediction of minimum surface roughness in end milling mold parts using neural network and genetic algorithm, Materials and Design 27 (2006) 735-744.

[3]　I. E. Kiryushin, D. E. Kiryushin and T. G. Nasad, Ensuring Surface Quality of Parts in High-Speed End Milling of Quenched Steel, Russian Engineering Research, 2008, Vol. 28, No. 12, pp. 1248-1251.

[4]　Azlan Mohd Zain, Habibollah Haron and Safian Sharif, Prediction of surface roughness in the end milling machining using Artificial Neural Network, Expert Systems with Applications 37 (2010) 1755-1768.

[5]　Babur Ozcelik and Mahmut Bayramoglu, The statistical modeling of surface roughness in high-speed flat end milling, International Journal of Machine Tools & Manufacture 46 (2006) 1395-1402.

[6]　D. E. Kiryushin, I. E. Kiryushin and T. G. Nasad, High-Speed End Milling of Titanium Alloys, Russian Engineering Research, 2008, Vol. 28, No. 10, pp. 1022-1025.

19

Hydro-Structure Analysis of Composite Marine Propeller under Pressure Hydrodynamic Loading

Hassan Ghassemi[*], Manouchehr Fadavie, Daniel Nematy

Department of Ocean Engineering, Amirkabir University of Technology, Tehran, Iran
*Corresponding author: gasemi@aut.ac.ir

Abstract This paper aims to predict the hydrodynamic characteristics and structural analysis of the marine propeller under pressure hydrodynamic loading. Because of the loading on the propeller blade, it goes under significant deformation that may affect the hydrodynamic performance of the propeller. Thus, the blade deformation of a propeller due to fluid pressure should be analyzed, considering hydro-elastic analysis. The propeller was made of anisotropic composite materials, and the geometry of the propeller is for one skew angle. First, the hydrodynamic pressure loading is obtained by FVM and then the deformation of the blade due to this pressure was calculated. Next, the pressure load for deformed propeller is achieved; it is again repeated to obtain the new deformed propeller. This procedure is repeated to converge the thrust, torque and efficiency. We present all results of the pressure distribution, hydrodynamic characteristics, stress and deformation of the propeller.

Keywords: *hydrodynamic characteristics, structural deformation, pressure and stress, composite material*

1. Introduction

Marine propellers are elements in which operate behind the vessel's hull and generate the desired thrust due to pressure loading on the blades. These propulsion systems are the most efficient driving equipment to drive the marine vessels. Meanwhile metal alloys were widely used from old days as conventional materials in manufacturing the propellers. Also, Nickel-Aluminum-Bronze and Manganese-Nickel-Aluminum-Bronze, because of their superior properties such as high strength, corrosion resisting and manufacturing, are alloys that mostly employed for propellers structure.

Nowadays composite materials applications are expanding in many industries such as aerospace, marines and many other vital vehicles. Composite materials have special properties such as light weight, strength-to-weight ratio, stiffens-to-weight ratio and easy manufacturing. Nowadays, extra enthusiasm for the use of composite materials in marine industries, in terms of improving properties and performance of structures, exists, especially in manufacturing the propeller blades as a result of anisotropic properties of these composite materials. Also, increase of efficiency and decrease of noise, fuel consumption, vibration, loading on blades, weight and pressure fluctuations are the common consequences of applying composite materials in marine propellers

According to increasing applications of composite materials, studying about these materials is needed more than ever. Due to complicated geometry of the propeller,

structure analyzing is difficult. Many researchers have employed analytical and numerical method to determine the fluid-structure interactions of the propeller [1,2]. Formerly, cantilever beam theory is used for majority of investigations. Furthermore because of complicated shape of the blade FEM is mostly used for structural analyzing of propellers.

For hydrodynamic analysis and calculation of the pressure distribution on the propeller blade several methods have been developed until now; such as blade element theory, lifting lines theory, vortex lattice method and boundary element method. Lin and Lin worked on nonlinear hydro-elastic behavior of propellers using a finite-element method and lifting surface theory [3]. Cho and Lee [4] published an article entitled propeller blade shape optimization for efficiency improvement. They studied an optimization method to generate the optimal shape of the propeller to increase the efficiency. Meanwhile lifting line theory and lifting surface theory are applied for hydrodynamic calculations.

A coupled BEM-FEM method for exploring the fluid structure interaction of flexible composite marine propellers is developed by Young [5,6]. Also, Young and Motely investigated influence of material and loading effects on the propellers made of advanced materials [7]. Lin et al. published nonlinear hydroelastic behavior of propellers using a finite-element method and lifting surface theory [8] and also experimental paper which presented optimization and experiments on a composite propeller [9]. A FSI (fluid structure interaction) analysis for a flexible propeller is done by Sun et al. [10]. Kulczyk and Tabaczek studied the propeller-hull interaction in the propulsion system [11]. The application of composite materials on the ships propeller performance is investigated by Takestani et al. [12]. Also, other works of the hydro-elastic analysis of marine propellers have been performed in [13,14].

The boundary element method (BEM) is widely used for analyzing the flow around the propellers. This method is applied in three-dimensional turbulence flow, based on potential theory. The BEM is developed for simulating the cavitation in face and back of the immersed, supercavitation and surface piercing propeller in wake flow by Kinnas and Fine [15]. Additionally, the accuracy of this method was so good for no vibration case, pressure and cavitation calculated by BEM was considerably close to experimental results. Ghassabzadeh and Ghassemi employed a hybrid BEM-FEM to determine the hydrodynamic characteristics of a composite propeller based on hydro-elastic analyzing [16]. Recently, investigation of different methods of noise reduction for submerged marine propeller s and their classification carried out by Feiz Chekab et al. [17].

In this paper, finite volume method is used to calculate the hydrodynamic pressure. On the other hand, the blade deformation is analyzed by FEM. Firstly, the hydrodynamic pressure for undeformed propeller is calculated by FVM and the results are exported to the FEM. Then the strain and stress of the blade is calculated. After that the pressure distribution on the propeller using the deformed shape of the blade is recalculated and the new pressure loading is exported to the FEM code and evaluate new deformation again. This process was repeated until the optimum shape and results are obtained.

2. FVM-FEM Coupled

The finite volume method with coupled finite element method employed to the present calculations. Pressure on the blade of the propeller and hydrodynamic characteristic of the propeller are first calculated by FVM. Deformed

blades and stress results may be determined by finite element method.

Here, the computational results of thrust and torque compared to experimental results [16] for B-series propeller are shown in Figure 1, which examine a very good agreement between them. In other word, the numerical results based on experimental data were proved.

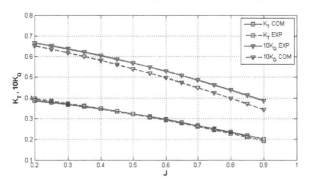

Figure 1. Comparison of computational and experimental hydrodynamic characteristic

Geometry and Computation Grid

In this paper, a marine propeller for B-series was chosen and modeled. The results of hydrodynamic and structural analyzing were presented. The process is according to the flowchart shown in Figure 2. The propeller geometric properties were also shown in Table 1, given from reference [16].

Table 1. Propeller Geometry Properties

parameter	Value
Number of Blades	5
Diameter	0.5 m
Hub Ratio	0.2
Pitch Ratio	Variable (at 0.7R=0.8)
Expanded Area Ratio (EAR)	0.65
Skew Angle	Variable (at 0.7R=25 degree)
Rake Angle	10 degree

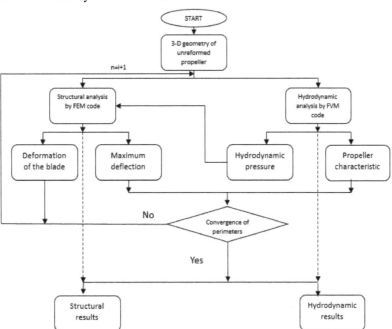

Figure 2. Calculations flowchart

At the beginning of the process, the propeller geometry and domain is generated as shown in Figure 3. The Cartesian coordinates system is used, where x,y and z denote downstream, starboard and upward direction respectively. The origin was located at the center of the hub, and positive directions are upstream, starboard and downstream. The domain dimensions were shown in Figure 4. The solution field was divided into global and sub-domain. The sub-domain frame simulates the propeller rotation and employs the Coriolis acceleration terms in the governing equations for the fluid. The global frame surrounds the sub-domain frame. The global frame is a circular cylinder with 4D diameter, where D is the propeller diameter. The distance between the sub-domain frame and inlet is nearly 4D, while it is nearly 6D for the outlet and dynamic frame. The sub-domain is sphere with a diameter of D.

bigger because the flow in this region is smoother, and smaller grids just cause longer calculation procedure. We employed different grid number and it is concluded that with number of elements 991640 are sufficient for the present calculations.

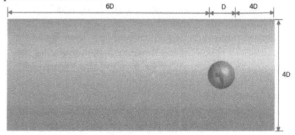

Figure 4. Domain dimensions

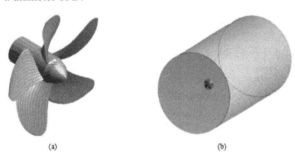

(a)   (b)

Figure 3. (a) 3D propeller (b) calculation domain

Discretized domain around propeller is illustrated in Figure 5. In the next step, boundary conditions are imposed. For this case we divided the solution field into two blocks. The rotating block was meshed with unstructured tetrahedral cells. These elements have a smaller size because the flow around the propeller blade is varying, and we need more accuracy there. The other block is meshed also with unstructured tetrahedral elements. In this block, the sizes of the elements are

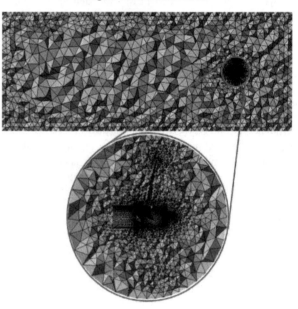

Figure 5. Domain descritization

Table 2. Mechanical Properties of Composite and Coper High Tensile Brase

Property Material	Modulus of elasticity (GPa)	Poisson ratio	Shear modulus elasticity (GPa)	Density (ton/m3)
Composite	E_x=132 E_y=10.8 E_z=10.8	v_{xy}=0.24 v_{yz}=0.49 v_{xz}=0.24	G_{xy}=5.65 G_{yz}=3.38 G_{xz}=5.65	ρ_P=1.5
Copper High Tensile Brass	E_z=102.97	v_{xz}=0.35	-	ρ_P=8.25

Boundary Conditions and Numerical Setups

In the present paper, we used CFX solver in order to evaluate the hydrodynamic performance of a B-series propeller. The governing equations of this problem were solved by the finite volume method based on the RANS equations. A SIMPLE algorithm was used for solving the pressure-velocity coupling equations. The k-epsilon model was used for turbulence model. The fluid (water) was assumed as an incompressible fluid, therefore; on the inlet boundary, uniform velocity is imposed and on the outlet boundary, pressure outlet was imposed. The axial velocity is 0.2m/s in the opposite direction. The no-slip wall condition was imposed on the propeller blades.

ANSYS structural 14 used for structural analyzing. The solver type was iterative. Propeller blade was made of linear elastic anisotropic composite with mechanical properties as shown in Table 2. We also define a fixed support at the end of the hub.

Hydrodynamic and Structural Results

The performance characteristics of the marine propeller can be defined using the non-dimensional coefficients such as the advance ratio, thrust coefficient, torque coefficient, and efficiency, which can be computed, respectively, as follows:

$$\begin{cases} J = \dfrac{V}{nD}, & K_T = \dfrac{T}{\rho n^2 D^4} \\ K_Q = \dfrac{Q}{\rho n^2 D^5}, & \eta = \dfrac{J}{2\pi}\dfrac{K_T}{K_Q} \end{cases} \quad (1)$$

After solving, the pressure contour, at J=0.2, in face and back of the deformed propeller is shown in Figure 6 and Figure 7, respectively. It is clear that the deformation of the blade affects the pressure distribution. The hydrodynamic pressure of the deformed blade is less than that of the undeformed blade. As can be seen the pressure in regions near to the root has its maximum amount and as we move toward the tip the pressure loading, due to structural reasons, decreases. However, areas near to mid-sections, between 0.5R and 0.7R generate the most thrust.

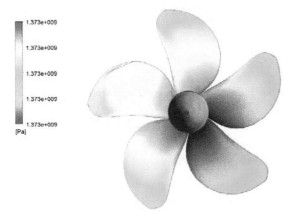

Figure 6. Pressure contour in face

Figure 7. Pressure contour is back

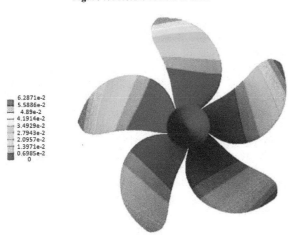

Figure 8. Deformation field of the propeller blade

Figure 8 shows the deflection of the blade along the radial at J=0.2. From the hydrodynamic point of view, applied loading on the blade is the heavy condition that may act to reach high deformation. As can be seen, deflections are distributed according to the nonlinear function. Blade is deformed higher at the tip. Maximum deflection is about 6 cm at the tip. The reason of this phenomenon is that a major thrust was generated in middle sections of the blade and also as we move toward the tip the bending moment about the root increases. Also, if deformation of the blade monitored carefully, the deformation of the trailing edge is more than of the leading edge. There is also an explanation for that, the stress, which illustrated in following Figures, is higher in the trailing edge.

The deformed and undeformed shape of the blade were shown with ten times enlargement at J=0.2, in Figure 9. As mentioned earlier, the deformation increases from root to tip of the blade, due to structural reasons mentioned earlier. The vectors of deformation or one blade is shown in Figure 10.

Figure 9. Deformed and undeformed blade of the propeller

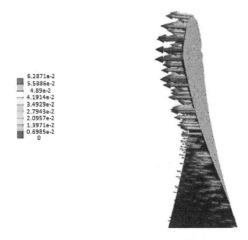

Figure 10. Deformation vectors of the blade

The equivalent von-misses stress in the face and back of the propeller was shown in Figs. 11 and 12 respectively. The stress in the trailing edge has its maximum value and as we move towards the other sections of the blade the equivalent stress decrease. The stress in tip is about 22% of maximum equivalent stress and in the leading edge in about 66.5% of maximum equivalent stress. In the back of the propeller in middle sections, we have the maximum equivalent stress and as we move to the other blade sections the stress decrease.

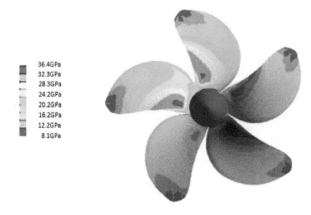

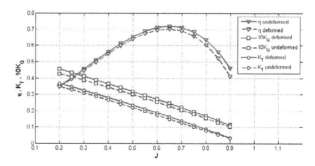

Figure 14. Hydrodynamic coefficients of deformed and undeformed propeller

Figure 11. Equivalent stress in the face of the propeller

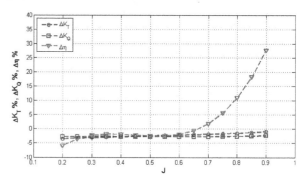

Figure 15. Relative error in hydrodynamic coefficients

3. Conclusions

Figure 12. Equivalent stress in the back of the propeller

Figure 13 shows the maximum deflection versus the advanced velocity ratio; it is clear that, an increase in advanced velocity ratio causes the maximum deflection of the blade to decrease. Therefore, the maximum deflection for the propeller occurs at J=0.2.

In the present paper, coupled FVM-FEM hydro-elastic procedure was used to investigate the hydrodynamic and structure behavior of a marine propeller. The solving procedure was an iterative process; the hydrodynamic pressure was calculated in FVM code and the structural analysis performed by FEM code. In the case study, a five blade composite propeller was chosen to study the deformations and the effect of deformation on the propeller performance. Admittedly, the deformation toward the blade tip increases and stress decreases. The stress in the face of the blade and leading edge is maximum and in back and in the middle of the blade, and as moves to the other areas decrease. Increasing in advanced velocity ratio causes the maximum deflection, the thrust coefficient and the torque coefficient decreases.

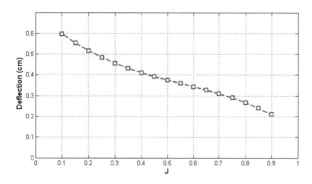

Figure 13. Deflection of the blade versus advanced coefficient

In Figure 14, the hydrodynamic coefficient of the propeller versus the advanced coefficient is presented for deformed and undeformed propellers. Deformed propeller gives lower efficiency relative to the undeformed propeller, especially at high advance ration.

In order to examine the effect of the advanced velocity ratio on the hydrodynamic characteristics precisely, the relative error of the propeller characteristics relative to those of the undeformed blade are shown in Figure 15 at all advanced velocity ratio. It is clear that relative error in the propeller characteristics are nearly constant, and so the performance of the propeller is independent of the rotational speed and the advance velocity.

We also determined the effect of this deformation on the propeller hydrodynamic performance. Obviously, in low advanced coefficients, when the propeller is working in heavy conditions, both thrust force coefficient and torque coefficient increase. While the propeller performance decrease. Hydrodynamic coefficients of the propeller are approximately constant. That means that the deformation of the propeller is independent of the advance velocity ratio.

Acknowledgment

This research was supported by the Marine Research Center (MRC) of Amirkabir University of Technology (AUT) whose works are greatly acknowledged.

References

[1] Blasques, J. P., Berggreen, C., and Andersen, P. (2010). Hydro-elastic analysis and optimization of a composite marine propelle. *Marine Structures, 23* (1), 22-38.

[2] Cho, J., and Lee, S.-C. (1998). Propeller Blade Shape Optimization for Efficiency Improvment. *Computers and Fluids, 27* (3), 407-419.

[3] Li, G., Li, W., You, Y., and Yang, C. (2013). Study on Fluid-Structure Interaction Characteristics of Composite Marine Propeller. *In The Twenty-third International Offshore and Polar Engineering Conference.* Anchorge: International Society of Offshore and Polar Engineers.

[4] Lee, H., Song, M. C., Suh, J. C., and Chang, B. J. (2014). Hydro-elastic analysis of marine propellers based on a BEM-FEM coupled FSI algorithm. *International Journal of Naval Architecture and Ocean Engineering, 6* (3), 562-577.

[5] Young, Y. L. (2007). Fluid–structure Interaction Analysis of Flexible Composite Marine Propellers. *Journal of fluids and structures.*

[6] Young, Y. L. (2006). Numerical and experimental investigations of composite marine propellers. *Proceedings of Twenty-Sixth Symposium on Naval Hydrodynamics.* Rome.

[7] Young, Y. L., and Motley, M. R. (2011). Influence of Material and Loading Uncertainties on the Hydroelastic Performance of Advanced Material Propellers. *Second International Symposium on Marine Propulsors,* (pp. 10-17). Hamburg.

[8] Lin, H. J., and Lin, J. J. (1996). Nonlinear hydroelastic behavior of propellers using a finite-element method and lifting surface theory. *Journal of Marine Science and Technology, 1* (2), 114-124.

[9] Lin, C. C., Lee, Y. J., and Hung, C. S. (2009). Optimization and experiment of composite marine propellers. *Composite Structures, 89* (2), 206-215.

[10] Sun, H. T., and Xiong, Y. (2012). Fluid-structure interaction analysis of flexible marine propellers. *Applied Mechanics and Materials, 226* (228), 479-482.

[11] Kulczyk J. and Tabaczek T. (2014) Coefficients of Propeller-hull Interaction in Propulsion System of Inland aterway Vessels with Stern Tunnels, *the International Journal on Marine Navigation and Safety of Sea Transportation,* Vol. 8 No. 3.

[12] Takestani, T., Kimura, K., Ando, S., and Yamamoto, K. (2013). Study on Performance of a Ship Propeller Using a Composite Material. *3rd International Symposium on Marine Propulsors.* Launceston, Tasmania.

[13] He, X. D., Hong, Y., and Wang, R. G. (2012). Hydroelastic optimisation of a composite marine propeller in a non-uniform wake. *Ocean Engineering, 39,* 14-21.

[14] Mulcahy, N. L., Prusty, B. G., and Gardiner, C. P. (2010). Hydroelastic tailoring of flexible composite propellers. *Ships and Offshore Structures, 5* (4), 359-370.

[15] Kinnas, S. A., and Fine, N. E. (1993). A Boundary Element Method for the Analysis of the Flow Around 3-d Cavitating Hydrofoils. *Journal of Ship Research, 3* (37), 213-224.

[16] Ghassabzadeh, M., Ghassemi, H., and Saryazdi, M. G. (2013). Detrmination of Hydrodynamics Characteristics of Marine Propeller Using Hydroelastic Analysis. *Brodogradnja, 64* (1), 40-45.

[17] Feizi Chekab M., Ghadimi P., Djeddi, S.R., Soroushan M., Investigation of Different Methods of Noise Reduction for Submerged Marine propeller s and Their Classification*, American Journal of Mechanical Engineering.* 2013 1 (2).

20

Kinematic Analysis of the Press Mechanism Using MSC Adams

Darina Hroncová, Peter Frankovský[*], Ivan Virgala, Ingrid Delyová

Technical University of Košice, Faculty of Mechanical Engineering, Letná 9, 042 00 Košice, Slovakia
*Corresponding author: peter.frankovsky@tuke.sk

Abstract The aim of the paper is to present the application of MSC Adams/View for kinematic analysis of a press mechanism. The press mechanism is simulated in MSC Adams/View software. MSC Adams and its modules Adams/View work with this module and its basic operation is dedicated to the solution of kinematics by means of numerical methods. Press mechanism works on the principle of converting rotational motion of a crank to translational motion of a slider block. This paper deals with model press mechanism in Adams/View, simulation running, plotting of the mechanisms points trajectory and kinematic parameters of mechanism members. The computer program shows displacement, velocity and acceleration, and angular velocity and angular acceleration. The paper presents the results with graphic display of parameters such as displacement, velocity, and acceleration.

Keywords: *MSC Adams/View, press mechanism, simulation, kinematic analysis*

1. Introduction

Current software simulation technologies make it easy to design mechanisms with complex kinematic structure. Computer programs significantly reduce time and facilitate the work when solving practical mechanisms. Applying software simulation model, we create a mechanism which corresponds to a real machine. With computer models we can analyze in detail the solution of real objects in practice. Using computer simulations, we can expect desired behavior of the model under loads that may occur because the elimination of problems in the real system is financially much more demanding and, of course, time-consuming.

Kinematic analysis is the process of measurement of kinematic quantities which is used to describe motion. In engineering, for instance, kinematic analysis may be used to find the range of movement for a given mechanism [3]. Kinematic synthesis designs a mechanism for a desired range of motion [4].

In this paper, we deal with kinematic analysis of the press mechanism in Figure 1 which is driven by the crank which rotates with constant angular velocity.

This work deals with the simulation program MSC Adams which was used to simulate the movement of the pump mechanism. The result is presented in a graphical representation of the movement of individual elements as well as respective members of the mechanism [1]. Displacement, velocity and acceleration of key points are plotted and the trajectory of chosen points is also plotted [2].

2. MSC Adams Main Characteristic

Computer software MSC Adams (Automatic Dynamic Analysis of Mechanical Systems) is one of the most widely used multi-function computing software. The program allows us to create dynamic, kinematic and static analysis of the proposed mechanical systems and helps us to optimize and improve their properties. It helps in simulations of mechanical systems consisting of rigid and flexible bodies connected by different types of kinematic links and joints [2,6,7].

3. Model of the Press Mechanism

The press mechanism works on the principle of converting rotational motion of the crank 2 to translational motion of the slider block 6. Driving link OA has a counterclockwise angular velocity of the 6 rad.s⁻¹. The task is to calculate the absolute speed of the member 6, the size of angular velocity of the member 3, angular velocity and angular accelerations of the member 4 and kinematic parameters of other press machine members [8].

The aim of the computer simulation by MSC Adams/View is to build the model of the press mechanism. Our goal is to determine kinematic variables of rotational motion, translational motion and general plane motion of the members.

A generalized diagram of the press mechanism is shown in Figure 1.

Member OA of the mechanism is formed by the crank and in ADAMS/View program it is a rigid body with geometry named Link. Parameters of the body 2 are as follows: length 0,15 m, width 0,04 m, depth 0,04 m.

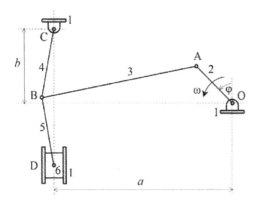

Figure 1. Model of the press mechanism

Member AB of the mechanism is formed by a connecting rod and in the program it is a rigid body with geometry named Link. Parameters of the body 3 are as follows: length 0,84 m, width 0,04 m, depth 0,04 m.

Member BC of the mechanism is formed by the crank and in the program it is a rigid body with geometry named Link. Parameters of the body 4 are as follows: length 0,5 m, width 0,04 m, depth 0,04 m.

Member BD of the mechanism is formed by the connecting rod and in the program it is a rigid body with geometry named Link. Parameters of the body 5 are as follows: length 0,7 m, width 0,04 m, depth 0,04 m.

The member 6 of the mechanism is formed by a piston and in the program it is a rigid body with geometry nemed Box. Parameters of the body 6 are as follows: length 0.1 m, width 0,16 m, depth 0,1 m.

Next parameters of the mechanism are angle $\varphi=45°$, a=1,05 m and b=0,5 m. The mechanism contains rotational joint, translational joint and fixed joint as shown in Figure 2b. Motion is in joint 1 in point O with angular velocity 6 rad. sec^{-1}.

Simulation parameters are: End Time 10 second and Steps 200.

3.1. Compilation of the Model in MSC Adams

Using basic building blocks, we compile a model of the press mechanism in MSC Adams/View (Figure 2). The model will be projected in several steps as described in the following sections [2].

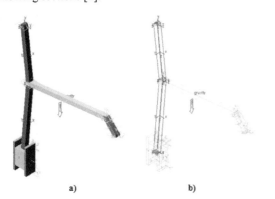

Figure 2. 3D Model of the press mechanism (a) full and (b) transparent

Trajectory of the mass center is plotted by function Trace Marker in Figure 3. It shows trajectory of point members 3, 5 and 6.

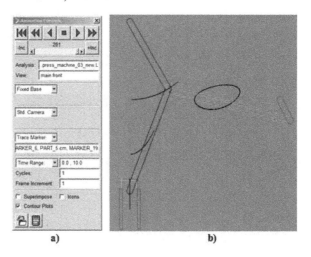

Figure 3. Illustration of trace marker determines (a) animation control and (b) trajectory of respective points

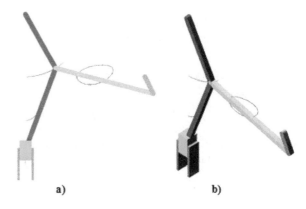

Figure 4. Model of the press mechanism and trajectories with animation control (a) front view of 2D model (b) 3D model

3.2. Creating Kinematic Variables of the Member 2

To determine the values of the parameters it is necessary to define several properties in measurement windows [1]. After opening the dialog box for the determined angle, velocity and acceleration, is necessary to define important parameters [3,4]. The time course angle of rotation member 2 is shown in Figure 5.

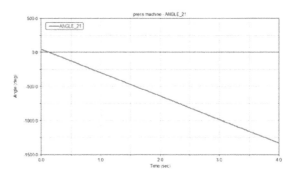

Figure 5. Rotation angle of the crank 2

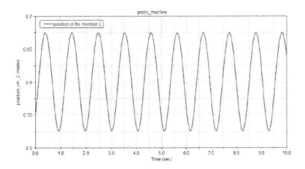

Figure 6. Position of the crank mass centre - PART_2

Angular velocity of the PART_2 is 343,77 rad.sec^{-1} (Figure 7).

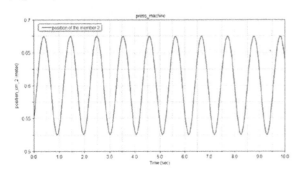

Figure 7. Angular velocity of the crank mass centre - PART_2

3.3. Creating Kinematic Variables of the Member 6

Figure 8 shows an example of measurement windows with kinematics parameter of the point D member 6.

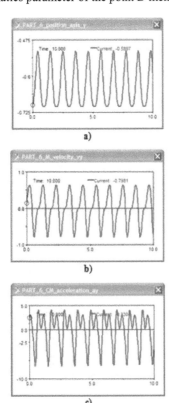

Figure 8. Measurements with kinematic parameters of the point D - PART_6 a) displacement y, b) velocity v_y, c) acceleration a_y

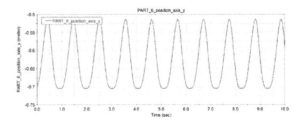

Figure 9. Kinematic parameters of the point D-position

Dependence of kinematic variable of velocity of the member 6 from time is shown in Figure 10.

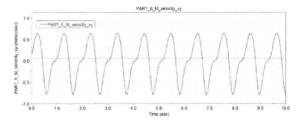

Figure 10. Kinematic parameters of the point D-acceleration

The following picture (Figure 11) demonstrates acceleration of point D in the member 6 in graphical form.

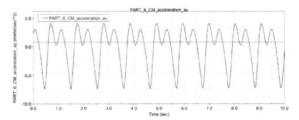

Figure 11. Kinematic parameters of the point D-velocity

3.4. Creating Kinematic Variables of the Member 6

Dependence of kinematic variables of angular velocity of the member 3 from time is shown in Figure 12.

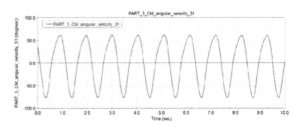

Figure 12. Angular velocity of the member 3 – PART_3

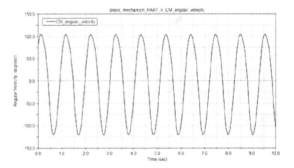

Figure 13. Angular velocity of the member 4 – PART_4

3.5. Creating Kinematic Variables of the Member 4

Dependence of kinematic variables of angular velocity of the member 4 from time are shown in Figure 13.

Dependence of kinematic variables of angular acceleration of the member 4 from time is shown in Figure 14.

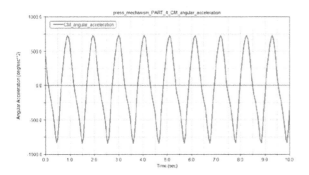

Figure 14. Angular acceleration of the member 4 – PART_4

3.6. Creating Kinematic Variables of the Member 4

The diagram in Figure 15 shows the dependence of angular velocity from time of the member 5.

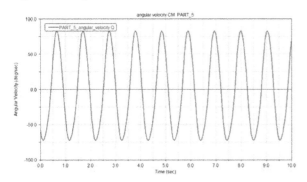

Figure 15. Angular velocity of the member 5 – PART_5

The diagram in Figure 16 shows the dependence of angular acceleration from time of the member 5.

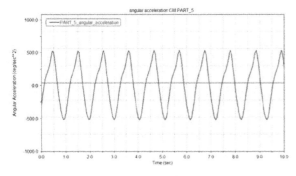

Figure 16. Angular acceleration of the member 5 – PART_5

Creating animation file of press mechanism is shown in Figure 17.

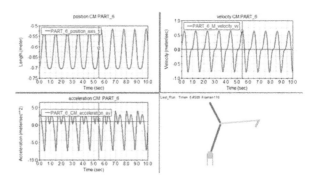

Figure 17. Kinematic parameters of the member 6 and animation in the same window

4. Summary

Great benefit of the study is the familiarity with the issues and new opportunities to learn how to work with the simulation program MSC ADAMS/View, which is used for modeling systems of more degrees of freedom, their static, kinematics and dynamic analysis [3-4]. The result of the simulation enables a graphical and numerical form [2]. Diagrams of kinematic variables were created in Adams/Postprocessor, one of the modules of the program MSC Adams/View. The program allows us to create a video file in *. avi format as shown in Figure 17.

In such form, the article may serve educational purposes to find out more about simulation in software MSC ADAMS/View.

Acknowledgement

This paper was supported in part by the Ministry of Education of the Slovak Foundation under KEGA projects No. 054 TUKE – 4/2014 "Using of modern numerical methods of mechanics as a base of scientific design to the development of knowledge base of students at the second and third level of university studies" and KEGA No. KEGA 004TUKE-4/2013.

References

[1] Delyová, I., Frankovský, P., Hroncová, D., "Kinematic analysis of movement of a point of a simple mechanism," *4th International Conference Modelling of mechanical and mechatronic systems*, Technical University Košice, Herľany, Slovakia, 2011.

[2] Hajžman, M., *Help text for an introduction to the basics of working with the system ADAMS*, Information on http://www.kme.zcu.cz/mhajzman/download/adams_zaklad.pdf.

[3] Juliš, K., Brepta, R., Mechanika I.díl, *Statika a Kinematika [Mechanics Part I, Statics and Kinematics]*, SNTL, Praha, 1987.

[4] Stejskal, V., Valášek, M., *Kinematics and dynamics of Machinery*, Marcel Dekker, Inc., New York, 1996.

[5] Kuryło, P., Papacz, W., "Wykorzystanie pakietu Matlab Simulink w modelowaniu zjawiska tarcia," *Tehnologiâ*, L. E. Švarcburg, Moskva: Moskovskij Gosudarstvennyj Tehnologičeskij Universitet Stankin, 2011, s. 207-218. ISBN: 978-5-8037-0420-1.

[6] Ángel, L., Pérez, M. P., Díaz-Quintero, C., & Mendoza, C., "ADAMS/MATLAB Co-Simulation: Dynamic Systems Analysis and Control Tool," *Applied Mechanics and Materials*, 232, 527-531.

[7] Božek, P., "Robot path optimization for spot welding applications in automotive industry," *Tehnički vjesnik*, 20 (5), 913-917.

[8] Hroncová, D., Delyová, I., Frankovský, P., "Kinematic Analysis of Mechanisms Using MSC Adams," In: *Applied Mechanics and Materials*. 2014. p. 83-89.

21

Evaluation of the Statistical Approach for the Simulation of a Swirling Turbulent Flow

D. Lalmi[1,*], R. Hadef[2]

[1]Faculty of exact sciences, natural sciences and life, Larbi Ben Mhidi University, Oeb Algeria
[2]Faculty of sciences and applid sciences, Larbi Ben Mhidi university, Oeb Algeria
*Corresponding author: Eldjemoui@gmail.com

Abstract This numerical study evaluates the performances of the statistical method of Reynolds realized (RANS) in the numerical simulation of a confined swirling turbulent flow. The confrontation of the computation results with real measurements confirmed the insufficiency of this approach. This handicap is related to the complexity of the structure of the flow (non stationary, three-dimensional, various scales of turbulence.....) revealed by hybrid approach DDES k-ω _SST.

Keywords: *Swirl, turbulence, simulation, RANS*

1. Introduction

Turbulent flows are largely used in engineering in particular within the turbojets and of the systems of combustion. They make it possible to increase the output of combustion by a better mixture of fuel with the air, to reinforce the stability of the flame and to reduce its length by the presence of the zone of central recirculation induced by the swirl [3]. In this area the flow is strongly non stationary with curved threads of current and presents a strong anisotropic turbulence [4]. To optimize the design (design) and to improve the performances of the burners, the properties of the swirling turbulent flows must be predicted perfectly. For that, several investigations based on a variety of methods were carried out [5,6,7].

2. Mathematical Formulation

The turbulent flow of an incompressible fluid is described by the realized equations of Navier Stocks expressed in a stationary regime by:

$$\frac{\partial U_l}{\partial x_l} = 0 \tag{1}$$

$$\frac{\partial U_i U_l}{\partial x_l} = -\frac{1}{\rho}\frac{\partial P}{\partial x_i} + \nu\frac{\partial^2 U_i}{\partial x_l \partial x_l} - \frac{\partial \overline{u'_i u'_l}}{\partial x_l} \tag{2}$$

Where: U_i and u'_i are the components average and fluctuating speed in the direction x_i, P is the pressure, ν is kinematic viscosity and ρ is the density of the fluid.

Additional equations must be derived for the terms from constraints of Reynolds $\overline{u'_i u'_j}$.

2.1. Model of Turbulence

In the model with constraints of Reynolds (RSM), the terms $\overline{u'_i u'_j}$ are calculated starting from their own transport equations written in the general form:

$$\frac{\partial \overline{u'_i u'_j} U_l}{\partial x_l} = \frac{\partial}{\partial x_l}\left[\left(\nu + \frac{\nu_t}{\sigma_k}\right)\frac{\partial \overline{u_i u_j}}{\partial x_l}\right]$$
$$+\left(-\overline{u'_i u'_l}\frac{\partial U_j}{\partial x_l} - \overline{u'_j u'_l}\frac{\partial U_i}{\partial x_l}\right) + \varphi_{ij} - \frac{2}{3}\varepsilon\delta_{ij} \tag{3}$$

In its version SSG [10], the term of correlation between the fluctuations in pressure and the deformation of the fluctuations speed $\varphi_{i,j}$ are expressed by:

$$\varphi_{ij} = -(C_1\varepsilon + C_1^* P_k)b_{ij} + C_2\varepsilon(b_{ik}b_{kj} - \frac{1}{3}b_{mn}b_{mn}\delta_{ij})$$
$$+(C_3 - C_3^*\sqrt{\Pi_{ij}})kS_{ij} + C_4 k(b_{ik}S_{jk} + b_{jk}S_{ik}) \tag{4}$$
$$-\frac{2}{3}b_{mn}S_{mn}\delta_{ij} + C_5 k(b_{ik}\Omega_{jk} + b_{jk}\Omega_{ik})$$

Where $S_{ij} = \frac{1}{2}\left(\frac{\partial U_i}{\partial x_j} + \frac{\partial U_j}{\partial x_i}\right)$ the tensor of the rate of average shearing is, $\Omega_{ij} = \frac{1}{2}\left(\frac{\partial U_i}{\partial x_j} - \frac{\partial U_j}{\partial x_i}\right)$ is the average tensor of vorticity, $b_{ij} = \frac{\overline{u'_i u'_j}}{2k} - \frac{1}{3}\delta_{ij}$ is the tensor of anisotropy and $\Pi_{ij} = b_{ij}b_{ij}$ its invariant. The turbulent kinetic energy is evaluated starting from its definition $k = \overline{u'_l u'_l}/2$, turbulent viscosity by its modeling in the model $v_t = 0.09 k^2/\varepsilon$ and the scalar ε is obtained by its transport equation of the model $k - \varepsilon$.

$$\frac{\partial \varepsilon}{\partial t} + U_l \frac{\partial \varepsilon}{\partial x_l} = \frac{\partial}{\partial x_l}\left\{ (v + v_t) \frac{\partial \varepsilon}{\partial x_l} \right\}$$
$$+ \frac{\varepsilon}{k}\left(-1.44 \overline{u_i u_j} \frac{\partial U_i}{\partial x_j} - 1.83\varepsilon \right) \quad (5)$$

The constants of the model are presented in Table 1.

Table 1. RSM_SSG constants model

C_1	C_1^*	C_2	C_3	C_3^*	C_4	$2C_5$
3.4	1.8	4.2	0.8	1.3	1.25	0.4

The model K ω SST is based on the general model K ω [11] whose transported variables are the turbulent kinetic energy K and the turbulent frequency ω. Its equations are as follows [12]:

$$\frac{\partial k}{\partial t} + U_l \frac{\partial k}{\partial x_l} = \frac{\partial}{\partial x_l}\left[(v + \sigma_k v_t) \frac{\partial k}{\partial x_l} \right] + \tilde{P}_k - \beta^* k\omega \quad (6)$$

$$\frac{\partial \omega}{\partial t} + U_l \frac{\partial \omega}{\partial x_l} = \frac{\partial}{\partial x_l}\left[(v + \sigma_\omega v_t) \frac{\partial \omega}{\partial x_l} \right] + \alpha_2 \frac{\omega}{k} P_k - \beta_2 \omega^2$$
$$+ 2(1 - F_1)\frac{\sigma_{\omega,2}}{\omega}\frac{\partial k}{\partial x_l}\frac{\partial \omega}{\partial x_l} \quad (7)$$

Where the function of F1 mixture (equal to the unit in close wall and null in the remote area) is defined by:

$$F_1 = \tanh\left\{ \left(\min\left[\max\left(\frac{\sqrt{k}}{\beta^* \omega y}, \frac{500v}{y^2 \omega} \right), \frac{4\sigma_{\omega,2} k}{CD_{k\omega} y^2} \right] \right)^4 \right\} \quad (8)$$

Where is *there* the normal distance to the wall nearest and the term $CD_{k\omega}$ equivalent to the positive portion to the term of cross diffusion of the equation (7). $CD_{k\omega}$ have a lower limit in order to avoid a division by 0 in the equation of F_1 and is defined by:

$$CD_{k\omega} = \max\left(2\sigma_{\omega,2}\frac{1}{\omega}\frac{\partial k}{\partial x_j}\frac{\partial \omega}{\partial x_j}, 10^{-10} \right) \quad (9)$$

The transition enters the two formulations, $K\omega$ and $K\varepsilon$, is done through the function F_1. Thus, when F_1 is 0 far from the walls, the formulation $k-\varepsilon$ is activated. Turbulent kinematic viscosity is given by:

$$v_t = \frac{\alpha_1 k}{\max(\alpha_1 \omega, SF_2)} \quad (10)$$

Where $S = \sqrt{S_{ij}S_{ij}}$ and F2 is related second to mixture defined by:

$$F_2 = \tanh\left\{ \left[\max\left(\frac{2\sqrt{k}}{\beta^* \omega y}, \frac{500v}{y^2 \omega} \right) \right]^2 \right\} \quad (11)$$

The Model SST contains also a limiting device in order to avoid the artificial construction of turbulence in the areas of stagnation:

$$P_k = v_t\left(\frac{\partial U_i}{\partial x_j} + \frac{\partial U_j}{\partial x_i} \right)\frac{\partial U_i}{\partial x_j} \rightarrow \tilde{P}_k = \min\left(P_k, 10\beta^* k\omega \right) \quad (12)$$

2.2. Hybrid Approach

It is known that the simulation of the great scales (Broad eddy simulation) has for principal advantage of producing results of great quality in such flows but however at a prohibitive price taking into account the smoothness of grid. From this consideration were born the hybrid methods of which we present here two approaches which were installation in this study, DES and Delayed DES. The hybrid approach of detached swirls type (OF for 'Detached-Eddy Simulation') is defined as a non stationary three-dimensional simulation using only one functioning model of turbulence as model of under mesh in the areas where the grid is sufficiently fine for a simulation on the great scales and as statistical model of Reynolds realized (URANS) in the areas where it is not it [13]. In theory, all models RANS can be modified to become models by introducing the dimension of the grid like a scale length of the model, which makes it possible to reduce the swirling level of viscosity of the model if the grid is refined.

In the case of the model Kω SST, a scale length is built starting from the variables of the basic model [14]:

$$L_t = \frac{\sqrt{k}}{\beta^* \omega} \quad (13)$$

and the hybrid model K-ω SST consists in comparing the scale L_t with a scale length $C_{DES}\Delta$, based on the dimension of the grid and the term of destruction of K in the equation (6) will be modified in the following way:

$$\beta^* k\omega = \frac{k^{3/2}}{L_t} \Rightarrow \frac{k^{3/2}}{C_{DES}\Delta} \quad pour \ C_{DES}\Delta < L_t$$

where C_{DES} is a constant of the model equal to 0.61 and Δ the width of the filter is to THEM the largest dimension of a given mesh: $\Delta = \max\left(\Delta_x; \Delta_y; \Delta_z \right)$.

That is translated, in the formulation, by the appearance of the function F_{DES} which causes an increase in the term of dissipation of K:

The formulation thus associates successfully the parietal areas modelling URANS and the areas where the grid fine with like is wished. It is notable that the approach OF uses for the two areas the same system of equations of statistical modeling with as difference the choice of the scale length of the turbulence which is carried out in each volume of control of the grid.

The original version of the model although very promising suffers nevertheless from some defects of which the release of the mode THEM in the boundary layer. No consideration on the thickness of boundary layer is included in the model OF, so that there is a risk of activation of the mode THEM in the boundary layers. The original formulation of rests on the assumption that grids RANS - for the applications in external aerodynamics generally have tangential refinements with the wall much larger than the thickness of boundary layer. If this assumption is not satisfied any more, the mode THEM can be activated in the boundary layer what results in the appearance of a false separation induced by the grid (grid-induced separation, GIS) or a reduction of the tensions of Reynolds (modelled stress depletion, MSD) [15]. To prevent this transition precise, it was proposed to preserve mode RANS in the boundary layer to use a parameter of correction [16]:

$$r_d = \frac{v_t + v}{\kappa^2 d_w^2 \left\{ \left[\sum_{i,j} \max \left(\partial u_i / \partial u_j \right)^2 \right]^{1/2}, 10^{-10} \right\}} \quad (14)$$

Where κ is the constant of Karman.

The quantity r_d is then used in the function:

$$f_d = 1 - \tanh \left[\left(20 r_d \right)^3 \right] \quad (15)$$

built in order to be worth 1 in the zones elsewhere, that is to say THEM and 0 in the boundary layer. Finally the scale is redefined such

$$L_{DDES} = L_t - f_d \max \left(0, L_t - C_{DES} \Delta \right) \quad (16)$$

Incorporated in the scale length of, this function delays the release of the mode THEM until apart from the boundary layer from where the name of the method: Delayed of (DDES).

2.3. Geometrical Configuration

The flow of air to be simulated is injected by an annular space of 27.5 mm and of ray external $R_o = 53.5\,mm$ in a cylindrical enclosure length 765 mm.

Its exit towards outside is ensured by a control of 600 mm (Figure 1). The fluid considered is at a temperature of 20°C and a volume through put of 426 m^3 / H corresponding at a speed retailer of U$_0$=17.23 m/s and a Reynolds number of 31500.

The intensity of the swirling movement of the flow is characterized by the value of the number of swirl at the entry expressing the relationship between flows of momentum tangential and axial and defined by:

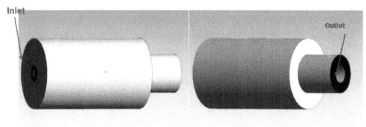

Figure 1. Global View of chamber

$$S_0 = \frac{\int_0^\infty U W r^2 \, dr}{R_o \int_0^\infty U^2 r \, dr} \quad (17)$$

where U and W are respectively the components axial and tangential mean velocity and $R_o = 53.5\,mm$ is the ray of annular space.

2.4. Solving Method

The resolution of the equations is carried out by the commercial code Fluent 13 in a grid of 1.1 cells million of the triangular type. The transport equations of all the required sizes are solved in steady operation, by diagram QUICK coupled with the SIMPLE algorithm for the determination of the field of pressure. The latter is estimated on the faces by diagram PRESTO. To accelerate convergence, the calculated sizes under-are released with each iteration. Two models RSM-ssg and DDES K ω SST are used in a grid of 2.3 million hexahedral cells. If first is supposed to be stationary, second is solved in a variable mode with a step of time of 0.05 ms.

3. Results

The computation results obtained are confronted with the real measurements carried out on the same configuration (geometry and conditions of entry) using a multiple probe of hot wire anemometer [17].

Figure 2 presents for the two statistical models, the radial profiles of the components mean velocity at various stations. The first remark is that, all the models corroborate measurements qualitatively and highlight the existence of the zone of central recirculation defined by a negative axial speed in the center of the enclosure. For z<4Ro, that the model K ω SST 2d is able to predict the sizes speed with an acceptable degree. Beyond, of this position all the models are unable. It has should be announced as the value the radial speed measured being non null on the axis beyond this station thus the flow cannot be axisymmetric. The three-dimensional configuration applied with Rsm-ssg did not improve the computation results, can be because of the low density of grid of the field. The turbulent kinetic energy profiles for each turbulent model are given in Figure 3. We have seen that the turbulent kinetic energy is correctly predicted by the turbulence model. It is seen that near the inlet region the comparison between the prediction calculated and experimental data is satisfactory, again approaching the exit of the domain of calculation comparison is not good agreement for the co and counter swirl, like in measurements, where the flow is may to established.

Taking into account this failure, one invested the expertise by the DDES to describe the flow beyond z=3R$_0$. The results presented qualitatively in Figure 4 and Figure 5 show that its structure is complex with scales of turbulence of various sizes and presents especially a bursting of the central swirl (PVC).

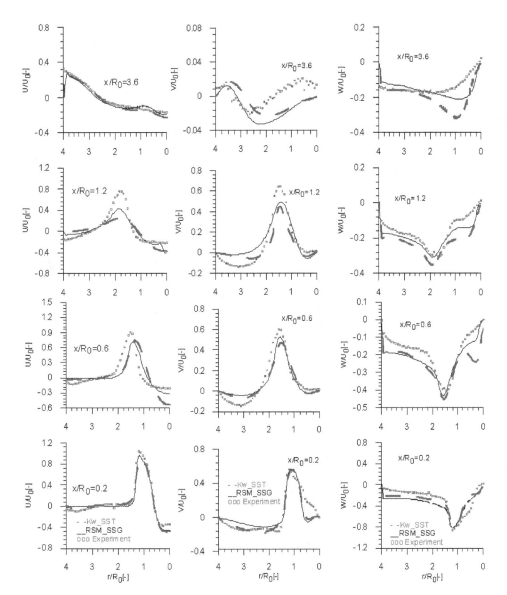

Figure 2. Radial Profiles of mean component velocity at diverse sections

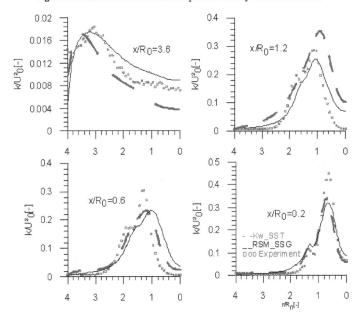

Figure 3. radial profiles of kinetic energy

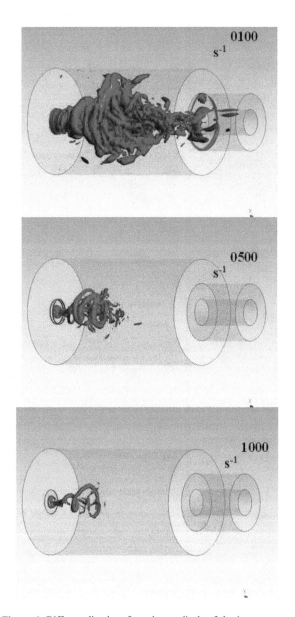

Figure 4. Different disvalues from the amplitude of the instantaneous vorticity in 3d

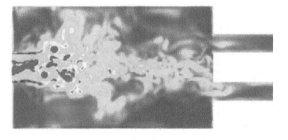

Figure 5. Contour of the amplitude of the instantaneous vorticity

4. Conclusion

The evaluation of the performance of the statistical method in simulation of a low swirling flow of air is investigated numerically using the commercial code Fluent 13.

The calculation results obtained were worn on the mean values of the flow confrontation. Theirs with actual measurements confirmed the unsufisensy of the statistical approach. This is due to the unsteady structure, three-dimensional with a broad spectrum of turbulence scales, confirmed by an extension of investigation by the method of calculation of large scales (DDES).

References

[1]   A. K. Gupta, D.G., Lilley Syrednd N., *Swirl Flows*. Abacus Press, Tunbridge Wells, 1984.

[2]   R. Hadef and B. Lenze, *Measurements of droplets characteristics in a swirl-stabilzed spray flame*, Exp. Therm. Fluid Sci. Journal, vol. 30, 117-130, 2005.

[3]   R. Hadef and B. Lenze, *Effects of co-and counter swirl on the droplets characteristics in a spray flame*, Chem. Eng. Proc. J., vol. 47, 2209-2217, 2008.

[4]   O. Lucca-Negro and T. O. O'Doherty, *Vortex breakdown: A review*. Prog. Ener. Comb. Sci. vol. 27, 431-481, 2001.

[5]   L. Chenzhou and L. C. Merkle, *Contrast between steady and time-averaged unsteady combustion simulations*, Computers & Fluids vol. 44, 328-3388, 2001.

[6]   B. Wegner, A. Maltsev, C. Schneider, A. Sadiki, A. Dreizler, and J. Janicka, *Assessment of unsteady RANS in predicting swirl flow instability based on LES and experiments*, Int. J. Heat and Fluid Flow, vol. 25, 528-536, 2004.

[7]   B. Wegner, A. Kempf, C. Schneider, A. Sadiki, A. Dreizler, J. Janicka, and M. Schäfer, *Large eddy simulation of combustion processes under gas turbine conditions*, Prog. Computational Fluid Dynamics, vol. 4, 257-263, 2004.

[8]   Li Zhuowei, N. Kharoua, R. Hadef, L. Khezzar, *RANS and LES simulation of a swirling flow in a combustion chamber with different swirl intensities*, ICHMT Intern. Symp. on Advances in Computational Heat Transfer Bath, England, 2012.

[9]   R. Hadef, D. Lalmi, *Prédiction d'un écoulement turbulent tourbillonnaire*, Congrès National de Mécanique des Fluides, USTHB Alger, 2012.

[10]   Speziale, C.S. Sarkar, S. and Gatski, T.B., *Modelling the Pressure-Strain Correlation of Turbulence: An Invariant Dynamical Systems Approach*, Journal of Fluid Mechanics, vol. 227, 245-272, 1991.

[11]   D.C. Wilcox. *Turbulence modeling for CFD*. Technical report, DCW Industries, 1993.

[12]   F.R. Menter. *Two-equation eddy-viscosity turbulence models for engineering applications*. AIAA-Journal, vol. 32-8, 1598-1605, 1994.

[13]   P.R. Spalart, W.-H. Jou, M. Strelets, and S.R. Allmaras. *Comments on the feasibility of LES for wings, and on a hybrid RANS/LES approach*. In Advances in DNS/LES., edited by C. Liu and Z. Liu (Greyden Press, Colombus, 1998.

[14]   M. Strelets. *Detached eddy simulation of massively separated flows*. AIAA Paper 2001-0870, Proceedings of the 39th AIAA Aerospace Sciences Meeting and Exhibit, Reno, Nevada, USA, 2001.

[15]   F.R. Menter and M. Kuntz, *Adapatation of Eddy-Viscosity Turbulence Models to Unsteady Separated Flow Behind Vehicles*. In The aerodynamics of Heavy Vehicles: Trucks, Busses and Trains, 2003.

[16]   P. R. Spalart, S. Deck, M. L. Shur, K. D. Squires, M. Kh. Strelets, A. Travin *A New Version of Detached-eddy Simulation, Resistant to Ambiguous Grid Densities*, Theoretical and Computational Fluid Dynamics, vol. 20-3, 181-195, 2006.

[17]   F. Holzäpfel, *Turbulence Structure in Confined and Free Swirling Flows*, PhD Thesis Universität Karlsruhe, Germany, 1996.

22

Computer Simulation of the Aerodynamic Structure of an Arched Roof Obstacle and Validation with Anterior Results

Slah Driss, Zied Driss[*], Imen Kallel Kammoun

Laboratory of Electro-Mechanic Systems (LASEM), National School of Engineers of Sfax (ENIS), University of Sfax (US), B.P. 1173, Road Soukra km 3.5, 3038 Sfax, TUNISIA
*Corresponding author: Zied.Driss@enis.rnu.tn

Abstract The present paper is dedicated to the numerical simulation of the aerodynamic structure of an arched roof obstacle with a height h=63 mm and a length L_2=118 mm. The governing equations of mass and momentum in conjunction with the standard k-ε turbulence model are solved using the software "Solid Works Flow Simulation". A finite volume discretization is used. The numerical results are validated with anterior results developed in a wind tunnel. The numerical simulations agreed reasonably with the experimental results and the numerical model was validated.

Keywords: CFD, turbulent flow, aerodynamic structure, arched roof

1. Introduction

Green buildings differ from conventional buildings due to the integration social and economic objectives. Environmental considerations roughly correspond direct and indirect environmental impacts, such as reduced emissions of greenhouse gases or reduced water consumption. Social considerations can be directly related to building itself like clean air and comfortable, natural light or beyond the framework Building. In this context, the study of geometry parameters effects on airflow patterns around numerous structures can be done inside wind tunnels, where airflow conditions are controlled. For example, Ntinas et al. [1] conducted an experiment inside a wind tunnel and the air velocity and turbulent kinetic energy profiles were measured around two small-scale obstacles with an arched-type and a pitched-type roof. Luo et al. [2] studied models of cuboid obstacles to characterize the three-dimensional responses of airflow behind obstacles with different shape ratios to variations in the incident flow in a wind-tunnel simulation. Tominaga and Stathopoulos [3] reviewed current modeling techniques in CFD simulation of near-field pollutant dispersion in urban environments and discussed the findings to give insight into future applications. Jiang et al. studied [4] three ventilation cases, single-sided ventilation with an opening in windward wall, single-sided ventilation with an opening in leeward wall, and cross ventilation. In the wind tunnel, a laser Doppler anemometry was used to provide accurate and detailed velocity data. Ahmad et al. [5] provided a comprehensive literature on wind tunnel simulation studies in urban street canyons/intersections including the effects of building configurations, canyon geometries, traffic induced turbulence and variable approaching wind directions on flow fields and exhaust dispersion. Smolarkiewicz et al. [6] performed large-eddy simulations (LES) of the flow past a scale model of a complex building. Calculations are accomplished using two different methods to represent the edifice. De Paepe et al. [7] simulated five different wind incidence angles using a turntable, in order to quantify their effect on indoor air velocities. The responses in local air velocities could largely be attributed to the relative position of the end walls of the scale models orientated towards the wind. This crucial position allows the measured air velocity trends to be explained. Lim et al. [8] presented a numerical simulation of flow around a surface mounted cube placed in a turbulent boundary layer. The simulations were carried at a Reynolds number, based on the velocity at the cube height, of 20,000 large enough that many aspects of the flow are effectively Reynolds number independent. Becker et al. [9] studied the structure of the flow field around three-dimensional obstacles of different aspect ratios, in two different types of boundary layers. The dimensions of the rectangular block obstacles were chosen to represent generic shapes of buildings. De Melo et al. [10] developed two Gaussian atmospheric dispersion models, AERMOD and CALPUFF. Both incorporating the PRIME algorithm for plume rise and building downwash, are inter-compared and validated using wind tunnel data on odour dispersion around a complex pig farm facility comprising of two attached buildings.

Richards et al. [11] used as input data for generating typical weather data required as input for building and heating, ventilation and air-conditioning (HVAC) system models in order to study the energy budgets of buildings and assess the performance of air-conditioning (A/C) systems. Gousseau et al. [12] used Large-Eddy Simulation (LES) to investigate the turbulent mass transport mechanism in the case of gas dispersion around an isolated cubical building. Close agreement is found between wind-tunnel measurements and the computed average and standard deviation of concentration in the wake of the building. Meslem et al. [13] observed changes in the prediction of local and global mean-flow quantities as a function of the considered turbulence model and by the lack of consensus in the literature on their performance to predict jet flows with significant three-dimensionality. The study reveals that none of the turbulence models is able to predict well all the jet characteristics in the same time. Cadirci et al. [14] studied the flow fields generated by a Java in a water channel. Detailed quantitative information about the performance of the Java on a flat-plate boundary layer is obtained. Java-induced boundary-layer profiles are clearly fuller at the wall and hence more resistant to flow separation. The 'positive' effects of the Java with different operating regimes on various boundary-layer flow characteristics such as displacement thickness, shape factor and the friction coefficient are presented. Kassiotis et al. [15] discussed a way to compute the impact of free-surface flow on nonlinear structures. The approach chosen relies on a partitioned strategy to solve the strongly coupled fluid-structure interaction problem. It is then possible to re-use the existing and validated strategy for each sub-problem. The structure is formulated in a Lagrangian way and solved by the finite element method. The free-surface flow approach considers a Volume-Of-Fluid (VOF) strategy formulated in an Arbitrary Lagrangian-Eulerian (ALE) framework, and the finite volume is used to discrete and solve this problem. In this paper, we are interested on the study of the aerodynamic structure of an arched-type roof obstacle placed in a wind tunnel.

2. Numerical Model

The vast majority of fluid flows encountered in aerodynamics are turbulent. The equations governing the air flow called the Navier-Stokes equations are obtained from the continuity equation, the equation of momentum, the transport equation of turbulent kinetic energy k and the transport equation of dissipation rate of turbulent kinetic energy ε. Several methods are available for including turbulence in the Navier-Stokes equations. Most of these involve a process of time-averaging of the conservation equations. When turbulence is included, the transported quantity is assumed to be the sum of equilibrium and a fluctuating component. The only term that remains positive definite is one containing the product of two fluctuating terms [16-23].

2.1. Computational Domain

The considered computational domain is shown in Figure 1. It is defined by the interior volume of the wind tunnel blocked by two planes: the first one is in the entry

and the second one is in the exit. The obstacle consists on an arched type roof with a height h=63mm and a width L_2=118 mm. The distance between the obstacle and the entry plane and that between the obstacle and the exits plane are defined by L_1=151 mm and L_3=529 mm respectively.

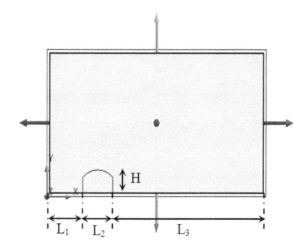

Figure 1. Computational domain for different highs

2.2. Boundary Conditions

The boundary conditions are required anywhere fluid enters or exits the system and can be set as a pressure, mass flow, volume flow or velocity. For the inlet velocity, we will take as a value U_{ref}=0.32 m.s^{-1}, and for the outlet pressure a value of 101325 Pa will be considered which means that at this opening the fluid exits the model to an area of static atmospheric pressure. A summary of the boundary conditions is given in the Figure 2.

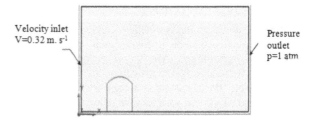

Figure 2. Boundary conditions in the test section

2.3. Mesh Resolution

In several ways, we can control the basic mesh by changing the number of the basic mesh cells along the X, Y and Z axes. The considered geometry can be resolved reasonably well. However, if we generate the mesh and zoom in the roof obstacle, we will see that it may still unresolve. In order to resolve these regions properly, we have used the Local Initial Mesh option (Figure 3). It allows us to specify an initial mesh in a local region of the computational domain to better resolve the model geometry and/or flow particularity in this region. The local region can be defined by a component of the assembly, specified by selecting the model face. All refinement levels are set with respect to the basic mesh cell [16,17,18].

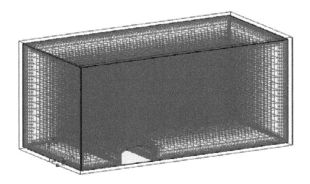

Figure 3. Representation of the basic mesh

3. Numerical Results

3.1. Velocity Field

Figure 4 presents the distribution of the velocity field in the longitudinal and the transverse planes defined by Z=0 mm and X=179 mm. According to these results, it has been noted that the velocity is weak in the inlet of the entry. It is indeed governed by the boundary condition value of the inlet velocity which is equal to 0.32 m.s^{-1}. In this region, the velocity field is found to be uniform and increases progressively downstream of the entry. While the position of the obstacle is characterized by the high velocity. Downstream of the arched type obstacle the velocity keeps increasing progress. Then, a decrease has been noted through the exit where the minimum velocity values are recorded in the lateral walls. The maximum velocity values are located in the top of the roof according to the distribution shown in the Z=0 mm plane. However, in the obstacle downstream a dead zone has been observed in the transverse plane shows a symmetric distribution.

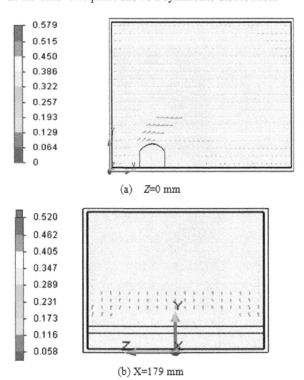

Figure 4. Distribution of the velocity field

3.2. Total Pressure

Figure 5 presents the distribution of the total pressure in the longitudinal and the transverse planes defined by Z=0 mm and X=179 mm. According to these results, it has been noted that a compression zone appears in the obstacle upstream of the obstacle. However, a depression zone has been observed in the downstream of the obstacle. In the top of the test section, the total pressure keeps a uniform distribution. The compression zones are located in the top of the roof. Downstream of the arched type obstacle, the total pressure keeps increasing progress.

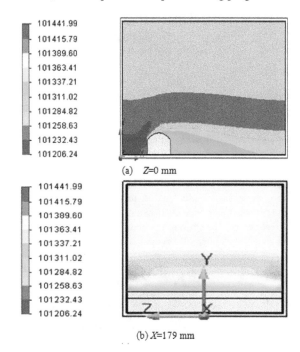

Figure 5. Distribution of the total pressure

3.3. Dynamic Pressure

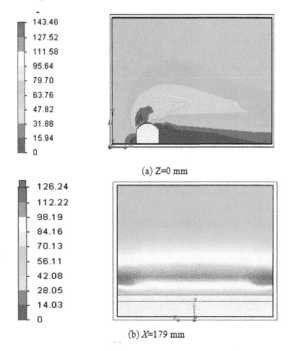

Figure 6. Distribution of the dynamic pressure

Figure 6 presents the distribution of the dynamic pressure in the longitudinal and the transverse planes defined by Z=0 mm and X=179 mm. According to these results, it has been noted that a compression zone appears in the obstacle upstream of the obstacle. However, a depression zone is located in the downstream of the obstacle. In the top of the test section, the dynamic pressure keeps a uniform distribution. Downstream of the arched type obstacle the pressure keeps increasing progress. The maximum pressure values are located in the top of the roof.

3.4. Turbulent Kinetic Energy

Figure 7 presents the distribution of the turbulent kinetic energy in the longitudinal and the transverse planes defined by Z=0 mm and X=179 mm. According to these results, it has been noted that the turbulent kinetic energy is weak in the section inlet. In this region, the turbulent kinetic energy is found to be uniform and increases progressively downstream of the test section. Downstream of the arched type obstacle, the turbulent kinetic energy keeps increasing progress. Then, a decrease has been noted through the exit where the minimum energy values are recorded in the lateral walls. The maximum turbulent kinetic energy values are located in the top of the roof. However, in the obstacle downstream a wake zone characteristic of the minimum values of the turbulent kinetic energy has been observed.

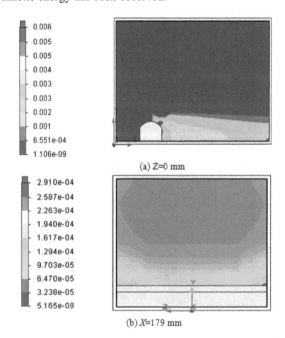

Figure 7. Distribution of the turbulent kinetic energy

3.5. Dissipation Rate of the Turbulent Kinetic Energy

Figure 8 presents the distribution of the turbulent dissipation rate of the turbulent kinetic energy in the longitudinal and the transverse planes defined by Z=0 mm and X=179 mm. From these results, it is clear that the dissipation rate of the turbulent kinetic energy is weak in the inlet of the test section. In this region, the dissipation rate of the turbulent kinetic energy found to be uniform

and increases progressively in the test section. Downstream of the arched type obstacle, the dissipation rate of the turbulent kinetic energy keeps increasing progress. Then, a decrease has been noted through the exit where the minimum values of the dissipation rate of the turbulent kinetic energy values are recorded in the lateral walls. The maximum values of the dissipation rate of the turbulent kinetic energy are located in the top of the roof. However, in the obstacle downstream a wake zone characteristic of the minimum value of the turbulent dissipation ratio has been observed.

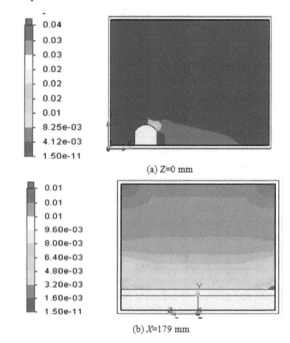

Figure 8. Distribution of the turbulent dissipation rate

3.6. Vorticity

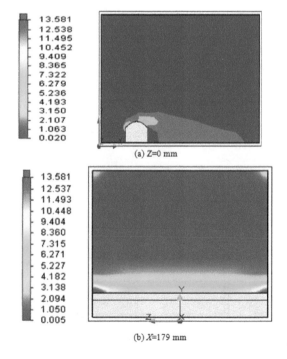

Figure 9. Distribution of the vorticity

Figure 9 presents the distribution of the vorticity in the longitudinal and the transverse planes defined by Z=0 mm and X=179 mm. From these results, it has been noted that the vorticity is weak in the inlet of the test section. In this region, the vorticity is found uniform and increases progressively in the test section. Downstream of the arched type obstacle, the vorticity keeps increasing progress. Then, a decrease has been noted through the exit where the minimum vorticity values are recorded in the lateral walls. The wake characteristic of the maximum vorticity values are located in the top of the roof according. However, in the obstacle downstream the vorticity values vanish.

4. Experimental Validation

The velocity profile around the arched obstacle is presented in Figure 10 for a height h=63 mm and a length L_2=118 mm. These results were predicted along the centre section of the obstacle defined by the plane Z=0 mm and the downstream edges defined by X=6.88 H. The average velocity values were deduced, at each point of the computational domain, by processing the computed values. According to these results, it's clear that the obstacle height has a direct effect on the velocity profile. In fact, the maximum value of the velocity field increases with the increase of the obstacle height. In these conditions, the maximum value of the velocity is equal to V=2.2 Uref. The predicted averaged velocity results are also compared with the measured data of Ntinas et al. [1]. Our numerical results present a good agreement with the experimental data founded from the literature. The gap between results is about 6%. These results confirm the validity of the numerical method.

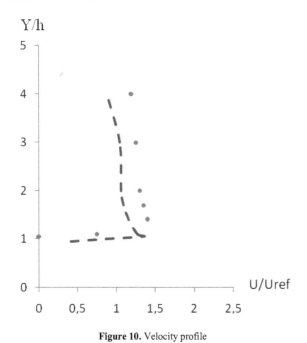

Figure 10. Velocity profile

5. Conclusion

A numerical simulation model was developed to study the aerodynamic structure of arched roof obstacle with a height h=63 mm and a length L_2=118 mm. Numerical results, such as velocity fields, pressure and turbulent characteristics are presented in the considered control volume. The satisfactory agreement between the numerical simulation and the experimental data, from a wind tunnel by anterior results, validated the mathematical model used in this study. According to the obtained results, the velocity keeps increasing progress downstream of the arched type obstacle. Indeed, a decrease has been noted through the exit where the minimum velocity values are recorded in the lateral walls. All these observations are useful for the building design. Particularly, we propose to use this knowledge to optimize ecological buildings.

Nomenclature

H	obstacle height, m
k	turbulent kinetic energy, $J.kg^{-1}$
l	length, m
P	pressure, Pa
Re	Reynolds number
V	magnitude velocity, $m.s^{-1}$
Xi	Cartesian coordinate, m
X	Cartesian coordinate, m
Y	Cartesian coordinate, m
Z	Cartesian coordinate, m
ε	dissipation rate of the turbulent kinetic energy, $W.kg^{-1}$
μ	dynamic viscosity, Pa.s
$μ_t$	turbulent viscosity, Pa.s
ρ	density, $kg.m^{-3}$

References

[1] G.K. Ntinas, G. Zhang, V.P. Fragos, D.D. Bochtis, Ch. Nikita-Martzopoulou, Airflow patterns around obstacles with arched and pitched roofs: Wind tunnel measurements and direct simulation, European Journal of Mechanics B/Fluids 43 (2014) 216-229.

[2] W. Luo, Z. Dong, G. Qian, J. Lu, Wind tunnel simulation of the three-dimensional airflow patterns behind cuboid obstacles at different angles of wind incidence, and their significance for the formation of sand shadows, Geomorphology 139-140 (2012) 258-270.

[3] Y. Tominaga, T. Stathopoulos, CFD simulation of near-field pollutant dispersion in the urban environment: A review of current modeling techniques, Atmospheric Environment 79 (2013) 716-730.

[4] Y. Jiang, D. Alexander, H. Jenkins, R. Arthur, Q. Chen, Natural ventilation in buildings: measurement in a wind tunnel and numerical simulation with large-eddy simulation, J. Wind Eng. Ind. Aerodyn. 91 (2003) 331-353.

[5] K. Ahmad, M. Khare, K.K. Chaudhry, Wind tunnel simulation studies on dispersion at urban street canyons and intersections- a review, J. Wind Eng. Ind. Aerodyn. 93 (2005) 697-717.

[6] P.K. Smolarkiewicz, R. Sharman, J. Weil, S.G. Perry, D. Heist, G. Bowker, Building resolving large-eddy simulation and comparison with wind tunnel experiments, journal of computational Physics 227 (2007) 633-653.

[7] M. De Paepe, J.G. Pieters, W.M.Cornelis, D. Gabriels, B. Merci, P. Demeyer, Airflow measurements in and around scale-model cattle barns in a wind tunnel: Effect of wind incidence angle, Biosystems Engineering 115 (2013) 211-219.

[8] H.C. lim, T.G. Thomas, Ian P. Castro, Flow around a cube in a turbulent boundary layer: LES and experiment /J. Wind Eng. Ind. Aerodyn. 97 (2009) 96-109.

[9] S. Becker, H. Lienhart, F. Durst, Flow around three-dimensional obstacles in boundary layers, J. Wind Eng. Ind. Aerodyn. 90 (2002) 265-279.

[10] A.M.V. De Melo, J.M. Santos, I. Mavrroidis, N. Costa Reis Junior, Modelling of odour dispersion around a pig farm building complex using AERMOD and CALPUFF. Comparison with wind tunnel results, A.M. Vieira de Melo et al. / Building and Environment 56 (2012) 8-20.

[11] K. Richards, M. Schatzmann, B. Leitl, Wind tunnel experiments modelling the thermal effects within the vicinity of a single block building with leeward wall heating, J. Wind Eng. Ind. Aerodyn. 94 (2006) 621-636.

[12] P. Gousseau, B. Blocken, G.J.F Heijst, Large-Eddy Simulation of pollutant dispersion around a cubical building: Analysis of the turbulent mass transport mechanism by unsteady concentration and velocity statistics, Environmental Pollution 167 (2012) 47-57.

[13] A. Meslema, F. Bodeb, C. Croitorub, I. Nastaseb, Comparison of turbulence models in simulating jet flow from a cross-shaped orifice : European Journal of Mechanics B/Fluids 44 (2014) 100-120.

[14] S. Cadircia, H. Gunesa, U. Ristb, Numerical investigation of a jet and vortex actuator in a cross flow boundary layer: European Journal of Mechanics B/Fluids 44 (2014) 42-59.

[15] C. Kassiotis, A. Ibrahimbegovica, H. Matthiesb, Partitioned solution to fluid–structure interaction problem in application to free-surface flows, European Journal of Mechanics B/Fluids 29 (2010) 510-521.

[16] Z. Driss, M. Ammar, W. Chtourou, MS. Abid, CFD Modelling of Stirred Tanks. Engineering Applications of Computational Fluid Dynamics 5 (2011) 145-258.

[17] Z. Driss, MS. Abid Use of the Navier-Stokes Equations to Study of the Flow Generated by Turbines Impellers. Navier-Stokes Equations: Properties, Description and Applications 3 (2012) 51-138.

[18] S. Driss, Z. Driss, I. Kallel Kammoun, Study of the Reynolds Number Effect on the Aerodynamic Structure around an Obstacle with Inclined Roof, Sustainable Energy, 2014, Vol. 2, No. 4, 126-133.

[19] Z. Driss, MS. Abid, Numerical and experimental study of an open circuit tunnel: aerodynamic characteristics. Science Academy Transactions on Renewable Energy Systems Engineering and Technology 2 (2012) 116-123.

[20] Z. Driss, A. Damak, S. Karray, MS. Abid, Experimental study of the internal recovery effect on the performance of a Savonius wind rotor. Research and Reviews: Journal of Engineering and Technology 1 (2012) 15-21.

[21] Z. Driss, G. Bouzgarrou, W. Chtourou, H. Kchaou, MS. Abid, Computational studies of the pitched blade turbines design effect on the stirred tank flow characteristics, European Journal of Mechanics B/Fluids 29 (2010) 236-245.

[22] M. Ammar, W. Chtourou, Z. Driss, MS. Abid, Numerical investigation of turbulent flow generated in baffled stirred vessels equipped with three different turbines in one and two-stage system, Energy 36 (2011) 5081-5093.

[23] Z. Driss, O. Mlayeh, D. Driss, M. Maaloul, M.S. Abid, Numerical simulation and experimental validation of the turbulent flow around a small incurved Savonius wind rotor, Energy, 74. 506-517, 2014.

23

Mechanical Model of Hydrogen Bonds in Protein Molecules

Zahra Shahbazi[*]

Department of Mechanical Engineering, Manhattan College, NY
*Corresponding author: Zahra.shahbazi@manhattan.edu

Abstract The unique properties of protein molecules have motivated researchers exploit them in the design and fabrication of bio-mimetic nano devices to perform a special task. Function of protein molecules is in turn dependent on their 3D structure and their ability to modify their shape for a specific task. To study and manipulate protein molecules we need to have knowledge of mechanical properties of these molecules. In this paper a multiscale model to predict stiffness of helical protein molecules has been developed. Hydrogen bonds as major contributing factor to proteins flexibility, are modeled as elastic springs based on their empirical potential energy. Such mechanical representation of hydrogen bonds enables us to obtain the stiffness ellipsoid of hydrogen bonds which leads to an understanding of the directional stiffness of protein molecules. The model has also been applied to three different protein molecules whose stiffness were reported in the literature. The comparison shows an agreement between the stiffness computed by the proposed model and that obtained through experiments and/or Molecular Dynamics (MD) simulations.

Keywords: protein molecule, hydrogen bond, stiffness, mechanical model

1. Introduction

Unique physical properties of biomaterials such as self-assembly, self-healing, adaptability and changeability have encouraged researchers to devote special attention to designing and fabricating synthetic materials that have similar properties [1,2,3]. These unique properties are facilitated by building blocks of biomaterials, proteins. In order to understand the properties of proteins, one has to initially understand their structure. This structure can be considered on several levels, all of which influence the final property and behavior of the material.

Previous efforts of finding the stiffness of protein molecules can be classified into two major categories of experimental studies and computational analysis. Regarding experimental efforts, recently Atomic Force Microscopy (AFM) has been used for studying mechanical properties of protein molecules. This technique measures the force required to stretch individual molecules and calculates its stiffness [4,5,6,7]. Laser tweezer is another experimental technique used to analyze material properties in molecular level [8]. Researchers also conducted computational simulations to predict mechanical properties of protein molecules [9,10,11].

Experimental studies tend to increase in difficulty as the size of the molecule becomes smaller. Existing computational efforts such as molecular dynamics, on the other hand, are more computationally expensive as the molecules become larger [12]. Such size dependency reveals the inherent difficulty of finding the mechanical properties of protein molecules.

Here we seek to establish the relation of the structure at its atomic length scale to the mechanical properties of protein molecules. To do this, we started with the interatomic bonds. In most protein molecules, hydrogen bonds are the key contributor to the molecules' stiffness. Therefore, we have developed a computational approach that models each hydrogen bond as a novel three-spring system. Using an already existing empirical interatomic potential for the hydrogen bonds, the stiffness ellipsoid is developed which can provide us with the stiffness of each bond at any given direction. As the next step, we have investigated the fragility of extending the analysis to the entire protein molecule to find its stiffness. To verify our proposed method, three protein molecules whose stiffness values have already been reported in literature using alternative methods described earlier, are investigated and the results show a reasonable agreement suggesting the feasibility of the technique for larger protein molecules.

A more comprehensive detail of part of the work presented here can be found in [13]. Additionally, a kinematic perspective of such modeling was also presented in 21st century kinematics workshop [14]. The foundational work in the mechanical modeling of protein molecules can be found in references [15-21].

2. Hydrogen Bonds

Atoms in proteins are held together by various types of bonds with different strength including very strong covalent bonds, moderate hydrogen bonds and weak van der Waals bonds. Hydrogen bonds contribute significantly to the folding of the proteins into stable conformations. Under external stimulations hydrogen bonds can change their length and corresponding angles. This gives them more flexibility than covalent bonds [22]. Therefore hydrogen bonds play an important role in determining mechanical properties of protein molecules and stiffness in particular.

2.1. Geometry

A hydrogen bond within a protein molecule is an interaction between an electronegative donor and a hydrogen that is covalently bonded to an acceptor atom from non-adjacent residues. The relative placement (position and orientation) of the acceptor atom with respect to the donor atom can be rigorously defined through a set of geometric parameters (angles and distances) [19]. A unified set of geometric parameters is illustrated in Figure 1 where a donor (nitrogen), an

acceptor (oxygen), acceptor antecedent, donor antecedent and hydrogen atoms are shown. The bond lengths are:

m: the covalent bond length between donor (nitrogen) and hydrogen

r: the hydrogen bond length between hydrogen and acceptor (oxygen)

d: the distance between donor and acceptor

and the angles are defined as:

α: between AA (acceptor antecedent), A (oxygen) and hydrogen (AA–A–H)

θ : between AA (acceptor antecedent), A (oxygen) and D (nitrogen) (AA–A–D)

λ : between D (nitrogen), A(oxygen) and hydrogen (D–A–H)

β: between D (nitrogen), hydrogen and A(oxygen) (D–H–A)

γ: between hydrogen, D (nitrogen) and A (oxygen) (H–D–A)

ε: between DD (donor antecedent), D (nitrogen) and A (oxygen) (DD–D–A).

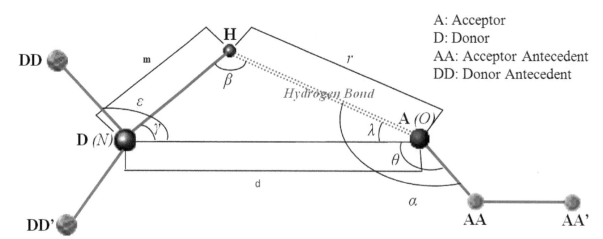

A: Acceptor
D: Donor
AA: Acceptor Antecedent
DD: Donor Antecedent

Figure 1. Geometric parameters

Different geometric formation criteria have been developed to predict hydrogen bonds occurrence in the protein molecules [19]. In this paper, in order to identify hydrogen bonds we used the suggested geometric criteria by [19] shown in Table 1.

Table 1. Geometric criteria [19]

	α^0	β^0	$r(Å)$	$d(Å)$
α-helices	[110,180]	[110,180]	<2.5	< 3.5
β-sheets	[120,180]	[110,180]	< 2.5	< 3.5
Overall main-chain	[90,180]	[100,180]	< 2.5	< 3.5
Overall main-chain, side-chain and mixed	[90,180]	[90,180]	< 2.5	< 3.5

2.2. Hydrogen Bond Energy

Hydrogen bond energy depends on the chemistry of the donor and acceptor atoms as well as their orientation. Based on the potential function used in "Dreiding" force field, ([23]) suggested a function (equation 1) to calculate hydrogen bond energy. In this energy function, constants are: $v_0 = 8$ kCal/mol^{-1} and $d_0 = 2.8$Å.

$$E_{HB} = v_0 \left[5(d_0/d)^{12} - 6(d_0/d)^{10} \right] g(\beta, \alpha, \phi) \quad (1)$$

$$g(\beta, \alpha, \phi) = \cos^2 \beta \cos^2 (\alpha - 190.5)$$

$$for sp^3 donor - sp^3 acceptor$$

$$g(\beta, \alpha, \phi) = \cos^2 \beta \cos^2 \alpha$$

$$for sp^3 donor - sp^2 acceptor$$

$$g(\beta, \alpha, \phi) = \cos^4 \beta$$

$$for sp^2 donor - sp^3 acceptor$$

$$g(\beta, \alpha, \phi) = \cos^2 \beta \cos^2 (\max[\alpha, \phi])$$

$$for sp^2 donor - sp^2 acceptor$$

Dahiyat energy function (equation 1) considers hydrogen bonds of protein molecules in any solvent. The energy values predicted by this function (due to both positional and angular coordinates) are commonly regarded in by other researchers in reasonable match with experimental data [24]. Although, we are using [19] geometric criteria and [23] energy function for prediction of the stiffness of hydrogen bonds, the proposed method remains valid for any other method of predicting the bond geometry and its energy.

3. Proposed Hydrogen Bond Mechanical Model

Under the application of external forces on the molecule, or internal forces due to shape changes in the functional proteins, hydrogen bonds exhibit flexibility. Hydrogen bonds can tolerate ±0.17 Å change in bond length in average before breaking [25,26,27]. Therefore we modeled hydrogen bonds as flexible mechanical elements (linear springs). Our simulation is valid for small displacements ranging ± 0.12 Å. Considering the energy function of these bonds shown in equation 1 we modeled each individual hydrogen bond with specific geometry as a combination of three linear springs accurately estimating the bond energy change for small changes in the bond geometry. Consequently our model calculates the mechanical properties of the spring such as spring stiffness and spring free length (in three directions) for each individual hydrogen bond. This is the major achievement of this study. Previous mechanical model of

hydrogen bonds considered the bond to be flexible solely in bond direction and assumed the same stiffness for all bonds with different geometry [28].

3.1. Equivalent Stiffens for Individual Hydrogen Bonds

To calculate stiffness of each bond the energy function of the bond (shown in equation 1) has been used. In this equation the energy is a function of three geometric parameters: donor-acceptor bond length (d), and the two angles (α&β) as shown in Figure 2-A. To simplify the model, angular coordinates (α&β) in energy function are replaced by the positional coordinates l and r, using geometric relations 2 and 3.

$$d^2 = (NH)^2 + r^2 - 2r(NH)\cos(\beta) \qquad (2)$$

$$l^2 = (OC)^2 + r^2 - 2r(OC)\cos(\alpha) \qquad (3)$$

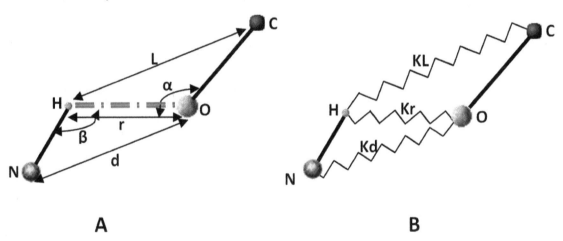

Figure 2. (A) The Geometry of hydrogen bonds. (B) The equivalent mechanical model of hydrogen bonds

Figure 2 B shows our proposed equivalent mechanical system to model hydrogen bonds. This model consists of three linear springs. The energy of this system should be the same as the energy of the bond itself computed by equation 1. Using Hook's law the equivalent energy of the spring system is shown in equation 4.

$$E = 0.5k_l(l_2 - l_0)^2 + 0.5k_r(r_2 - r_0)^2 + 0.5k_d(d_2 - d_0)^2 \quad (4)$$

To fully define the model, we need to calculate the values of the six constants (k_r, r_0, k_l, l_0, k_d, d_0) in the right hand side of the equation 4. In order to obtain these constants, we have set up an optimization problem with the aim of minimizing the fitting error; that is the difference between the bond energy obtained by empirical studies (equation 1) and our proposed model (equation 4).

The optimization problem has the following constraints: The energy computed at the configuration prior to the small displacements by both methods (equations 1 and 4) should be equal. This condition is implemented in equation 6.

The three spring constants (for the springs in Figure 2) are the slope of graph describing each spring force versus interatomic separation distance [29]. These conditions are implemented in equations 7 to 9.

Stiffness and the free lengths of springs should be positive numbers (equations 10 to 15).

The optimization problem (minimizing the difference between the bond energy obtained by empirical studies and our proposed model after small displacement) is therefore summarized as follows:

Minimize M (the objective function)

$$M = 0.5k_l(l_2 - l_0)^2 + 0.5k_r(r_2 - r_0)^2$$
$$+ 0.5k_d(d_2 - d_0)^2 - E(d2, \alpha2, \beta2) \qquad (5)$$

Subject to the following constraints:

$$0.5k_l(l_2 - l_0)^2 + 0.5k_r(r_2 - r_0)^2$$
$$+ 0.5k_d(d_2 - d_0)^2 - E(d1, \alpha1, \beta1) = 0 \qquad (6)$$

$$k_l - dE_l/(l_1 - l_0) = 0 \qquad (7)$$

$$k_r - dE_r/(r_1 - r_0) = 0 \qquad (8)$$

$$k_d - dE_d/(d_1 - d_0) = 0 \qquad (9)$$

$$-k_d + s^2 \le 0 \qquad (10)$$

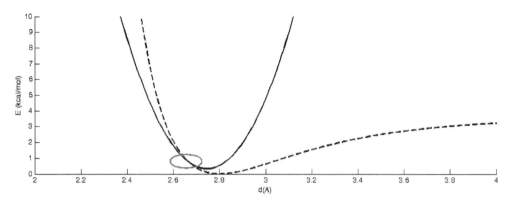

Figure 3. The new energy function calculated verses the LJ function

$$-k_d + p^2 \leq 0 \qquad (11)$$

$$-k_l + q^2 \leq 0 \qquad (12)$$

$$-l_0 + m^2 \leq 0 \qquad (13)$$

$$-r_0 + n^2 \leq 0 \qquad (14)$$

$$-d_0 + o^2 \leq 0 \qquad (15)$$

This optimization problem is then solved using Lagrange Multipliers method [30]. Figure 3 shows the energy of the hydrogen bond versus the bond length (r) obtained by empirical studies (dashed curve) and our optimized model (solid curve). As shown in Figure 3, the

proposed function can properly estimate the empirical energy function for small variations of the bond length.

3.2. Equivalent Stiffness Ellipsoid

Equivalent stiffness ellipsoid Once we have the stiffness of three springs, modeling individual hydrogen bonds, we can calculate bond stiffness at any given direction. To do this, stiffness matrix and consequently stiffness ellipsoid (ellipsoid equation) for each bond is calculated using static equilibrium equations (16, 17 and 18) for the mechanical system shown in Figure 4. In this system link NH is assumed to be fixed and link OC is free to move (from O1C1 to O2C2). The free body diagram of the system is shown in Figure 4-B.

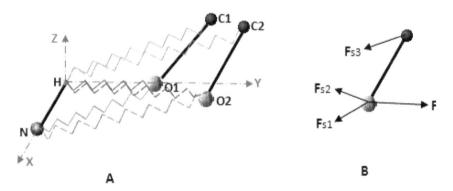

Figure 4. A: Small displacement of atoms result in small change in the length of springs B: Free body diagram of link OC

$$F_x = k_1 dx_0 \cos\alpha_1^2 + k_2 dx_0 \cos\alpha_2^2 + k_3 dx_c \cos\alpha_3^2$$
$$+ k_1 dy_0 \cos\alpha_1 \cos\beta_1 + k_2 dy_0 \cos\alpha_2 \cos\beta_2$$
$$+ k_3 dy_c \cos\alpha_3 \cos\beta_3 + k_1 dz_0 \cos\alpha_1 \cos\gamma_1$$
$$+ k_2 dz_0 \cos\alpha_2 \cos\gamma_2 + k_3 dz_c \cos\alpha_3 \cos\gamma_3 \qquad (16)$$

$$F_y = k_1 dx_0 \cos\alpha_1 \cos\beta_1 + k_2 dx_0 \cos\alpha_2 \cos\beta_2$$
$$+ k_3 dx_c \cos\alpha_2 \cos\beta_2 + k_1 dy_0 \cos\beta_1^2 + k_2 dy_0 \cos\beta_2^2$$
$$+ k_3 dy_c \cos\beta_3^2 + k_1 dz_0 \cos\beta_1 \cos\gamma_1$$
$$+ k_2 dz_0 \cos\beta_2 \cos\gamma_2 + k_3 dz_c \cos\beta_3 \cos\gamma_3 \qquad (17)$$

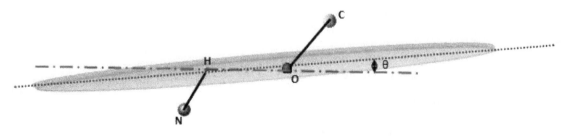

Figure 5. A schematic presentation of stiffness ellipsoid and hydrogen bond

$$F_z = k_1 dx_0 \cos\alpha_1 \cos\gamma_1 + k_2 dx_0 \cos\alpha_2 \cos\gamma_2$$
$$+k_3 dx_c \cos\alpha_3 \cos\gamma_3 + k_1 dy_0 \cos\gamma_1 \cos\beta_1$$
$$+k_2 dy_0 \cos\gamma_2 \cos\beta_2 + k_3 dy_c \cos\gamma_3 \cos\beta_3 \quad (18)$$
$$+k_1 dz_0 \cos\gamma_1{}^2 + k_2 dz_0 \cos\gamma_2{}^2 + k_3 dz_c \cos\gamma_3{}^2$$

Rearranging these equations, the stiffness matrix can be determined from equation 19.

$$\begin{pmatrix} F_x \\ F_y \\ F_z \end{pmatrix} = \begin{pmatrix} k_{11} & k_{12} & k_{13} & k_4 & k_{15} & k_{16} \\ k_{21} & k_{22} & k_{24} & k_{24} & k_{25} & k_{26} \\ k_{31} & k_{32} & k_{33} & k_{34} & k_{35} & k_{36} \end{pmatrix} * \begin{pmatrix} dx_0 \\ dy_0 \\ dz_0 \\ dx_c \\ dy_c \\ dz_c \end{pmatrix} \quad (19)$$

$$F = KX \quad (20)$$

The maximum and minimum stiffness of the system at point C, and their corresponding directions can be computed from stiffness matrix. The process is as follows. A unit displacement vector at point C is assumed. The corresponding force that is required to cause this displacement is calculated from equation 20. The process of finding the minimum and maximum possible force vectors is then setup as an optimization problem:

Minimize(Maximize) $F = KX$

Subject to unit displacement vector $XX^T = I$.

It can be shown that these maximum and minimum forces are the eigenvectors of the $K^T K$ matrix, with magnitudes of these forces being the square root of the corresponding eigenvalues. The three obtained eigenvalues are the corresponding diameters of the stiffness ellipsoid, and the eigen vectors are ellipsoid's principal directions.

The stiffness of the bond at any other direction, Ut, can be computed by calculating the intersection of the stiffness ellipsoid and the line passing through the center of the ellipsoid in the Ut direction. Figure 5 shows a schematic presentation of stiffness ellipsoid and its orientation with respect to bond geometry.

4. Results and Model Validations

4.1. For Single Hydrogen Bond

The developed methodology in previous section has been applied on several sample hydrogen bonds. Table 1 provides an overview of the results, detailing the following parameters: (E) the bond energy (kCal/mol), $(k_l, k_r \& k_d)$ stiffness (N/m) of each spring, $(l_0, r_0 \& d_0)$ free length of springs (Å), eigen values of stiffness matrix which are the maximum, mean and minimum stiffness of the bonds (N/m), (θ^*) the angle between maximum stiffness direction and hydrogen bond direction (degrees), (k^{**}) the hydrogen bond stiffness in bond (OH) direction (N/m). Table 3 lists stiffness for various bonds obtained from [31]. Our reported stiffness for hydrogen bonds in Table 2, ranging from 1.19-5.74 N/m with an average of 3.43N/m and standard deviation of 1.3N/m, are in agreement with the empirical results reported by [31] (3-6N/m). It has been shown in Table 2, column 11 which hydrogen bonds' maximum stiffness direction needs to be rotated 5 to 10 degrees to be aligned with bond direction (OH). On the other hand as shown in columns 8 and 10 of Table 2 the stiffness ellipsoid is so thin because of very small amount of minimum stiffness comparing to the maximum one. These observations lead us to an important conclusion: "hydrogen bonds are much stiffer in the bond direction than in perpendicular direction to the bond". This achievement combined with the analysis of particular arrangement of hydrogen bonds in α helices (studied in next section) provides mathematical prove for the observation of α helices being stiffer in axial direction than in lateral direction [9].

Table 2. Hydrogen bond stiffness for some sample bonds

E(energy)	k_l	k_r	k_d	l_0	r_0	d_0	max k	mean k	min k	θ^*	k^{**}
-2.303	51	26.3	4.8	2.9	2.2	3	54.53	1.82	0.14	6.9	3.36
-2.29	40.9	8	1.7	2.8	2	2.9	34.16	0.49	0.07	10	1.19
-3.576	68.3	23.6	3	2.8	2	2.9	64.68	1.61	0.07	5.4	2.1
-3.232	111.2	37.9	6.9	2.8	1.9	2.8	104.51	2.03	0.14	7.9	4.83
-4.791	87.9	26.6	3.3	2.8	2	3	80.43	1.19	0.14	5.5	2.31
-4.843	127.7	28.8	3.9	2.8	1.8	2.9	109.69	1.26	0.21	6.4	2.73
-3.676	69.6	22.7	3.5	2.8	2	3	64.82	1.19	0.14	6.8	2.45
-2.238	57.1	41.9	6.5	2.9	2.2	3	70.56	3.15	0.14	5.1	4.55
-2.043	45.3	22.1	4.6	2.9	2.2	3	47.39	1.33	0.14	8.2	3.22
-3.04	62.1	26.4	4.4	2.9	2.1	3	62.23	1.61	0.21	6.7	3.08
-1.37	46.9	49.6	8.2	3.1	2.4	3.1	69.51	4.2	0.14	4.6	5.74
-2.856	60.1	27.7	4.8	2.9	2.1	3	61.81	1.61	0.14	6.9	3.36
-3.765	109.3	29.8	5.3	2.8	1.9	2.8	97.44	1.33	0.14	8	3.71
-1.172	39	37.3	7.7	3	2.4	3.1	54.53	2.87	0.28	6.1	5.39

Table 3. Different bonds' stiffness [31]

Bond Type	Bond stiffness (N/m)
Covalent	50 - 180
Metallic	15 - 75
Ionic	8 - 24
Hydrogen bond	6 - 3
Van der Walls	0.5 - 1

4.2. Application to Protein Molecules

The secondary structures of protein molecules (α helices and β sheets) are stabilized by hydrogen bonds [32]. Based on this fact, one can infer that the stiffness of protein molecules, as well as their strength in different directions, is greatly determined by hydrogen bonds. Therefore to predict the stiffness of protein molecules the individual stiffness of hydrogen bonds has been used.

Figure 6-A illustrates a schematic arrangement of hydrogen bonds in a helices. It is shown that in each turn of a helix there are 3 to 4 hydrogen bonds [33]. Each turn

has 3.6 residues with 5.4Å pitch along the helical axis [32]. The equivalent mechanical model is shown in figure 6-B. In this model, hydrogen bonds in each turn are assumed to be parallel and turns are connected in a serial form. Considering the Hook's law for springs, the stiffness of the entire a helix can be obtained using equation 21.

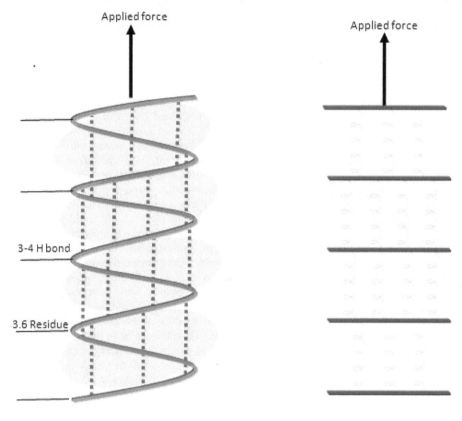

A: Schematic Biological System   B: Mechanical Equivalent Model

Figure 6. A schematic arrangement of hydrogen bonds in α helices

$$K_{eq} = 1/(1/k_{turn1} + 1/k_{turn2} + \ldots + 1/k_{turn-end}). \quad (21)$$

4.2.1. Calculating stiffness of some sample protein molecules

Now we can use the approach discussed in previous section to calculate the stiffness of several protein molecules. The proteins in this numerical experiment were chosen such that empirical or computed data on their stiffness is readily available in the literature. Secondly, we chose proteins with helical structures because of their specific geometry and arrangement of hydrogen bonds. The helical motif can greatly simplify the evaluation of protein stiffness since hydrogen bonds are placed almost parallel to the helix axes. Table 4 lists three sample proteins with the required specification, their PDB (Protein Data Bank) code, the number of amino acids and the number of detected hydrogen bonds (based on the geometric and energetic criteria reported in [21]). The first molecule is a synthetic peptide, cysteine - lysine - cysteine (C3K30C) specifically designed to study hydrogen bonding by [6]. Under the experimental conditions [6] this synthetic peptide adopts the a helix structure as a result of hydrogen bonding (31 bonds) within the molecule. The

peptide has a length of 53Å and a diameter of approximately 15Å. The stiffness of this protein molecule was measured using atomic force microscopy. Each molecule was stretched from the a-helical state into a linear chain. The stiffness was found to vary with molecule displacement. In order to conduct a meaningful comparison between our calculated stiffness of this protein molecule and the empirical measurement reported in [6], only small displacements are considered. In particular, we concentrate on deformation of 0.12Å or smaller for the subject protein. With such deformation protein molecule will keep its helical shape and none of the hydrogen bonds breaks. The reported stiffness of this protein molecule varies between 0.3-0.4N/m. Results of our computational analysis applied on this protein molecule are summarized as follows. 31 hydrogen bonds are detected within this protein molecule.

The stiffness of the studied molecule (0.38N/m) is in agreement with the expected range of the stiffness of this molecule ((0.3-0.4N/m)) from empirical data.

The elastic properties of characterized myosin II S2 sub domain using molecular dynamics and normal mode analysis has been studied [9]. Tis protein molecule with 1NKN PDB code, is 87 residue long a-helix. We predicted 146 hydrogen bonds within this molecule structure and

0.083N/m stiffness. This results are in good agreement with the stiffness predicted by [9] for this protein molecule (0.06-0.08N/m).

Table 4. Sample proteins

PDB code	# of amino acids	# Detected hydrogen bonds
Synthetic peptide	35	31
1gk6	53	42
1nkn	150	146

To study elasticity and strength of secondary structures of protein molecules [1] compared the force-extension behavior of α-helices, β-sheets and tropocollagen domains in protein molecules. Their results are provided by using atomistic modeling of nano-mechanical response of the protein molecules at ultra-slow deformation rates. The stiffness predicted for a molecule with helical structure (1GK6) in this study is 0.571N/m. The predicted stiffness for this molecule using our proposed method is 0.52N/m.

Table 5. Stiffness for different protein molecules (N/m)

PDB code	K from simulation	k reported in litrature
Synthetic peptide	0.380	0.300-0.400 [6]
1gk6	0.520	0.571 [10]
1nkn	0.083	0.060-0.080 [9]

5. Discussion and Conclusion

This paper presents a novel methodology to study mechanical properties of protein molecules in atomic level. Two major results of this research are: (1) estimating the stiffness of the hydrogen bonds at any given direction and; (2) predicting the stiffness of protein molecules (helical structure). We achieve this through developing equivalent system of mechanical elements (three linear springs) that mimic physical behavior of the actual bonding arrangements. The empirical value of energy is obtained through the energy function of hydrogen bonds reported in literature. Then the stiffness of each spring is calculated based on this empirical data. To obtain these results we solve an optimization problem for each individual hydrogen bond to make sure that both mechanical model and the actual bond have similar potential energies, and change of potential energies through small displacements. Once the mechanical properties of each spring are obtained, the stiffness ellipsoid of the bond is calculated. Having the stiffness ellipsoid one can calculate the stiffness on hydrogen bonds at any given direction. The comparison between stiffness of some sample hydrogen bonds (average of 3.43N/m and standard deviation of 1.3N/m) versus the expected value from literature (3-6N/m) showed that our results are in the expected range.

Furthermore, to obtain the stiffness of helical protein molecules a mechanical model based on hydrogen bonds particular geometry and arrangement is proposed. This methodology is employed on three different helical protein molecules with reported stiffness in literature. The obtained stiffness for these protein molecules are in good agreement with empirical and computational values reported in literature.

In addition further analysis on bond's directional stiffness combined with the proposed mechanical model

of α helices, suggest that the α helices are stiffer in axial direction than in the lateral one. This provides direct mathematical proof to this behavior of helical protein molecules.

This mechanical model to calculate protein molecules stiffness is developed for the helical molecules. Developing a more elaborate mechanical model of the protein molecules will result in the prediction of the stiffness of protein molecules with other structure (e.i. β-sheets).

To the best of our knowledge the analysis reported here is the first direct mechanical analysis of the directional stiffness of individual hydrogen bonds. This analysis, combined with a mechanical model of the arrangement of hydrogen bonds in protein molecules, provides us with a multiscale model to calculate mechanical properties of biological materials starting from atomic levels.

Acknowledgements

The support provided by National Science Foundation, University of Connecticut and Manhattan College are greatly acknowledged.

References

[1]   Buehler, M. J., and Keten, S., 2008. "Elasticity, strength and resilience: a comparative study on mechanical signatures of a-helix, -sheet and tropocollagen domains". Nano Research, 1, pp. 63-71.

[2]   Kazerounian, K., 2004. "From mechanisms and robotics to protein conformation and drug design". Journal of Mechanical Design, 126, pp. 40-45.

[3]   Hu, X., Cebe, P., Weiss, A. S., Omenetto, F., and Kaplan, D. L., 2012. "Protein-based composite materials". Materials Today, 15(5), pp. 208-215.

[4]   Rief, M., Gautel, M., Oesterhelt, F., Fernandez, J. M., and Gaub, H. E., 1997. "Reversible unfolding of individual titin immunoglobulin domains by afm". Science, 276, p. 1109.

[5]   Gabovich, A. M., and Li, M. S., 2009. "Mechanical stability of proteins". Journal of Chemical Physics, 131, p. 024121.

[6]   Lantz, M. A., Jarvis, S. P., Tokumoto, H., Martynski, T., Kusumi, T., Chikashi, N., and Miyake, J., 1999. "Stretching the a-helix: A direct measure of the hydrogen bond energy of a single peptide molecule". Chemical Physics Letters, 315, pp. 61-68.

[7]   Tskhovrebova, L., Trinick, K., Sleep, J., and Simmons, M., 1997. "Elasticity and unfolding of single molecules of the giant muscle protein titin". Nature, 387, p. 308.

[8]   Kellermayera, M. S. Z., Smithb, S. B., Bustamanteb, C., and Granzierc, H. L., 1998. "Complete unfolding of the titin molecule under external force". Journal of Structural Biology, 122, pp. 197-205.

[9]   Adamovic, I., Mijailovich, S. M., and Karplus, M., 2008. "The elastic properties of the structurally characterized myosin ii s2 subdomain: A molecular dynamics and normal mode analysis". Biophysical Journal, 94, pp. 3779-3789.

[10] Buehler, M., and Keten, S., 2008. "Elasticity, strength and resilience: A comparative study on mechanical signatures of a helix, b-sheet and tropocollagen domains". Nano Res, 1, pp. 63-71.

[11] Hamdia, M., Ferreiraa, A., Sharmab, G., and Mavroidis, C., 2008. "Prototyping bio-nanorobots using molecular dynamics simulation and virtual reality". Microelectronics Journal, 39, p. 190201.

[12] Poursina, M., Bhalerao, K. D., Anderson, K. S., Flores, S., and Laederach, A. "Strategies for articulated multibodybased adaptive coarse grain simulation of rna". Methods in Enzymology, 487(31), pp. 73-98.

[13] Shahbazi, Z., 2011. "Role of hydrogen bonds in kinematic mobility and elasticity analysis of protein molecules". PhD thesis, University of Connecticut.

[14] Kazerounian, K., and H., I., 2012. 21st Century Kinematics. Springer, ch. Protein Molecules: Evolution's Design for Kinematic Machines, pp. 217-244.

[15] Kazerounian, K., Latif, K., and Alvarado, C., 2005. "Protofold: A successive kinetostatic compliance method for protein conformation prediction". Journal of Mechanical Design, 127(4), pp. 712-717.

[16] Kazerounian, K., Latif, K., Rodriguez, K., and Alvarado, C., 2005. "Nano-kinematics for analysis of protein molecules". Journal of Mechanical Design, 127(4), pp. 699-711.

[17] Subramanian, R., and Kazerounian, K., 2007. "Kinematic mobility analysis of peptide based nano-linkages". Mechanism and Machine Theory, 42(8), pp. 903-918.

[18] Subramanian, R., and Kazerounian, K., 2007. "Improved molecular model of a peptide unit for proteins". Journal of Mechanical Design, 129(11), pp. 1130-1136.

[19] Shahbazi, Z., Ilies, H., and Kazerounian, K., 2010. "Hydrogen bonds and kinematic mobility of protein molecules". Journal of Mechanisms and Robotics, 2, pp. 021009-1,9.

[20] Shahbazi, Z., and Demirtas, A., 2015. "Rigidity analysis of protein molecules". Journal of Computing and Information Science in Engineering.

[21] Z., S., F., P. T. A. P., H., I., K., K., and P., B., 2010. Advances in Robot Kinematics, Issue on Motion in Man and Machine. Springer, ch. A Kinematic Observation and Conjecture for Stable Construct of a Peptide Nanoparticle, pp. 203-210.

[22] Wales, T. E., and Fitzgerald, M. C., 2001. "The energetic contribution of backbone–backbone hydrogen bonds to the thermodynamic stability of a hyperstable p22 arc repressor mutant". J Am Chem Soc, 123(31), pp. 7709-10.

[23] Dahiyat, B. I., Gordon, B., and Mayo, S. L., 1997. "automated design of the surface positions of protein helices". Protein Science, 6, pp. 1333-1337.

[24] Jacobs, D., Rader, A. J., Kuhn, L. A., and Thorpe, M. F., 2001. "Protein flexibility predictions using graph theory". Proteins: Structure, Function, and Genetics, 44(2), pp. 150-165.

[25] Baker, E. N., and Hubbard, R. E., 1984. "Hydrogen bonding in globular proteins". Prog Biophys Mol Biol, 44(2), pp. 97-179.

[26] Artymiuk, P. J., and Blake, C. C., 1981. "Refinement of human lysozyme at 1.5 a resolution analysis of non-bonded and hydrogen-bond interactions". J Mol Biol, 152(4), pp. 737-62.

[27] Xu, D., Tsai, C. J., and Nussinov, R., 1997. "Hydrogen bonds and salt bridges across protein-protein interfaces". Protein Engineering, 10(9), pp. 999-1012.

[28] Brown, I. D., 2006. The chemical bond in inorganic chemistry: the bond valence model. Oxforf University Press.

[29] Mitchel, B. S., 2003. An introduction to materials engineering and science. John Wiley & Sons, INC.

[30] Arora, J. S., 2004. Introduction to optimum design. Elesevier Academic Press.

[31] Ashby, M., Shercliff, H., and Cebon, D., 2007. Materials: engineering, science, processing and design. Butterworth-Heinemann.

[32] Branden, C., and Tooze, J., 1999. Introduction to protein structure, second ed. Garland publishing.

[33] Ackbarow, T., Chen, X., Keten, S., and Buehler, M. J., 2007. "Hierarchies, multiple energy barriers, and robustness govern the fracture mechanics of alpha-helical and beta-sheet protein domains". Proc Natl Acad Sci U S A, 104(42), pp. 16410-5.

24

Production of Biogas by Anaerobic Digestion of Food Waste and Process Simulation

Faisal Kader[*]**, Abdullah Hil Baky, Muhammad Nazmul Hassan Khan, Habibullah Amin Chowdhury**

Islamic University of Technology, Gazipur, Bangladesh
*Corresponding author: mfaisaliut@yahoo.com

Abstract Anaerobic Digestion is a biological process that takes place naturally when microorganisms break down organic matter in the absence of oxygen. In an enclosed chamber, controlled anaerobic digestion of organic matter produces biogas which is predominantly methane. The produced methane then can be directly used for rural cooking; or after certain conditioning, can be used in onsite power generation, heating homes or as vehicular fuel. Besides, food waste is increasingly becoming a major problem in every society imposing serious economic and environmental concerns. For this reason, many contemporary researches are emphasizing in finding sustainable solutions to recycle and produce energy from such waste. In this context, this paper aims to study and optimize the production of biogas from food waste (rice). For the experiment, an existing wet digestion biogas plant installed in Islamic University of Technology was used. The food waste (rice) for the research was collected from the cafeteria of Islamic University of Technology. Furthermore, a process simulation was performed by PROII software to estimate the methane production rate. Eventually, the simulated and experimental results were compared. The duration of the study period was 120 days. The experimental results showed that an average specific gas production of 14.4 kg-mol/hr can be obtained for 0.05 kg-mol/hr of starch loading rate. In case of the simulated results, the gas production was found to be 19.82 kg-mol/hr for the same loading rate of starch. The percentage of methane and CO_2 obtained in the biogas plant was 69% and 29% respectively.

Keywords: *Anaerobic Digestion, food waste, process simulation, biogas*

1. Introduction

Concerning the treatment of solid waste, the anaerobic digestion of solid waste has been studied in recent decades, trying to develop a technology that sum up advantages for volume and mass reduction as well as for energy and sources recovering. Anaerobic digestion, besides aerobic composting can be an alternative strategy for the reduction of MSW. In contrast to aerobic composting, anaerobic digestion of solid waste does not require air and produce biogas with high volumetric fraction of methane (50-70%). Furthermore the anaerobic digestion processes are best applicable for wet waste and the area requirement are satisfactory [1,2]. Until 1970's the anaerobic digestion (AD) was commonly used only for wastewater treatment [3]. Due to continuous increase of generated solid waste and the large environmental impacts of its improper treatment, its management has become an environmental and social concern. In Dhaka the main source of municipal solid waste are domestic, streets, market places, commercial establishment clinics and hospitals. At present, Dhaka City generates about 3500-4000 tons of solid waste per day, the per capita generation being 0.5kg/day. The density of solid waste is reported to be 600 kg/m³(JICA 2005). According to a study conducted by Japan International Corporation Agency (JICA 2005), as of 2004, the total waste from domestic source was estimated to be 1945 t/d out of a population of 5.728 million with average generation rate of 0.34 kg/day per person. The total solid waste from business sources was estimated to be 1035 t/d and by cleaning the streets, the amount of waste generation was predicted to be 200 t/d (0.365 t/kmx550 km=201 t/d), making the total estimated waste generation of about 3200 t/d.

By 2025 the demand of energy is expected to increase by 50%. Hence there is an ongoing search to develop renewable energy sources which gives sustainable, affordable, environmentally friendly energy [4,5]. Biofuels are renewable and environmentally clean which significantly decreases the fossil fuel consumption [6].

For anaerobic digestion food waste is a highly degradable substrate because of its biodegradability and high nutrient content. A typical food waste contains 7-31 wt. % of total solid and the biochemical methane potential of the food waste is estimated to be 0.44-0.48 m² CH4/kg of the added volatile solid [7,8,9].

Therefore the aim of this study was to investigate the potential of anaerobic digestion for biogas production using only rice as food waste. The production of methane was simulated using process simulation software PROII.

Then the simulated results were verified by the experimental data.

2. Materials and Method

2.1. Sample Collection

The food waste (rice) for the research was collected from the cafeteria of Islamic University of Technology. The water was collected from the main water supply line of the university. The rice waste was fed into the digester by diluting with water. The waste was manually mixed with water without shredding or any pretreatment.

2.2. Experimental Setup

A fixed dome wet-digestion plant was constructed in the IUT campus. The schematic diagram of the plant is shown below. The digester consisted of three major parts: inlet chamber, fermentation chamber and a hydraulic chamber. The inlet chamber was used for charging the feed stock into the fermentation chamber where the fermentation of the feedstock takes place. The fermentation chamber had a fixed non movable gas space. The gas was stored in the upper part of the chamber and collected through the gas-outlet pipe. When the production commences, the pressure of the gas displaced the slurry into the hydraulic chamber. The hydraulic chamber had a volume of 3 m^3 which indicated the maximum gas holding capacity of the digester.

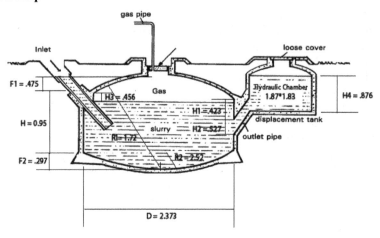

Figure 1. Schematic diagram of a wet digestion plant

3. Analytical Method of Biogas Production

Generally three main reactions occur during the entire process of the anaerobic digestion to methane: hydrolysis, acid forming and methanogenesis. Although AD can be considered to take place in three stages all reactions occur simultaneously and are interdependent.

Hydrolysis is a reaction that breaks down the complex organic molecules into soluble monomers. Hydrolysis reaction of the organic fraction of the MSW can be represented by the following reaction: [10]

$$C_6H_{10}O_4 + 2H_2O \rightarrow C_6H_{12}O_6 + 2H_2 \quad (1)$$

Acid forming stage comprises two reactions – fermentation and the acetogenesis reactions. Typical reactions occurring during this stage are the conversion of the glucose to ethanol and the conversion of the glucose to propionate. Important reactions during the acetogenesis stage are the conversion of glucose to acetate, ethanol to acetate, propionate to acetate and bicarbonate to acetate.

The reactions that occur during methanogenesis stage are as follows [10]:

Acetate conversion:

$$2CH_3CH_2OH + CO_2 \rightarrow CH_3COOH + CH_3 \quad (2)$$

Followed by:

$$CH_3COOH \rightarrow CH_4 + CO_2 \quad (3)$$

Methanol conversion:

$$CH_3OH + H_2 \rightarrow CH_4 + H_2O \quad (4)$$

Carbon dioxide reduction by hydrogen:

$$CO_2 + 4H_2 \rightarrow CH_4 + H_2O \quad (5)$$

4. Process Simulation of Methane Production

For process simulation in the PRO II software at first the components needs to be identified. Since ethanol, H_2O, CO_2 were present in the library they were added from the library components and hypothetical components Glucose and Starch were filled up according to the UNIFAC structures since they were not in the system.

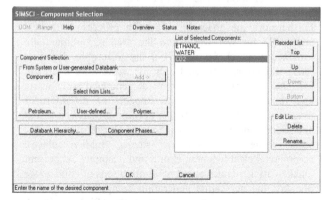

Figure 2. Component selection

Filling up Glucose and Starch:

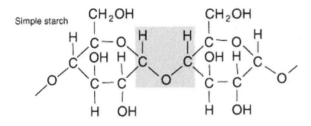

Figure 3. Structure of starch

Table 1. Group and number of the components of starch

Category	Group	Number
Alcohols	0200	1600
Alcohols	0277	800
Ethers	0602	1600
Paraffins	0902	2400

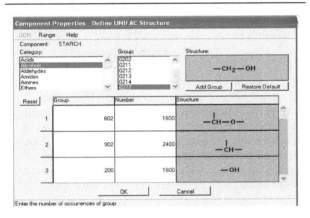

Figure 4. Component properties for starch

Table 2. Group and number for glucose

Category	Group	Number
Alcohols	0200	4
Alcohols	0214	1
Aldehyde	0300	1
Paraffins	0902	3

SimSci Group ID	Number	Structure
1	200	4	—OH
2	214	1	—CH—CH₂OH
3	300	1	—CH≡O
4	902	3	—CH—

Figure 5. Component properties for glucose

After that the reactions were defined. The reaction set names were provided with the reaction data where the reaction was balanced (Figure 6). All the reactions were balanced likewise.

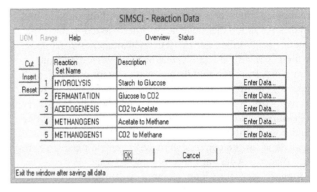

Figure 6. Reaction set names

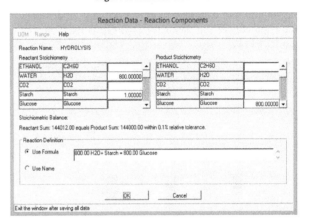

Figure 7. Reaction data

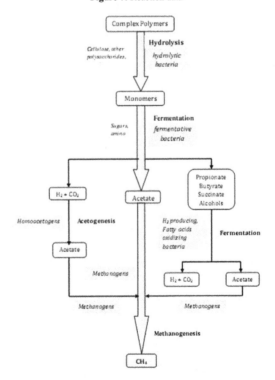

Figure 8. Overall process diagram of AD system

Then the PFD was drawn. It was drawn according to the overall process diagram of the AD system as shown in Figure 8. The production of methane is divided into three streams; three different line of production. The PFD was drawn for each of the three streams from which methane can be obtained separately as shown in Figure 9, Figure 10 & Figure 11. To be explained, in the left stream the monomer glucose is converted into carbon dioxide and hydrogen which then with acidogenesis reaction turns into acetic acid followed by the methanogenesis reaction, which convert it into methane. Likewise methane is produced from the middle and right stream of the overall process diagram. Each of the PFD consists of some conversion reactors and input and output lines. By clicking on the input lines the conditions like the flow rate and composition for the input substance is defined. Same was done for the conversion reactors for the parameters to be placed. After all the parameters are defined the program was run. The reactors will turn blue if the conditions are set perfectly in the input/output lines and in the reactors. Here, water and starch flow rate was taken as 0.5kg-mol/hr and 0.05 kg-mol/hr respectively. In different reactor the conversion factor for Hydrolysis reaction was set as 0.89 [11].

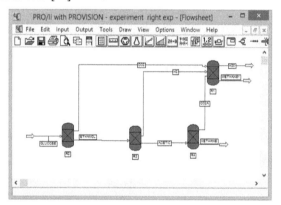

Figure 9. Flow diagram for right stream of the process diagram

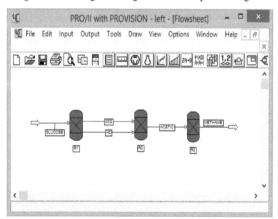

Figure 10. Flow diagram for left stream of the process diagram

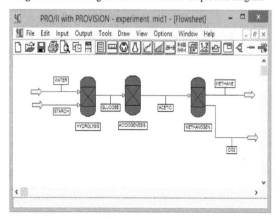

Figure 11. Flow diagram for the middle stream of the process diagram

The flow of methane from three different streams was obtained from the output lines after the simulation was run. From the simulated results, the total flow rate of methane from three different streams was found to be 19.82 kg-mol/hr.

5. Experimental Result

The sample calculation of methane flow rate for starch loading rate 0.05 kg-mol/hr is done using the continuity equation. The density of methane is 0.668 kg/m^3. The output pipe used to collect the methane was of 18 mm diameter. Using the anemometer the velocity of the output methane was calculated. Using these values the mass flow rate for the methane was calculated.

Density of Methane, $\rho = 0.668$ kg/m^3
Diameter of Pipe, $D = 18$ mm
Cross sectional area of pipe, $A = \pi/4 D^2 = \pi/4 * 0.018^2$
                                $= 2.5 * 10^{-4}$ m^2
Velocity of flow, $V = 24$ m/s
Mass flow rate, $m = \rho A V = .668 * 2.5 * 10^{-4} * 24$
                                $= 4.00 * 10^{-4}$ kg-mol/s
                                $= 14.4$ kg-mol/hr

The comparison between experimental and simulated flow rate of methane with respect to starch loading rate is shown below. The observed deviation of the experimental result might be happened due to conversion factors used in the conversion reactors. A good approximation of the conversion factors might give a better solution. Also no necessary steps were taken to control the pH value in the digester which might have a great effect on the output.

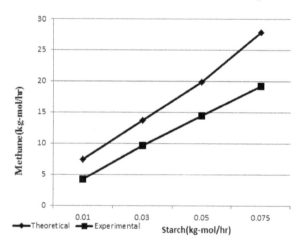

Figure 12. Comparison between starch loading rate vs. Methane flow rate of theoretical and experimental result

In wet digestion plant, gas volume was measured daily by measuring the water displacement in the hydraulic chamber. Each day, gas was released at the prescheduled time and the water level of the chamber was measured before and after the release. The difference between two readings was multiplied with the length and width of the chamber to obtain the gas volume each day in m^3.

The percentage of methane in the biogas obtained from the plant is shown below. Considering the cost of the test, the gas was tested only 3 times during the study period. Each test was done in approximately 3 weeks interval. As the degradation of starch particle continuously increased with the increase of the methanogenic bacteria, the percentage of methane in the outlet gas also increased. Also, as the food waste is mainly starch, which is a hydrocarbon, the gas obtained from the anaerobic digestion of food waste contains an incredibly high amount of methane. Furthermore, it was observed that gradually the amount of methane in the biogas came to an almost constant value of 69%. This is in accordance with a finding of Bangladesh Council of Scientific and Industrial Research (BCSIR).

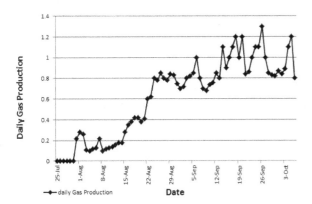

Figure 13. Daily gas production in, m^3

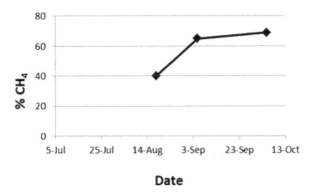

Figure 14. Percentage of methane, %

6. Discussion

In this research, PROII process simulation software was used to optimize the methane flow rate and comparisons have been done between the experimental and the simulated results. The reason of the observed deviation between the experimental and simulated results was predicted to be the lack of controlled environment and leakage of the gas and rain water out and in to the system respectively which affected the production as the plant was setup in an open space. Since both the flow rate and the percentage of methane in the biogas is important for its different useful usage this experiment was carried out with a view to maximizing both the flow rate and the methane percentage. And using wasted rice was of great response. Its reason remains in rice being a source of starch which is hydrocarbon.

Acknowledgments

We express gratitude to prof. Dr Abdur Razzak Akanda, Special thanks to Bizon Shahriar of Electrical and Electronics Engineering Department for making the collaboration possible with GTZ and A.N.M Jubayer of GTZ for his continuous guidance and help. Thanks to Sajjad Alam & Kazi Zayed Hasan of Mechanical and Chemical Engineering Department for their continuous assistance in the project. Moreover instructors and technicians of machine shop, Thermodynamics lab of mechanical engineering department are acknowledged for their help and suggestions. Most importantly, we acknowledge Tabassum Aziz for his earlier project related to biogas that helped us in many ways.

Nomenclature

AD- Anaerobic Digestion
IUT- Islamic University of Technology
GTZ- German Technical Corporation
JICA- Japan International corporation Agency
MCE- Mechanical and Chemical Engineering
MSW- Municipal Solid Waste
PFD- Process Flow Diagram

References

[1] Baldasano, J.M. and Soriano, 2000, "Emission of greenhouse gases from anaerobic digestion processes: comparison with other municipal solid waste treatments. Water science and technology," Vol. 41 (3); 275-282.

[2] Hartmann, H. and Ahring, B.K., 2006, "Strategy for the anaerobic digestion of the organic fraction of municipal solid waste: and overview. Water science & technology." Vol. 53 (8):7-22.

[3] Palmisano, A.C. & Barlaz, Morton A., 1996. "Microbiology of solid waste."

[4] Deublein, D. and Steinhauser, A. (2008). "Biogas from waste and renewable resources: An introduction." Wiley-V CH, Weinheim, Germany.

[5] Khanal, S.K. (2008). "Bioenergy generation from residues of biofuel industries. Anaerobic Biotechnology for Bioenergy Production: Principles and Applications." Wiley-Blackwell, Oxford, UK; John Wiley & Sons, New York, NY, USA, 161-188.

[6] Ersahin, M.V, Gomec, C.V., Dereli, R.K., Arikan, O. and Ozturk, I. (2011), "Bio methane production as an alternative: Bioenergy source from codigesters treating municipal sludge and organic fraction of municipal solid wastes." Journal of Biomedicine and Biotechnology, 8:1-8.

[7] Heo, N., Park, S., Lee, J., Kang, H. and Park, D. (2003). "Single-stage anaerobic codigestion for mixture wastes of simulated Korean food waste and waste activated sludge." Applied Biochemical and Biotechnology, 107: 567-579.

[8] Han, S.-K.and Shin, H.-S. (2004). "Biohydrogen production by anaerobic fermentation of food waste."

[9] Zhang, R., El-Mashad, H.M., Hartman, K., Wang, F., Liu, G., Choate, C. and Gamble, P. (2007). "Characterization of food waste as feedstock for anaerobic digestion." Bioresource Technology, 98: 929-935.

[10] Ostrem, K. & Themelis, Nickolas J., 2004. "Greening waste : anaerobic digestion for treating the organic fraction of municipal solid wastes". http://www.seas.columbia.edu/earth/wtert/sofos/Ostrem_Thesis_final.pdf.

[11] Rebecca Anne Devis, 2008. "Parameter Estimation for Simultaneous Saccharification and Fermentation of Food Waste into Ethanol Using Matlab Simulink".

25

Evaluation of Energy Losses in Pipes

LAHIOUEL Yasmina[1,*], LAHIOUEL Rachid[2]

[1]Laboratoire d'Analyse Industrielle et de Génie des Matériaux (LAIGM), Faculté des sciences et de la Technologie, Université de Guelma, B.P. 401, Guelma
[2]Laboratoire de Physique de Guelma (GPL), Université de Guelma, B.P. 401, Guelma, Algérie
*Corresponding author: ya_lahiouel@yahoo.fr

Abstract Energy losses in pipes used for the transportation of fluids (water, petroleum, gas, etc.) are essentially due to friction, as well as to the diverse singularities encountered. These losses are usually converted into head reductions in the direction of the flow. The knowledge of data of such transformation allows the determination of the necessary power needed for the transportation of the fluid between two points. It constitutes the necessary calculation basis necessary for the design and analysis of transport and distribution networks. The review of the different relationships allowing the determination of these losses and their comparison to the experimental results obtained by the authors constitute the object of this study.

Keywords: *head losses, frictional head losses, minor head losses, singular head losses, water distribution network*

1. Introduction

Pipe technology is based on the universal principles of fluid flow. When a real (viscous) fluid flows through a pipe, part of its energy is spent through maintaining the flow. Due to internal friction and turbulence, this energy is converted into thermal energy. Such a conversion leads to the expression of the energy loss in terms of the fluid height termed as the head loss and usually classified into two categories. Essentially due to friction, the first type is called linear or major head loss. It is present throughout the length of the pipe. The second category called minor or singular head loss is due to the minor appurtenances and accessories present in a pipe network. The appurtenance encountered by the fluid flow which is a sudden or gradual change of the boundaries results in a change in magnitude, direction or distribution of the velocity of the flow. This classification into major and minor head losses is rather relative. For a pipeline of small length having many minor appurtenances, the total minor head loss can be greater than the frictional head loss. In petroleum and water distribution networks, the pipelines are of considerable length and therefore the terms major head loss and minor head loss can be used without any confusion.

A great number of studies were carried out in order to achieve a general and precise formulation of the diverse types of head losses. [25] was the first to have come out with a relation for the head loss. As brought up by [2], Darcy contributed greatly to the application of the derived relation, thus associating his name with that of Weisbach. The relation is therefore most commonly known as the Darcy-Weisbach formula. It essentially depends on the friction coefficient and the relative roughness.

The friction coefficient is a function of the flow regime characterised by the Reynolds number. Several explicit and implicit relationships were proposed for the friction coefficient. [1] performed extensive experimentations involving smooth and artificially roughened pipes achieved using sand particles of uniform size. The Nikuradse diagram also known as the Stanton diagram or the Stanton-Pannel diagram, is the result of these investigations. Comparing the results included in Nikuradse's diagram, [8] found that its curves do not match with those of actual pipes. However, by introducing the concept of equivalent surface roughness, it is possible to use Nikuradse's results for commercial pipes.

Several other investigators provided the literature with diverse diagrams. [15] presented a diagram for commercial pipes using several non-dimensional groups. [21] plotted the friction coefficient represented by $\left(1/\sqrt{f}\right)$ against the Karman number represented by:

$$K = R_e \sqrt{f} = D^{\frac{3}{2}} \sqrt{2g \frac{S}{v}}$$. He produced the necessary

curves for the Colebrook transition zone. L. F. Moody suggested to Rouse converting his diagram by plotting the friction coefficient against the Reynolds number which he refused. [17] then plotted it himself, producing the actual universally known Moody diagram allowing the determination of the friction coefficient as a function of the Reynolds number and the ratio (ε/D).

Other investigators provided the literature with relationships. [3] suggested a relationship between the friction coefficient and the Reynolds number applicable solely for smooth turbulent flows. For the same regime, [9] suggested a relationship similar to that of Blasius. For a transitional turbulent flow, [8] derived a relationship which is presently commonly known as the Colebrook-White equation. For a rough turbulent flow, Prandtl [2] suggested a relationship expressing the friction coefficient as a function of the ratio (ε/D). It is now widely known as the Karman-Prandtl equation.

All these relationships are implicit implying the use of a try-and-error procedure in order to achieve a solution leading to a value of the friction coefficient. Explicit relations expressing the friction coefficient for all regimes of flow are available. For a smooth turbulent flow, [24] suggested a simple explicit relation based on the Karman-Prandtl equation while [5] proposed a simpler one. For a transitional turbulent flow, [17] enriched the specialised literature with a highly accurate relationship, on which the relations of [1,5,10,13,23,29,30] linking the friction coefficient f to the Reynolds number R_e and the relative pipe roughness (ε/D) have been based and led to an appreciable accuracy.

Relationships applicable for all regimes are also available. [7] proposed a friction factor equation which should be applicable for all fluid flow regimes whereas [6] derived explicit solutions for the Prandtl and Colebrook-White equations. One of the widely used formulas used is that developed by [27]. It expresses the head loss as a function of diameter, flow rate and length using an empirical coefficient. The Hazen-Williams coefficient as it has since been known depends on the pipe material and the flow velocity. Values of the Hazen-Williams coefficient applicable for diverse common pipe materials are recommended by [16] who tabled them after performing a great amount of experimentations. The Hazen-Williams formula was a subject of interest for many investigators. Based on his own tests, [26] confirmed it while [Janna et al, 1978] tried to modify it by introducing a new coefficient which vary with the Reynolds number, the relative roughness and the flow velocity. Earlier in 1923, as reported by [18] Strickler had suggested a simpler relationship based on a fixed coefficient. Later on, other empirical relationships were proposed. Hence, [22] suggested his relationship as early as 1930. It is similar to that of Hazen-Williams except for the use of a fixed coefficient. In 1965 as reported by [18], Calmon and Lechapt, suggested a relationship including three coefficients which vary with the pipe roughness. The Manning formula known as such because it has firstly been derived by Manning in 1891 [2] has, as reported by [20] and [28], since received a lot of contributions and is still being used.

The various experimentations carried out by the authors in order to determine the linear and singular head losses taking place in a variety of pipes of different dimensions and roughness are presented. Water is used at different flow regimes. The obtained linear results are compared to those computed using the most common pipe flow formulas presented earlier (Darcy-Weisback, Hazen-Williams, Manning, Strickler, Scobey and Calmon & Lechapt) leading to the domain of application and accuracy of these relationships.

Eight singularities widely used in distribution networks have also been investigated. They include sudden enlargements and contractions, 45° and 90° bends of different radiuses of curvature, Venturi meters, orifice meters, gate valves and ball valves. The results obtained are compared to those available in the literature.

2. Experimental Method

The experimental investigations were carried out on a test bench mainly constituted by a hydraulic bench providing the necessary flow discharge, a network of pipes of different dimensions and roughness able to simulate both frictional (major) and singular (minor) head losses.

The hydraulic bench used to distribute the water at the required flow rates, is principally constituted by a reservoir and two pumps which may be used in series or in parallel [4]. The water is pumped from reservoir to the pipes network through a closed circuit. A regulation valve, a direct rotameter, and water and mercury manometers are used to regulate the fluid flow and to determine the pressure in terms of head difference respectively.

The network is constituted by PVC pipes of different diameters and roughness. Four are pipes used for frictional head losses. Their diameter vary from 13 mm to 25 mm, and their surface roughness from smooth to $\varepsilon = 0.02mm$. The network of conduits also includes diverse appurtenances simulating singular head losses. It includes a Venturi meter, an orifice meter, a 45° bend, a ball valve, a sudden enlargement, a sudden contraction, two 90° bends of large and small radius of curvature respectively, a strainer, a gate valve and an open valve.

Measurements are carried out in terms of height using water and mercury manometers. They represent the difference of pressure between the respective positions where they are realized. The necessary flow discharges are obtained through two centrifugal pumps, and the flow rate is insured by a rotameter.

3. Theoretical Approach

3.1. Frictional or Major Head Losses

Frictional head losses are mainly due to the fluid viscosity and the flow regime. Their influence may be resented throughout the length of the pipe. In a long pipe, the frictional head losses are relatively important, and they cannot be neglected. A relationship expressing this loss is proposed by [25]. Known as the Darcy-Weisbach equation, it links the head loss the friction coefficient, the flow velocity and the pipe dimensions:

$$h_f = f \frac{L}{D} \frac{V^2}{2g} \qquad (1)$$

By expressing the velocity as a function of flow pipe section $(V = 4Q/\pi D^2)$ and by replacing the known parameters by their respective numerous values, equation (1) becomes:

$$j = \frac{f}{12.1D^5} Q^2 \qquad (2)$$

Dimensional analysis of the Darcy-Weisbach equation leads to:

$$h_f = \frac{L\,V^2}{D\;g}\Phi\!\left[\frac{\varepsilon}{D},R_e\right] \qquad (3)$$

Comparing equations (1) and (3) shows that the Darcy friction coefficient f is a function of both relative roughness and Reynolds number. [19] led the way in trying to express the friction coefficient by carrying out extensive experimentations leading to the measurement of the velocity distribution and head losses throughout the length of smooth and artificially roughened pipes. He applied the well-known Prandtl mixing length theory, and for smooth turbulent flow, developed a relationship for f which is unfortunately of implicit type requiring a recursive solution. For such a flow filed and for values of the Reynolds number less than 10^5, [3] reached a much more simple relation:

$$f = \frac{0.316}{R_e^{0.25}} \qquad (4)$$

However, almost all pipe flow regimes correspond to a transition zone where the frictional coefficient depends both on the Reynolds number and the relative roughness. [8] suggested a relationship in this direction since termed the Colebrook-White equation. It also had an implicit form, but was interesting since almost all the explicit solutions proposed later are approximations of this relation. [30] derived a complex relationship which has the advantage of being explicit and highly accurate for $4\times10^3 \le R_e \le 10^8$ and $0 \le \varepsilon/D \le 5\times10^{-2}$:

$$\frac{1}{\sqrt{f}} = -2\log\left\{\frac{\varepsilon}{3.7D} - \frac{5.02}{R_e}\log\left[\frac{\varepsilon}{3.7D} - \frac{5.02}{R_e}\log\left(\frac{\dfrac{\varepsilon}{3.7D}}{\dfrac{13}{R_e}}\right)\right]\right\} \qquad (5)$$

Other empirical relationships proposed later have the merit of expressing the head loss without the complexity introduced by the friction coefficient. They are mostly based on polynomial representation. Mostly used in the U.S.A., the relationship developed by [27] may be written as:

$$j = \frac{10.68}{C_{HW}^{1.852}\,D^{4.87}}\,Q^{1.852} \qquad (6)$$

The value of the Hazen-Williams coefficient C_{HW} is a function of the material and its dimensions. Its value is found to be 145, 146, 148 and 150 for PVC-roughened-14 mm-diameter pipes, PVC-roughened-25 mm -diameter pipes, PVC-smooth-13.3 mm -diameter pipes, and PVC-smooth-23.5 mm -diameter pipes respectively [2].

Another often used relationship is that known as the Manning equation, although its attribution to this later is contested as reported by [20] and [28]. It is expressed as:

$$j = \frac{10.29\,N^2}{D^{16/3}}\,Q^2 \qquad (7)$$

The Manning roughness coefficient is solely dependant on the pipe material. Its value is found to be 0095. 0 for the type of PVC used in the experimentations [2].

The other relationships investigated in this study are those due to:

Strickler [18]:

$$j = \frac{10.29}{K^2\,D^{16/3}}\,Q^2 \qquad (8)$$

Scobey, 1966:

$$j = \frac{1.58}{C_S^{1.886}\,D^{4.87}}\,Q^{1.887} \qquad (9)$$

Calmon and Lechapt [18]:

$$j = a\,\frac{Q^n}{D^m} \qquad (10)$$

where: $K = 0.95$ for the type of pipes and fittings used.

$C_S = 37$ for the new pipes used.

$a = 0.916.10^{-3}$; $n = 1.78$; $m = 4.78$ for the smooth pipes used.

$a = 1.01.10^{-3}$; $n = 1.84$; $m = 4.88$ for the roughned pipes used.

3.2. Singular Or Minor Head Losses

Head losses due to singularities or accessories are commonly termed minor head losses. This is due to the fact that for pipes of important length, their value can be neglected comparatively to that due to friction. However, their effect can be significant for short pipes. Minor head losses are expressed as:

$$h_m = K_m\,\frac{V^2}{2\,g} = \frac{8\,K_m}{\pi^2\,g\,D^4}\,Q^2 \qquad (11)$$

The constant K_m is a characteristic of the type of singularity, and its value varies consequently.

4. Results and Comparison

Figure 1 to Figure 4 present the results obtained concerning the frictional head losses for the four pipes investigated. They are presented in terms of the linear head loss $\left(h_f/L\right)$ versus the Reynolds number. All flow regimes were found to be turbulent.

For the two smooth pipes for which the results are presented in figure 1 and figure 2, the Hazen-Williams and Darcy approaches seem to be the more able to predicting the head loss. The integration of complex expressions for the friction coefficient with their large spectrum of application, like the one expressed in equation (5) [30], seems to be appropriate and ends up with satisfactory results.

It is nevertheless remarkable that the increase in the Reynolds number leads to higher errors. This is probably due to the fact that the flow field moves from the laminar/transitional regime to the turbulent regime. This difference can be corrected by integrating appropriate coefficients considering the change of regime, and this shows the complexity and the sensibility of the problem particularly when it is applied for the solution of distribution networks which may contain a great number of pipes with distinct diameters and roughness [12]. This problem is not met when one deals with the energy losses encountered in pipelines transporting petroleum since these have usually the same characteristics [11].

The same comments can be made for the case of the roughened pipes for which the results are presented in figure 3 and figure 4. The experimental results show head losses greater than those computed by the different empirical formulas used. The authors comment on this difference by the fact that the roughness coefficient which is the most important parameter might have not been suitably predicted. Its value has in fact been chosen equal to that suggested by the manufacturer, thus neglecting the possible presence of deposits which generally tends to increase the roughness of pipe walls and consequently the head loss.

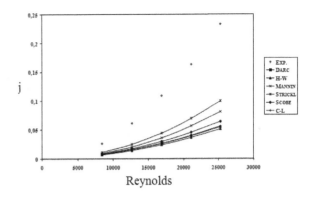

Figure 3. Head loss for a roughened pipe (D=25 mm-ε = 0.015 mm)

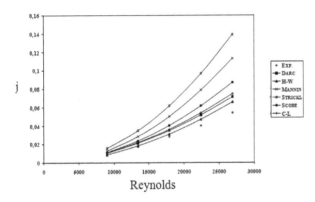

Figure 1. Head loss for a smooth pipe (D=23.5 mm)

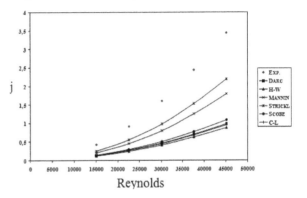

Figure 4. Head loss for a roughened pipe (D=14 mm-ε = 0.02 mm)

The capacity of transportation of pipelines decreases with time (age). This loss is mainly caused either by a diminution in the cross section area due to an accumulation of deposits inside the pipe, or an increase of the ruggedness, or both. The biological growth, the obstruction and the encrustation are the most common forms of such deposits which can vary from 1 mm to 10 mm in thickness. For the Hazen-Williams formula, the value of 140 is the most commonly used for C_{HW}. It is however understood that for an old pipe in good condition, a value comprised in the interval 120 100 - should be acceptable, while for a used pipe 80 40 - is mostly used. This situation shows the great difficulties faced by the authors in achieving an 'acceptable' value for the coefficients (C_{HW} as well as f and N) used in the developed relationships.

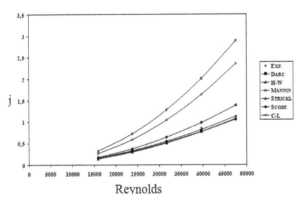

Figure 2. Head loss for a smooth pipe (D=13.3 mm)

Table 1. Head loss for Singularities

	Sudden Enlargement	Sudden Contraction	Sudden Bend (90°)	Progressive Bend (45°)	Venturi meter
	$\begin{cases} D_{upstream} = 13.3\,mm \\ D_{downstream} = 23.5\,mm \end{cases}$	$\begin{cases} D_{upstream} = 23.5\,mm \\ D_{downstream} = 13.3\,mm \end{cases}$	RC - median	-Long radius-	-Short pipe-
Experimental Result	0.21	0.39	0.70	0.197	0.61
	$V = 0.6\,m/s \div 1.3\,m/s$ $D = 1 \div 1.8$	$V = 0.6\,m/s \div 1.3\,m/s$ $D = 1.8 \div 1$	$V = 0.6\,m/s \div 1.3\,m/s$	With edge	$D : 1 \div 2$
Empirical results [réf]	0.40	0.34	0.75	0.18	0.72

The results obtained through the different singularities investigated are resumed in Table 1. Great difficulties have been faced by the authors when dealing with minor losses. Indeed, the measurements through such geometries have been found to be complex and, the literature dealing with such phenomenon is poor compared to that interested

in the friction losses. In the case of a sudden enlargement for example, the re-establishment of the velocity field takes a distance of approximately 100 the pipe diameter. Within this intermediate region, the flow is complex involving both friction and turbulence, and it is difficult to separate the effects of the latter from that due to friction.

Furthermore, the results published suggest values for the minor head loss coefficients generally neglecting important parameters such as material which may have a great influence upon the result.

The results proposed for the minor losses seem to be acceptable, except in the case of the sudden enlargement where there is a clear divergence. The authors feel that more interest should be given to this kind of local losses, and that future investigations should try to take into account all the factors involved in such head losses.

5. Conclusions

The present study investigated the different approaches developed for the determination of energy losses inside fluid transportation pipes. These losses may be caused either by friction or appurtenances.

The experimental results obtained and their comparison to those computed using the relationships developed by diverse investigators over the last decades show a preponderance of the relations proposed by Darcy and Hazen-Williams. The errors found seem to be mainly due to the difficulty to determining the friction coefficient which is a function of the Reynolds number and wall pipe roughness. The main difficulty which arises when trying to determine the pipe roughness is due to the fact that such pipes are subject to age effect resulting in erosion, corrosion, deposits, etc. The coefficient value proposed by the manufacturer is no more valid. Further research might well then be devoted to the phenomenon of aging of pipes.

More complexity is faced when head losses through singularities are investigated. The intermediate region downstream of any appurtenance is a mixture of friction and turbulence phenomena, and it is difficult to separate the effects of each one. Further research should be directed towards defining precise minor coefficients by taking into account all the factors involved in such head losses.

The results presented however do form a set of benchmark data for possible improvement and application in similar cases involving flow transportation inside pipes.

Notations

a,m,n: Calmon-Lechapt coefficients
C_{HW}: Hazen-Williams coefficients
C_S: Scobey coefficient
D: Pipe diameter (m)
f: Pipe frictional coefficient
g: Gravitational acceleration (m/s^2)
h_f: Pipe frictional head loss (m)
h_m: Pipe singular head loss (m)
j: Head loss per unit length (m/m)
K: Strickler coefficient
Ka: Karman number ($R_e \sqrt{f}$)
K_m: Singularity coefficient
L: Pipe length (m)
N: Manning roughness coefficient
Q: Fluid flow rate (m^3/s)
R: Radius (m)
R_e: Reynolds number
V: Pipe mean flow velocity (m/s)

ε: Pipe roughness
Φ: Function

References

[1]   Barr, D.I.H., 1975, "New Forms of Equation for the correlation of Resistance Data", *Proc. Inst. of Civil Engrs.*, 59 (2), pp 827-835.

[2]   Bhave, P.R., 1991, "Analysis of flow in water distribution networks", Technomic Pub. Co., Inc., USA.

[3]   Blasius, H., 1913, "DasAhnlichkeitsgesetz in Flussigkeiten", *Verein Deutscher Ingenieure, Forschungsheft*, p 131.

[4]   Boudebza C. and Bouachari, A., 1999, "Etude des approches de calcul des pertes de charge par frottement et singulières et comparaison avec l'expérimentation", Guelma Univ., Algeria.

[5]   Chen J.J.J., 1984, "A Simple Explicit Formula for the Estimation of Pipe Friction Factor", *Proc. Inst. of Civil Engrs.*, 77 (2), pp 49-55.

[6]   Chen, J.J.J., 1985, "Systematic Explicit Solutions of the Prandtl and Colebrook-White Equation for Pipe Flow", *Proc. Inst. of Civil Engrs.*, pp 383-389.

[7]   Churchill, S.W., 1977, "Friction Factor Equation Spans All Fluid Flow Regimes", *Chemical Engineering*, 84, pp 91-92.

[8]   Colebrook, C.F., 1939, "Turbulent Flow in Pipes with Particular reference to the Transition Region between the Smooth and Rough Pipe Laws", *J. Inst. of Civil Engrs.*, I1, pp 133-156.

[9]   Ger, A.M. a*nd Holly, E.R., 1976, "Comparison of single point injections in Pipe Flow"*, J. Hydraul. Div., *Am. Soc. of Civil Engrs.*, 102 (HY6), pp 731-746.

[10] Haaland, S.E., 1983, "Simple and Explicit Formulas for the Friction Factor in Turbulent Pipe Flow", *J. Fluids Engrs.*, pp 89-90.

[11] Haddad, A., Bouachari, A. and Boudebza, C., 1999, "Pertes de charge régulières et singulières le long du tronçon Oléoduc SP-3", Internal report, Guelma Univ., Algeria.

[12] Haddad A. and Lahiouel Y., 1999, "Analyse des réseaux de distribution d'eau par la méthode de Cross et l'approche linéaire", *Trans. of the 4th Mech. Engg.Cong.*, vol. 2, p 89, Mohammadia, Morocco.

[13] Jain, A.K., 1976, "Accurate Explicit Equation for Friction Factor", J. Hydraul. Div., *Am. Soc. of Civil Engrs.*, pp 674-677.

[14] Jain, A.K., Mohan, D.M. and Khanna, P., 1978, "Modified Hazen-Williams Formula", J.Environ. Engrg. Div., *Am. Soc. of Civil Engrs.*, pp 137-146.

[15] Johnson, S.P., 1934, "A Survey of Flow Calculation Methods", Preprinted Paper for Summer Meeting, Am. Soc. of Mech. Engrs.

[16] Lamont, P., 1969, "The Choice of the Pipe Flow Laws for Practical Use", *Water and Water Engineering*, pp 55-63.

[17] Moody, L.F., 1944, "Friction Factors for Pipe Flow", *Trans. American Society of Mechanical. Engineers.*, 66, pp 671-684.

[18] Morel, M.A. and Laborde, J.P., 1994, "Exercices de Mécanique des Fluides", Volume 1, Chihab-Eyrolles Editions, Algeria.

[19] Nikuradse, J., 1933, "Strmungsgesetze in Rauben Rohren", *Verein Deutsher Ingenieure, Forschungsheft*, p 361.

[20] Powell, R.W., 1968, "The Origin of Manning Formula", J. Hydraul. Div., *American Society of Civil Engrs.*, pp 1179-1181.

[21] Rouse, H., 1943, "Evaluation of Boundary Roughness", *Proc. 2nd Hydraul. Conf.*, Univ. of Lowa.

[22] Scobey, F.C., 1966, "The flow of water in commercially smooth pipes", Water Ressources Center Archives, Series report 17, Univ. of California, Berkeley, USA.

[23] Swamee, P.K. and Jain, A.K., 1976, "Explicit Equations for Pipe Flow Problems", *J.Hydraul. Div.*, Am. Soc. of Civil Engrs., 102 (HY11), pp 1707-1709.

[24] Techo, R., Tichner, R.R. and James, R.E., 1965, "An Accurate Equation for the Computation of Friction Factor for Smooth Pipes from the Reynolds Number", *J. of Applied Mechanics*, 32, pp 443.

[25] Weisbach J., 1855, "Die Experimental Hydraulik", Freiberg, Germany: Engelhardt.

[26] White, W.R., 1974, "The Hydraulic Characteristics of Clayware Pipes", Hydraulic Research Station, Report No. Int. 133, Wallingford, Berkshire, England.

[27] Williams, G.S. and Hazen, A., 1933, "Hydraulic tables", 3rd Edition, John Wiley & Sons nc., USA.

[28] Williams G.P., 1970, "Manning Formula-A Misnomer?", J. Hydraul. Div., *American Society of Civil Engrs.*, pp 193-200.

[29] Wood, D.J., 1972, "An Explicit Friction Factor Relationship", *Civil Engineering*, Am. Soc. of Civil Engrs. pp 383-390.

[30] Zigrang, D.J. and Sylvester, N.D., 1982, "Explicit Approximations to the Solution of Colebrook's Friction Factor Equation", *J. Am. Inst. of Chemical Engrs.*, pp 514-515.

26

Numerical Study of an Unconventional Savonius Wind Rotor with a 75° Bucket Arc Angle

Zied Driss[*], Olfa Mlayeh, Slah Driss, Makram Maaloul, Mohamed Salah Abid

Laboratory of Electro-Mechanic Systems (LASEM), National School of Engineers of Sfax (ENIS), University of Sfax (US), B.P. 1173, Road Soukra km 3.5, 3038 Sfax, TUNISIA
*Corresponding author: Zied.Driss@enis.rnu.tn

Abstract In this paper, computer simulation has been conducted to study the aerodynamic structure around an unconventional Savonius wind rotor with a 75° bucket arc angle. The numerical model is based on the resolution of the Navier-Stokes equations in conjunction with the standard k-ε turbulence model. These equations were solved by a finite volume discretization method. The software "Solidworks Flow simulation" has been used to characterise the flow characteristics in different transverse and longitudinal planes. The good comparison between numerical and experimental results confirms the validity of the numerical method.

Keywords: *unconventional Savonius wind rotor, bucket arc angle, wind tunnel, aerodynamic structure, CFD*

1. Introduction

The Savonius wind rotor has aroused a large credit, not only in research and academic communities but also in industrial appliances. In comparison to that of other kinds, the efficiency of Savonius rotor is lower. The reason of low efficiency mainly rests on the fact that one bucket moves against wind when another one moves in the direction of wind [2,3]. In this context, numerical and experimental investigations have been conducted to improve the Savonius wind rotor performance. For example, Khan et al. [4] tested Savonius rotor both in tunnel and natural wind conditions with the provision of variation of overlap. Research conducted by Grinspan et al. [5] in this direction led to the development of a new blade shape with a twist for the Savonius rotor. They obtained a maximum power coefficient of 0.5 for its model. Saha and Rajkumar [6] performed work on twist bladed metallic Savonius rotor and compared the performance with conventional semi-circular blades having no twist. They obtained an efficiency of 0.14. Their rotor also produced good starting torque and larger rotational speeds. Saha et al. [7] conducted wind tunnel tests to assess the aerodynamic performance of single, two and three-stage Savonius rotor systems. Both semicircular and twisted blades have been used. Aldos [8] studied power augmentation of Savonius rotor by allowing the rotor blades to swing back when on the upwind side. He reported a power augmentation of the order of 11.25% with the increase in Cp. He further concluded that different basic rotors configuration might produce different power augmentation. Sabzevari [9] examined the

effects of several ducting, concentrators and diffusers on the performance improvements of a split Savonius rotor. A circularly ducted Savonius rotor equipped with a number of identical wind concentrators and diffusers along the periphery of circular housing produced efficiency of the order of 40%. In order to eliminate the low aerodynamic performance of Savonius wind rotors, Mohamed et al. [10] studied several shapes of obstacles and deflectors placed in front of two and three blades Savonius turbine. A rounded deflector structure was placed in front of two counter-rotating turbines. An experimental investigation was carried out by Golecha et al. [11] to identify the position of the deflector plate to yield higher coefficient of power for single stage modified Savonius rotor. Akwa et al. [13] discussed the influence of the buckets overlap ratio of a Savonius wind rotor on the averaged moment and power coefficients, over complete cycles of operation. Kamoji et al. [14] compared the helical Savonius rotor with the conventional Savonius rotor. The results indicate that the helical Savonius rotors have positive coefficient of static torque. D'Alessandro et al. [15] developed a mathematical model of the interaction between the flow field and the rotor blades. The aim of their research was to gain an insight into the complex flow field developed around a Savonius wind rotor and to evaluate its performance. Irabu and Roy [16] improved and adjusted the output power of Savonius rotor under various wind power and suggests the method of prevention the rotor from strong wind disaster.

On the basis of the previous studies, it appears important to propose a new design to improve the performance of the conventional Savonius wind rotor. For this purpose, computer simulations have been conducted to study the aerodynamic structure around an unconventional Savonius wind rotor with a 75° bucket arc angle.

2. Geometrical Arrangements

Figure 1 proposes different designs of the Savonius wind rotor. This rotor is constituted by two half buckets characterized by the height H=300 mm, the diameter c=100 mm, the bucket arc angle ψ=75°, and the thickness e=6 mm. The two buckets are collected on a common axis, with a shaft diameter equal to s=10 mm, and they are fixed within screws to make an angle equal to 180°.

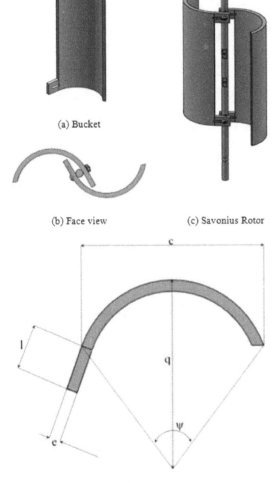

(a) Bucket

(b) Face view   (c) Savonius Rotor

(d) Geometrical arrangments

Parameter	Value
c	100 mm
q	100 mm
H	300 mm
l	24 mm
e	6 mm
ψ	75°

Figure 1. Modified Savonius wind trotor

3. Numerical Model

In this work, the software "Solidworks Flow Simulation" has been used to study the turbulent flow around unconventional Savonius wind rotors. This code is based on solving Navier-Stokes equations with a finite volume discretization method. The technique consists in dividing the computational domain into elementary volumes around each node in the grid; it ensures continuity of flow between nodes. The spatial discretization is obtained by following a procedure for tetrahedral interpolation scheme. As for the temporal discretization, the implicit formulation is adopted. The transport equation is integrated over the control volume [21-25]. The computational domain is defined by the interior volume of the wind tunnel blocked by two planes. The first one is in the tranquillization chamber entry and the second one is in the exit of the diffuser [18]. A boundary condition is required anywhere fluid enters or exits the system and can be set as a pressure and velocity. For the inlet velocity, we take as a value V=3 m.s-1, and for the outlet pressure a value of p=101325 Pa is considered which means that at this opening the fluid exits the model to an area of static atmospheric pressure.

4. Results and Discussions

In this section, we are interested to study the aerodynamic characteristics of the flow such as the velocity fields, the average velocity, the total pressure, the dynamic pressure, the turbulent kinetic energy, the dissipation rate of the turbulent kinetic energy, the turbulent viscosity and the vorticity. For thus, three planes were considered defined by z=0 mm, x=0 mm, and y=0 mm (Figure 2). According to the air flow direction, the first is a transverse plane; however, the two others are longitudinal planes.

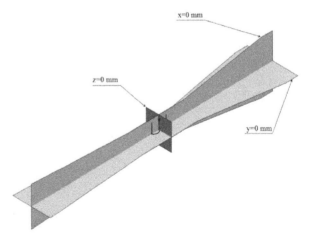

Figure 2. Visualization planes

4.1. Velocity Fields

Figure 3 shows the distribution of the velocity field in the planes defined by x=0 mm, y=0 mm and z=0 mm. While examining these results, it can easily be noted that the velocity is weak in the inlet of the collector. In fact, it is governed by the boundary condition value of the inlet velocity which is equal to V=3 m.s-1. In this region, the velocity field is found to be uniform and increases progressively downstream of the collector. At the test vein, an important increase of the velocity value has been noted due to the reduction of the tunnel section that causes the throttling of the flow. While the upstream of the rotor is characterized by the high velocity, a brutal drop is located in the concave surface of the two buckets. Downstream of the rotor, the velocity value keeps increasing till the out of the test section. Then, a sharp decrease has been noted through the diffuser where the minimum velocity values are recorded in the lateral walls of the diffuser. The maximum velocity values are located in the convex surface of the two buckets according to the distribution shown on the y=0 mm plane. The velocity fields on the longitudinal plane, defined by y=0 mm, show a symmetric distribution. Also, a deceleration of the velocity field around the shaft and in the concave surface of the buckets has been observed. Another zone is located in the top and the left wall of the test vein which corresponds to the zone around the advancing bucket.

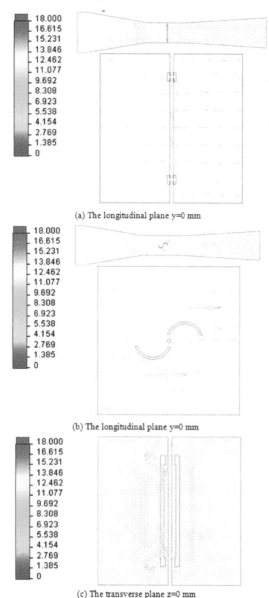

(a) The longitudinal plane y=0 mm

(b) The longitudinal plane y=0 mm

(c) The transverse plane z=0 mm

Figure 3. Distribution of the velocity field

4.2. Average Velocity

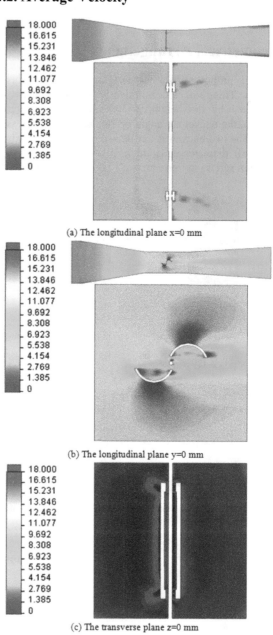

(a) The longitudinal plane x=0 mm

(b) The longitudinal plane y=0 mm

(c) The transverse plane z=0 mm

Figure 4. Distribution of the Magnitude velocity

Figure 4 shows the distribution of the magnitude velocity in the planes defined by x=0 mm, y=0 mm and z=0 mm. According to these results, it has been noted that the velocity is weak in the inlet of the collector. In this region, the average velocity field is found to be uniform and increases progressively downstream of the collector. At the test vein, an important increase of the velocity value has been noted due to the reduction of the tunnel section that causes the throttling of the flow. While the upstream of the rotor is characterized by the high velocity, a brutal drop is located in the concave surface of the two buckets. Downstream of the rotor, the average velocity value keeps increasing till the out of the test section. Then, a sharp decrease has been noted through the diffuser where the minimum average velocity values are recorded in the lateral walls of the diffuser. The wakes characteristics of the maximum average velocity values are located in the convex surface of the two buckets according to the distribution shown in the longitudinal plane defined by y=0 mm. However, the wakes characteristics of the minimum average velocity values are located in the concave surface of the buckets and in the bucket sides.

4.3. Total Pressure

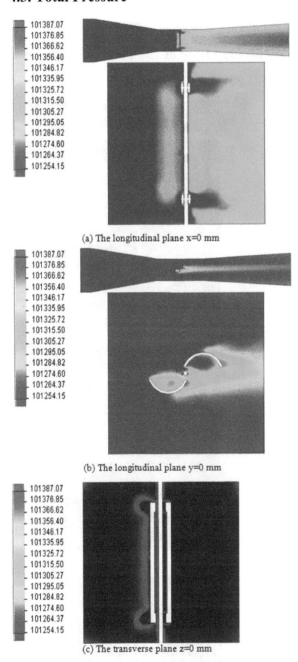

(a) The longitudinal plane x=0 mm

(b) The longitudinal plane y=0 mm

(c) The transverse plane z=0 mm

Figure 5. Distribution of the total pressure

Figure 5 shows the distribution of the total pressure in the planes defined by x=0 mm, y=0 mm and z=0 mm. According to these results, it has been noted that the total pressure is on its maximum in the intake and is globally uniform in the collector and the upstream of the rotor in the test vein. A brutal drop of the total pressure has been noted in the concave surface of the rotor, downstream of the advancing bucket and around the axis of the rotor. The distribution of the total pressure in the longitudinal planes

defined by x=0 cm and y=0 cm shows that the depression zones are located in the downstream of the concave surface of the returning bucket, the convex surface of the advancing bucket and the extremities of the rotor axis. Downstream of the rotor, the total pressure starts to increase gradually in the way out of the test vein and keeps increasing through the diffuser. The transverse planes show the formation of depression zones near the rotor axis and the buckets. Around the rotor, the total pressure is found to be relatively high and is increasing in the way out of the test vein.

4.4. Dynamic Pressure

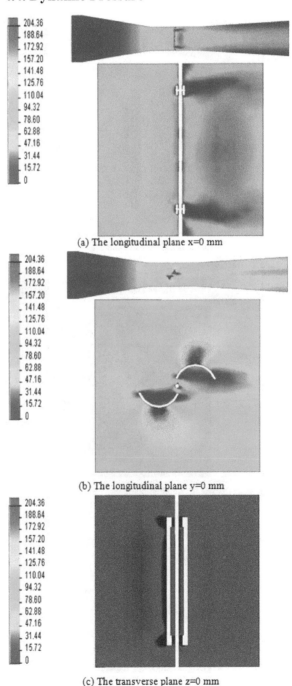

(a) The longitudinal plane x=0 mm

(b) The longitudinal plane y=0 mm

(c) The transverse plane z=0 mm

Figure 6. Distribution of the dynamic pressure

Figure 6 shows the distribution of the dynamic pressure in the planes defined by x=0 mm, y=0 mm and z=0 mm.

According to these results, the dynamic pressure is found to be weak in the collector inlet and increases gradually through the collector as long as the tunnel section gets smaller. When it gets to the test section, the dynamic pressure keeps increasing in the upstream of the rotor and around it. It reaches its maximum in the convex surface of both the advancing and returning buckets. A depression zone is recorded in the concave surface of the rotor buckets and around the shaft. Downstream of the rotor, the dynamic pressure remains relatively low in the test section and through the diffuser.

4.5. Turbulent Kinetic Energy

mm. According to these results, it is clear that the turbulent kinetic energy is found to be very weak in the wind tunnel except in the area surrounding the rotor and the diffuser exit. The distribution of the turbulent kinetic energy shows the increase of the turbulent kinetic energy near the rotor. The wake characteristic of the maximum value of the turbulent kinetic energy is recorded around the axis and along the concave surface especially in the advancing bucket as shown in the distribution in the longitudinal plane defined by y=0 mm.

4.6. Dissipation Rate of the Turbulent Kinetic Energy

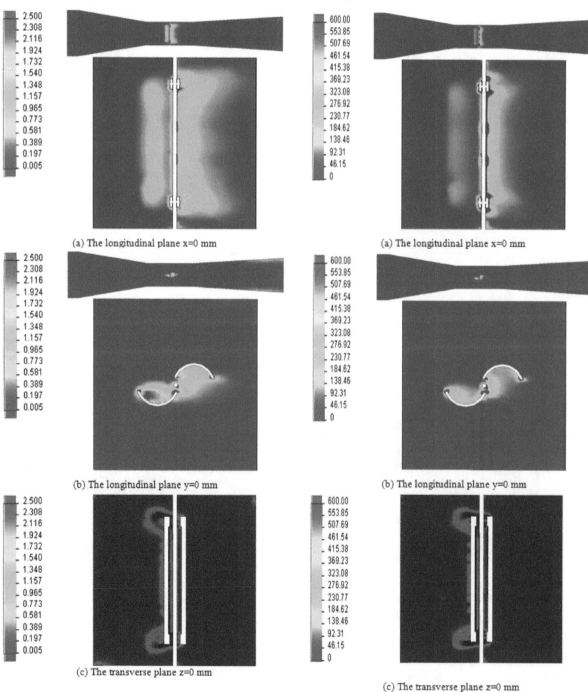

(a) The longitudinal plane x=0 mm

(b) The longitudinal plane y=0 mm

(c) The transverse plane z=0 mm

Figure 7. Distribution of the turbulent kinetic energy

(a) The longitudinal plane x=0 mm

(b) The longitudinal plane y=0 mm

(c) The transverse plane z=0 mm

Figure 8. Distribution of the dissipation rate of the turbulent kinetic energy

Figure 7 shows the distribution of the turbulent kinetic energy in the planes defined by x=0 mm, y=0 mm and z=0

Figure 8 shows the distribution of the turbulent dissipation rate in the planes defined by x=0 mm, y=0 mm and z=0 mm. According to these results, it has been noted that the turbulent dissipation rate is very weak in the wind tunnel except in the area surrounding the Savonius rotor. The wake characteristic of the maximum values of the turbulent dissipation rate of the turbulent kinetic energy appears around the rotor axis and in the end plates of the buckets.

4.7. Turbulent Viscosity

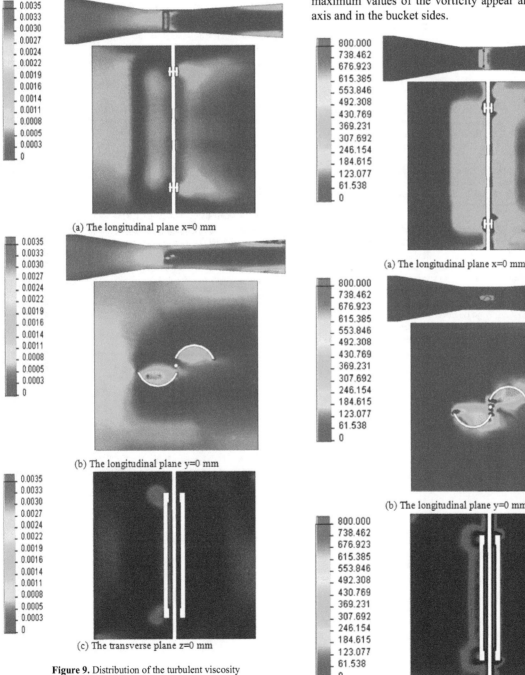

Figure 9. Distribution of the turbulent viscosity

Figure 9 shows the distribution of the turbulent viscosity in the planes defined by x=0 mm, y=0 mm and z=0 mm. According to these results, it has been noted that the viscosity is weak in the collector. It increases upstream and near the rotor and when leaving the collector. The

wake characteristic of the maximum values of the turbulent viscosity in the test section is recorded in the concave surface of the buckets. In the rotor downstream, a rapid decrease has been observed.

4.8. Vorticity

Figure 10 shows the distribution of the vorticity in the planes defined by x=0 mm, y=0 mm and z=0 mm. According to these results, it has been noted that the vorticity is very weak in the wind tunnel except in the region around the rotor. The wakes characteristics of the maximum values of the vorticity appear around the rotor axis and in the bucket sides.

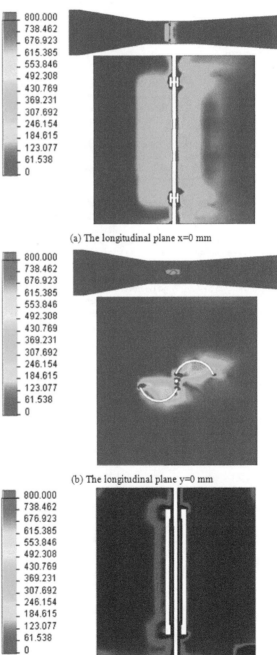

Figure 10. Distribution of the Vorticity

5. Conclusions

Computer simulations have been conducted to study the aerodynamic structure around an unconventional Savonius wind rotor with a 75° bucket arc angle. In this work, it has been observed that the bucket design has a direct effect on the local characteristics. Particularly, it has been noted that the depression zones are located in the concave surface of the bucket and downstream of the rotor. The acceleration zone, where the maximum velocity values are recorded, is formed in the convex surface of the rotor bucket.

In the future, we propose to use the particle image velocimetry laser (PIV) system for a finer analysis of the local air flow characteristics around unconventional Savonius wind rotors.

References

[1]   Twidwell JW, Weir AD. Renewable Energy Resources, The University Press Cambridge, Britain (1985) 1-411.

[2]   Khan N, Tariq Iqbal M, Hinchey M, Masek V. Performance of Savonius Rotor as Water Current Turbine. Journal of Ocean Technology 4 (2009) 27-29.

[3]   Grinspan AS, Kumar PS, Saha UK, Mahanta P, Ratnarao DV, Veda Bhanu G. Design, development & testing of Savonius wind turbine rotor with twisted blades. Proceedings of national conference on fluid mechanics & fluid power, India, 28 (2001) 28-31.

[4]   Saha UK, M. Jaya Rajkumar, On the performance analysis of Savonius rotor with twisted blades, Renewable Energy 31 (2006) 1776-1788.

[5]   Saha UK, Thotla S, Maity D. Optimum design configuration of Savonius rotor through wind tunnel experiments. Journal of Wind Engineering and Industrial Aerodynamics 96 (2008) 1359-1375.

[6]   Aldos TK. Savonius Rotor Using Swinging Blades as an Augmentation System. Wind Engineer 8 (1984) 214-220.

[7]   Sabzevari A. Power augmentation in a ducted Savonius rotor, Second international symposium on wind energy systems (1978) 1-6.

[8]   Mohamed MH, Janiga G, Pap E, Thévenin D. Optimization of Savonius turbines using an obstacle shielding there turning blade, Renewable Energy 35 (2010) 2618-2626.

[9]   Golecha K, Eldho TI, Prabhu SV. Influence of the deflector plate on the performance of modified Savonius water turbine. Applied Energy 88 (2011) 3207-3217.

[10]   Akwa JV, Júnior GA, Petry AP. Discussion on the verification of the overlap ratio influence on performance coefficients of a Savonius wind rotor using computational fluid dynamics. Renewable Energy 38 (2012) 141-149.

[11]   Kamoji MA, Kedare SB, Prabhu SV. Performance tests on helical Savonius rotors. Renewable Energy 34 (2009) 521-529.

[12]   D'Alessandro V, Montelpare S, Ricci R, Secchiaroli A. Unsteady Aerodynamics of a Savonius wind rotor: a new computational approach for the simulation of energy performance. Energy 35 (2010) 3349-3363.

[13]   Irabu K, Roy JN. Characteristics of wind power on Savonius rotor using a guide-box tunnel, Experimental Thermal and Fluid Science 32 (2007) 580-586.

[14]   Driss Z, Abid MS. Numerical and experimental study of an open circuit tunnel: aerodynamic characteristics. Science Academy Transactions on Renewable Energy Systems Engineering and Technology 2 (2012) 116-123.

[15]   Driss Z, Damak A, Karray S, Abid MS. Experimental study of the internal recovery effect on the performance of a Savonius wind rotor. Research and Reviews: Journal of Engineering and Technology 1 (2012) 15-21.

[16]   Damak A, Driss Z, Abid MS. Experimental investigation of helical Savonius rotor with a twist of 180°, Renewable Energy 52 (2013) 136-142.

[17]   Driss Z, Mlayeh O, Driss D, Maaloul M, Abid MS. Numerical simulation and experimental validation of the turbulent flow around a small incurved Savonius wind rotor, Energy 74 (2014) 506-517.

[18]   Driss Z, Ammar M, Chtourou W, Abid MS. CFD Modelling of Stirred Tanks. Engineering Applications of Computational Fluid Dynamics 5 (2011) 145-258.

[19]   Driss Z, Abid MS. Use of the Navier-Stokes Equations to Study of the Flow Generated by Turbines Impellers. Navier-Stokes Equations: Properties, Description and Applications 3 (2012) 51-138.

27

Mechanical Stress Analysis of Tree Branches

Zahra Shahbazi[1,*], Allison Kaminski[1], Lance Evans[2]

[1]Manhattan College, Mechanical engineering department, Riverdale, NY, United States
[2]Manhattan College, Biology department, Riverdale, NY, United States
*Corresponding author: Zahra.shahbazi@manhattan.edu

Abstract Various models have been developed to calculate stresses due to weight along tree branches. Most studies have assumed a uniform modulus of elasticity and others have assumed that branches are tapered cantilever beams orientated horizontally or at an angle. Astress model was evaluated in which branches are curved and that the modulus of elasticity may vary along the branch. For this model, the cross-sectional areasof branches were divided into concentric rings in which the modulus of elasticity may vary. Next, areas of rings were transformed according to their modulus of elasticity. Branches with curved shapes were also considered and best fit lines for branch diameters were developed. A generated diameter equation was used in the stress calculations to provide realistic results. From these equations, a Graphical User Interface (GUI) in Matlab, was developed to calculate stress within tree branches. Moreover, a Finite Element Model (FEM) was created in Abaqus to compare with the models.

Keywords: *stress analysis, tree branches, Finite Element*

1. Introduction

A tool used to accurately calculate the stress on tree branches from measured dimensions and properties can have multiple applications. This tool will be beneficial to understand relationships between tree morphology and branch stress [1].

Previous studies calculated tree branch stresses by examining them as tapered cantilever beams of either an elliptical or circular cross sections. These stress calculations assumed the tree branch was of a uniform material with a uniform Modulus of Elasticity (MOE) value [1]. The MOE is a measure of the stiffness of a material. Materials with a small MOE bend more easily while materials with a large MOE are stiffer. This study accounts for varying MOE values within tree branches. Multiple studies have shown that MOE depends on the age of the wood and its location within branches and among branches. As a tree branch grows, new wood layers are produced toward the outside [2]. The innermost (oldest) wood has a small MOE value, and has little mechanical significance to resist bending. The outermost wood resists a majority of the bending in a tree branch [3]. Therefore at a given branch cross section, MOE of the branch varies in the radial direction - as the radius increases, MOE values increase.

This study considers radial variances in MOE when calculating stress. Lower tree branches have a greater variance of MOE in the radial direction than branches located nearer the tops of trees [2]. Cases that include differences in MOE will provide a better characterization of all tree branches.

As mentioned above, in addition to modeling tree branches as being composed of a uniform material, previous studies modeled tree branches as tapered cantilevered beams [1,2]. Branches in nature do not resemble straight lines, they are curved. In this study we also analyzed stress for curved branches.

Here, we propose six cases, the first considering the varying MOE of branches in the radial direction and the second considering the curviness of branches. Four additional cases are used for branches with less complicated geometries and for comparison purposes. These cases include a cantilever beam of fixed circular and elliptical cross sections, as well as a tapered cantilever beam of circular and elliptical cross sections. For all of the six cases we developed analytical equations to calculate stress. Also, aGraphical User Interface (GUI) is developed using Matlab where users can simply select one of the six above cases and perform analytical stress analysis.

In addition, a Finite Element (FE) simulation is developed using Abaqus. This FE model enables the user to analyze more complicated geometries in three dimensions. To make the FE modelmore readily usable by a wide audience, another GUI in Matlab is developed in which users enter geometry and mechanical properties (as an excel file) and receive a python file which is simply executable in Abaqus. Therefore, any user with limited knowledge of FE and Abaqus will be able to perform stress analysis on tree branches.

Both developed analyticaland numerical methods are used to study stress along tree branches.

2. Method

2.1. Analytical Analysis

Six tree branch cases were created. Case 1 is a fixed circular cross section. Case 2 is a fixed elliptical cross section. Case 3 is a tapered circular cross section. Case 4 is a tapered elliptical cross section. Case 5 is a fixed circular cross section with a non-uniform material where the MOE varies in the radial direction. Case 6 is a curved branch of uniform material and a circular cross section.

Equations to calculate stress were derived for each cases. These equations were used to write a code in Matlab that allowed measurable dimensions of each modeled branch. Stress analysis at specified locations was determined. The Matlab code was able to calculate stress at the top, bottom and sides of any desired cross section at all locations along the length of the branches.

The model only considered stress due to the weight of the branch. No external loads were applied. The weight of the branch was analyzed as a distributed load that acts vertically downward. To perform the analysis, the weight was separated into two components, one parallel to the axis of the branch and one perpendicular to the branch (Figure 1).

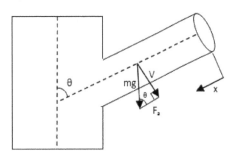

Figure 1. Tree branch model showing the force due to weight acting on the branch along with its components

The component of the distributed weight that acts perpendicular to the branch and parallel to the cross sectional area is the shearing distributed load, $w_s(x)$. The component of the distributed weight that acts parallel to the branch and perpendicular to the cross sectional area is the axial distributed load, $w_a(x)$. The shearing distributed load contributes to the normal stress due to bending, while the axial distributed load contributes to the normal compressive stress on the branch. Both stresses were needed to be considered in order to calculate the total normal stress acting on a given cross section.

General equations were derived to calculate stress for each of the six cases. All of the equations derived assume a uniform MOE throughout the branch. To calculate the axial compressive stress acting on a cross section of a tree branch, the general equation for the axial distributed load, the load per unit length Equation (1) was

$$w_a(x) = \frac{mg\cos\theta}{L} = \frac{\rho V(x)g\cos\theta}{x} = \frac{\rho A(x)xg\cos\theta}{x} \quad (1)$$

where m is the total mass of the branch, L is the total length of the branch, g is the gravitational constant, x is the distance from the tip of the branch to the where the stress analysis is desired, ρ is the density, $V(x)$ is the volume of the branch up to the specified point x, $A(x)$ is the branch cross sectional area and θ is the angle of the tree branch with respect to vertical. If the branch is perfectly horizontal then θ will be 90 degrees, making the

axial distributed load zero. Using Equation (1) the axial force acting at any location from the tip of the branch, $F_a(x)$, was determined from Equation (2) [4].

$$F_a(x) = -\int w_a(x)dx \quad (2)$$

The normal axial stress due to the axial component of the weight is [4]

$$\sigma_a(x) = \frac{F_a(x)}{A(x)} \quad (3)$$

The normal stress due to bending is caused by the component of weight acting perpendicular to the branch. Shearing distributed load, $w_s(x)$, which acts perpendicular to the length of branch was calculated using

$$w_s(x) = \frac{mg\sin\theta}{L} = \frac{\rho V(x)g\sin\theta}{x} = \frac{\rho A(x)xg\sin\theta}{x} \quad (4)$$

The variables are the same as for the distributed axial load. Shear force $V_s(x)$, is the component of weight that acts perpendicular to the length of the branch and parallel to the cross sectional area. Shear force was determined from the shearing distributed load [4].

$$V_s(x) = -\int w_s(x)dx \quad (5)$$

The bending moment in terms of distance from the tip of the branch (Equation 6) was obtained by integrated the shear force, $V_s(x)$, with respect to distance from the tip of the branch, x [4].

$$M(x) = \int V(x)dx \quad (6)$$

The normal stress due to bending was expressed as [4]

$$\sigma_b(x) = \pm\frac{M(x)c(x)}{I(x)} \quad (7)$$

where $c(x)$ is the radial distance from the center of the cross section to the location in the radial direction where the stress analysis is desired. For instance, to obtain the bending stress at the top of the cross section $c(x)$ would be the radial distance from the center of the cross section to the top of the cross section. $I(x)$ is the moment of inertia of the cross section. Normal stress due to bending is compressive (negative) or tensile (positive) depending on the location of the cross section being examined. Tops of branches will be in tension and bottoms will be in compression. The neutral axis is where the stress passes from positive to negative and does not experience the effects of bending. The left and right sides of the cross section lie along the neutral axis and therefore have a bending stress of zero.

Total normal stress acting on a cross section is the sum of the stress due to bending and axial stresses. When adding these two stresses together the sign convention of each stress must be considered.

Equations (1) through (7) were general equations used to perform the stress analysis for each of the cases. For some of the cases additional analyses were needed. For case 1 the generic procedure described above can be followed exactly.

Case 2 was a fixed elliptical cross section. The equations described above were used by substituting in the cross sectional area of an ellipse, $A = \pi R_V R_H$, and the

moment of inertia for an ellipse, $I = \frac{\pi}{4}R_H R_V^3$ [4].

Cases 3 and 4 were tapered; therefore additional calculations have to be considered. Case 3 was tapered and had a circular cross sectional area (Figure 2).

The angle of taper was calculated using [1]

$$\tan\alpha = \frac{R(x)}{x} = \frac{R_o}{L} \qquad (8)$$

where R_o is the radius at the base of the branch. Rearranging for the radius of the branch in terms of distance from the tip of the branch, $R(x)$, gives Equation (9) [1].

$$R(x) = \frac{R_o x}{L} \qquad (9)$$

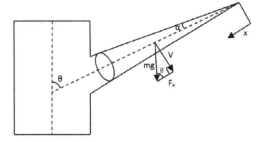

Figure 2. Case 3 tree branch model, tapered with a circular cross section. Where α is the angle of taper

The cross sectional area in terms of distance from the tip of the branch then became [1]

$$A(x) = \pi R(x)^2 = \pi\left(\frac{R_o x}{L}\right)^2 \qquad (10)$$

The moment of inertia also varied with x and had to be calculated using $R(x)$. By making these adjustments to the generic equations the stress analysis was performed for case 3.

Case 4 modeled the branch as a tapered elliptical beam. For case there were two angles of taper, one in the x-y plane and another in the x-z plane (Figure 3).

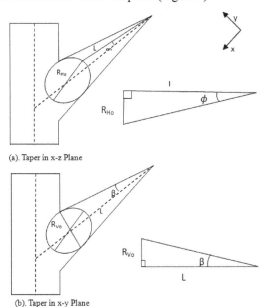

(a). Taper in x-z Plane

(b). Taper in x-y Plane

Figure 3. Case 4 taper in the x-y planes and x-z planes. R_{Ho} and R_{Vo} are the radii in the horizontal and vertical directions respectively at the base of the tree branch. φ is the angle of taper in the horizontal direction. β is the angle of taper in the vertical direction

The angle of taper in the x-z plane was determined from the horizontal radius of the ellipse and was calculated using Equation (11).

$$\tan\phi = \frac{R_H(x)}{x} = \frac{R_{Ho}}{L} \qquad (11)$$

Where ϕ is the angle of taper in the x-z plane, R_{Ho} is the horizontal radius at the base of the branch, and $R_H(x)$ is the horizontal radius with respect to distance from the tip of the branch. Equation (12) provides the horizontal radius at any location, x, along the branch.

$$R_H(x) = \frac{R_{Ho} x}{L} \qquad (12)$$

The angle of taper in the x-y plane was calculated using

$$\tan\beta = \frac{R_V(x)}{x} = \frac{R_{Vo}}{L} \qquad (13)$$

Where β is the angle of taper in the x-y plane, R_{Vo} is the vertical radius at the base of the branch, and $R_V(x)$ is the vertical radius with respect to distance from the tip of the branch.

$$R_V(x) = \frac{R_{V0}(x)}{L} \qquad (14)$$

Radii of the branch in the horizontal and vertical directions varied along the length of the branch; therefore, the cross sectional area of the branch needed to be expressed in terms of x, the distance from the tip of the branch as

$$A(x) = \pi R_V(x) R_H(x)$$
$$= \pi\left(\frac{R_{V0}(x)}{L}\right)\left(\frac{R_{Ho}(x)}{L}\right) = \frac{\pi R_{V0} R_{Ho}}{L^2} x^2 \qquad (15)$$

The moment of inertia equation for the branch became

$$I = \frac{\pi}{4} R_H(x) R_V(x)^3 \qquad (16)$$

Substituting these equations into the generic equations the stress analysis was performed for case 4.

Case 5 was a fixed circular cross section with a non-uniform material. The MOE varied in the radial direction (Figure 4).

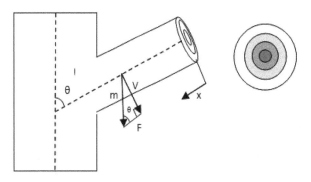

Figure 4. Case 5, a tree branch model of a fixed circular cross section and varying MOE in the radial direction

The cross section of the branch was separated into concentric rings of equal width in which the MOE may vary. The derived equations assumed the branch was made of a uniform material. In order to use these equations the moment of inertia of each ring had to be transformed based on MOE values. Then each branch was analyzed as being composed of a uniform material and the generic equations for the stress analysis was used.

The Matlab program for case 5 was written so that the branch was subdivided into many rings. Therefore based on the outer diameter of the branch and the desired number of rings as inputs, the radius of each ring was calculated. The outer radius of each concentric ring is

$$r_h = h \times \frac{R}{m} \qquad (17)$$

where m represents the total number of rings in the cross section and h represents the number assigned to each ring. The inner-most ring is $h=1$, while for the outermost ring $h=m$. R represents the outer radius of the branch. The area of each ring was calculated.

The innermost ring had the smallest MOE value and therefore its moment of inertia was not transformed. All the other rings hadlarger MOE values than the innermost ring. To account for larger MOE values, a ratio of each ring's MOE value to the MOE value of the innermost ring were used to transform the moment of inertia of each section. For example, if a ring had a MOE that was two times greater than that of the innermost ring, then the moment of inertia of the outer ring was needed to be two times greater as well. After transforming the moment of inertia values, the branch was analyzed as having a uniform material. The ratio of the MOE of an outer ring to the MOE of the innermost section was calculated as [4]

$$n_h = \frac{E_h}{E_1} \qquad (18)$$

Note for $h=1$ the ratio became 1. The transformed moment of inertia became [4]

$$I_h = n_h \frac{\pi}{4}\left(r_h{}^4 - r_{h-1}{}^4\right) \qquad (19)$$

The new moment of inertia of the entire cross section is the sum of the transformed moments of inertia for each ring.

$$I_{new} = I_1 + I_2 + \ldots + I_m \qquad (20)$$

The stress due to the bending moment on each ring was calculated using the new moment of inertia and the MOE ratio [4].

$$\sigma_{b_h}(x) = \pm \frac{n_h M(x) r_h}{I_{new}} \qquad (21)$$

The axial stress on each ring was calculated by considering the various MOE values of each ring. First the distributed axial load, $w_a(x)$, and the total axial force, $F_a(x)$, were calculated using the total area of the cross section, $A_{total}(x)$. These variables are independent of the MOE of the branch.

The area and MOE of each ring was considered in all calculations of the axial force acting on its section. The sum of the axial forces acting on each ring was equivalent to the total axial force (Equation 22).

$$F_1(x) + F_2(x) + \ldots + F_m(x) = F_a(x) \qquad (22)$$

where $F_a(x)$ is the total axial force due to the weight of the branch. Forces $F_1(x)$ through $F_m(x)$ were the compressive

forces acting on each ring. The forces on each ring caused the branch to deform along the axis of the branch, making the branch shorter. This deformation was calculated as [4]

$$\delta_h = \frac{F_h L}{A_h E_h} \qquad (23)$$

where δ_h is the deformation of the ring, F_h is the axial force acting on the ring, A_h is the area of the ring, and E_h is the MOE of the ring. Since all of the rings were connected to one another they allhad the same deformation.

$$\delta_1 = \delta_2 = \ldots = \delta_m$$
$$\frac{F_1 L}{A_1 E_1} = \frac{F_2 L}{A_2 E_2} = \ldots = \frac{F_m L}{A_m E_m} \qquad (24)$$

Using equation 24, the axial force on each ring was determined. The axial force on any of the rings in terms of the axial force on ring 1 is

$$F_h = \frac{A_h E_h}{A_1 E_1} F_1 \qquad (25)$$

The total axial force on the branch in terms of the force on ring 1 became

$$F_a = F_1 + \frac{A_2 E_2}{A_1 E_1} F_1 + \ldots + \frac{A_m E_m}{A_1 E_1} F_1 \qquad (26)$$

Equation 26was rearranged to solve for the axial force acting on ring 1 (Equation 27).

$$F_1 = \frac{F_a}{1 + \frac{A_2 E_2}{A_1 E_1} + \ldots + \frac{A_m E_m}{A_1 E_1}}$$
$$F_1 = \frac{F_a}{\frac{1}{A_1 E_1}\left(A_1 E_1 + A_2 E_2 + \ldots + A_m E_m\right)} \qquad (27)$$

This procedure was repeated to determine the axial forces acting on each ring. Using the axial force acting on each ring the axial stress on each ring was calculated as

$$\sigma_{a\,h}(x) = -\frac{F_h(x)}{A_h} \qquad (28)$$

The total stress for case 5 is the sum of the axial and bending stresses.

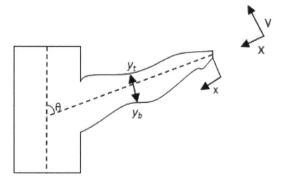

Figure 5. Case 6: Curvy tree branch model, where y_t is the height of the branch on top of the axis,y_b is the height below the axis

Case 6 accounted for the curves of a branch in the x-y plane, seen from the side view. The curviness of the branch was determined by measuring the height of the branch above and below the axis of the branch at incremental distances along its length. The axis line was the line that connects the center of the cross section at the base of the branch to the center of the cross section at the tip of the branch (Figure 5).

The distance from the axis to the top of the branch is y_t. The distance from the axis to the bottom of the branch is y_b. The diameter of the branch is the sum of y_t and y_b. The branch was assumed to have a circular cross section in this case because the varying heights of the branch were only considered in the x-y plane (side view), not the x-z plane (top view).

To perform a stress analysis for case 6, stresses were calculated at the base of the branch, where $x=L$. At this location the axis of the branch coincided with the center of the branch cross section. The axis was defined earlier as the line that connects the centers of the cross sections at the base of the branch and at the tip of the branch. Performing the stress analysis at the base provided the maximum stresses experienced by the branch.

To account for the curviness of the branch a best fit polynomial equation was determined for y_t and y_b in terms of x. Because the branch is curved, the centroid of the branch may not lie on the branch axis. Therefore the axial component of weight applied at the centroid must be moved to the axis to perform the stress analysis. To account for moving the forces an additional bending moment must be added.

The centroid of the area in the x-y plane was determined by first calculating the centroid of the areas above and below the axis separately. Then the two centroids were combined to obtain the centroid of the entire branch. The areas of the top and bottom parts of the branch must be calculated separately, by integrating the y_t and y_b equations to find the area under the curve. Using the areas the x-coordinate of the centroid for the top (Equation 29) and bottom (equation 30) portions of the branch were calculated using [4]

$$\bar{x}_t = \frac{1}{A_t} \int_A x f_t(x) dx \qquad (29)$$

$$\bar{x}_b = \frac{1}{A_b} \int_A x f_b(x) dx \qquad (30)$$

where y_t and y_b are both functions of x, $y_i = f_i(x)$. To determine the y-coordinate of the centroid, two additional best fit polynomial equations were determined for x in terms of y_t and y_b, giving $x_i = f_i(y)$. The y-coordinates of the

centroids above (Equation 31) and below (Equation 32) to the axis line are [4]

$$\bar{y}_t = \frac{1}{A_t} \int_A y f_t(y) dy \qquad (31)$$

$$\bar{y}_b = \frac{1}{A_b} \int_A y f_b(y) dy \qquad (32)$$

Next the x and y coordinates of centroid, relative to the branch axis, were calculated for the entire branch (Equations 33 and 34) [4].

$$\bar{x} = \frac{\sum_i A_i \bar{x}_i}{\sum_i A_i} = \frac{(A_t \times \bar{x}_t) + (A_b \times \bar{x}_b)}{(A_t + A_b)} \qquad (33)$$

$$\bar{y} = \frac{\sum_i A_i \bar{y}_i}{\sum_i A_i} = \frac{(A_t \times \bar{y}_t) + (A_b \times \bar{y}_b)}{(A_t + A_b)} \qquad (34)$$

To perform the stress analysis, the axial force, F_a, was moved from the centroid to the center of the branch's cross section. The perpendicular distance that F_a must be moved was $\bar{y}$. Therefore the additional bending moment that must be added to account for moving the axial force was

$$M_{addtional} = F_a \times \bar{y} \qquad (35)$$

The bending moment that was due to the shear force (Equation 36) was the product of the shear force and the perpendicular distance between the x-coordinate for the centroid and the base of the branch.

$$M_{shear} = V_s \times (L - \bar{x}) \qquad (36)$$

Recall that $\bar{x}$ was measured from the tip of the branch, therefore it was subtracted by the length to get the distance between the base and the centroid.

The total bending moment was the sum of these two bending moments where their signs were considered. After the total bending moment was calculated, the stress due to bending was determined. The axial stress was calculated using Equation 3 after the force was moved to the center of the cross section. The total normal stress acting on the cross section at the base was the sum of normal stress due to bending and the axial stress in which the signs were considered.

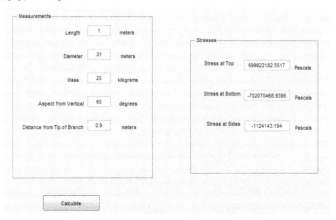

Figure 6. Image of the Graphical User interface

The equations derived for each of the cases were used to write codes in Matlab that can perform a stress analysis on any branch. The Matlab code provided an analytical solution for the stress analysis. To make the code more user friendly a Graphical User Interface (GUI) was generated to determine the stresses (Figure 6).

2.2. Finite Element Analysis

Using an approximate solution would allow branches with a wide variety of geometries to be analyzed, the software Abaqus was selected to perform the stress analysis so that branches with various geometries could be modeled. Abaqus uses Finite element analysis (FEA) to compute an approximate solution for stress. Therefore the method selected to model the branch in the program would have an effect on the final stress calculated values. Two separate methods for modeling tree branches were considered, a 2-D wireframe and a 3-D wireframe. Stress values for each option were compared.

A simple branch with a constant cross section, and material was modeled using each of the two methods(2D and 3D wireframe) described above in Abaqus. For each branch, the base end was fixed; preventing deflection and rotation in all directions, and gravity was applied. Once the model was completed, Abaqus performed the stress analysis and provided a contour image displaying stress, strain or displacement throughout the entire branch (Figure 7 and Figure 8).

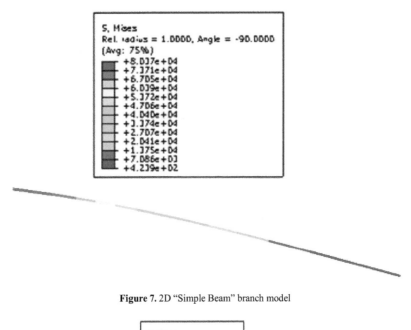

Figure 7. 2D "Simple Beam" branch model

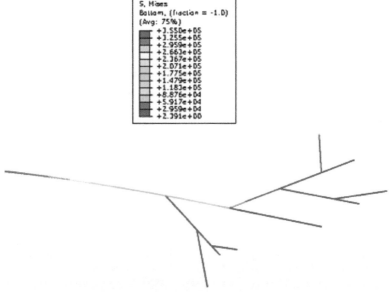

Figure 8. 2D Model containing secondary branches

The results from Abaqus and our analytical solution were identical (zero error). The 2D wireframe and the 3D wireframe both had the same approximate solutions. We selected 3D wireframe for all subsequent simulations. The 2D wireframe alternative was possible but the 3D wireframe was more appropriate since branches are three dimensional.

After verifying FE simulation, a more comprehensive simulation was developed. The model simulated 3D geometry for branches with side branches and considered the presence of leaves and fruits. Each branch was modeled in Abaqus as a 3D wire frame. The geometry was modeled as connected datum points with 3D wires. Each branch segment (between adjacent points) was assigned a material and a diameter. The material assignment was based on whether the segment was categorized as old, medium or new growth. The diameter of the segment was the average diameter of the two ends of the segment.

After the branch was created, loads and a boundary condition were applied to the branch. A load was applied to each point where a leaf or a fruit was present, and a gravitational force was applied over the entire branch. As stated above, a boundary condition was applied to the base of the branch. The boundary condition was fixed, preventing deflection and rotation in any direction. The stress analysis was then performed. Figure 9 shows a sample simulation on a real tree branch.

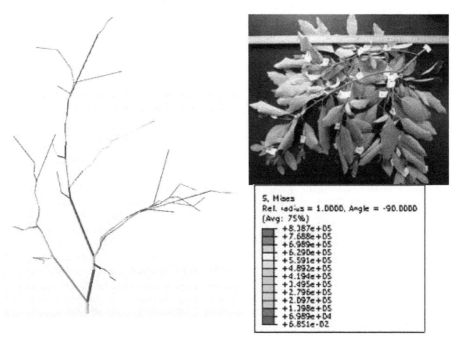

Figure 9. 3D Finite Element analysis on a real tree branch (*Sassafras albidium*)

Once the model was created, the GUI is developed which required data of geometry and mechanical properties within an excel file and outputted an executable file in Abaqus. Thereafter, the program was initiated in Abaqus to perform stress analysis. This process enables users, even with limited knowledge on using Abaqus, to perform a 3D stress analysis on tree branches.

Next, the developed analytical analysis on the 6 cases examined trends. For cases 1 through 5 all of the input variables were kept constant except for one in order to examine how changing that single variable affects the results. The variables considered included branch length, diameter, density and aspect (or angle) with respect to the vertical. For case 5 changes in MOE were also considered. We also analyzed the results from FE model.

3. Results and Discussion

For each of the cases tested, all of the cases produced similar trends. With an increase in branch length, mass increased since density was constant. However the ratio of length to mass remained constant since they both increased at the same rate. Therefore, as branch length increased, maximum stress increased (Figure 10).

As the length increased, maximum stress increased at a greater than linear rate possibly due to an increased bending arm. As for the tree branch models, cases 1, 2, and 5 (non-tapered branches), all produced nearly identical results. Note that the data points for case 1 in

Figure 10 coincide with those of case 2. According to the stress analysis, the benefits of elliptical model versus the circular cross sections were negligible. Tapered branches (cases 4 and 5) had smaller stress values than non-tapered branches. As branch length increased, differences between tapered and non-tapered branches were apparent. Non-Tapered branches were larger, with larger stress values.

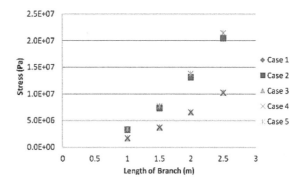

Figure 10. Length of branch versus maximum stress

When branch diameters were varied among samples, masses also varied since densities were constant. However, diameters and masses did not increase at the same rates. When examining changes in diameters solely, increases in diameters produced decreases in stress (Figure 11). However, increases in diameters per unit mass lead to higher stresses (Figure 12).

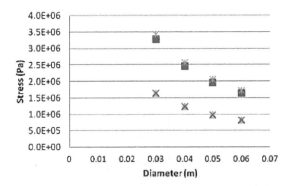

Figure 11. Diameter versus stress

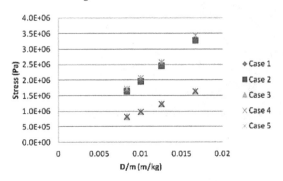

Figure 12. Diameter to mass ratio versus maximum stress

In the above figures cases 1 and 2 yielded nearly identical results. For all of the cases as branch diameters increased, stresses decreased. Therefore, when branch mass was not considered, smaller diameter branches had less stress. Among the branches tested, larger diameter branches had more stress. Relationships among tree branch models were the same in which the tapered models had less stress than the fixed models, and the shape of the cross section had a negligible effect on the stress. Thicker branches resisted more stress.

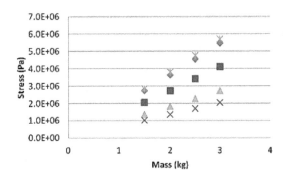

Figure 13. Mass versus maximum stress

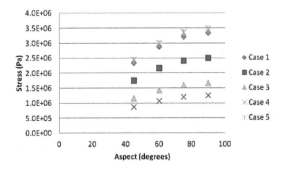

Figure 14. Aspect from vertical versus maximum stress

For each of the cases the branch density and branch aspect relative to vertical were also analyzed. The density was varied by keeping the dimensions of the branch constant and varying mass. As branch density increased, the stress increased (Figure 13) and as branch aspect increased (became more horizontal), stress increased (Figure 14).

There were noticeable differences among the various cases. Cases 1 and 5, the two cases with fixed circular cross sections, produced higher stress than the other cases. Cases 1 and 5 also produced very similar results. Case 2 had lower stress values than case 1, which indicated that branch aspect and mass the elliptical shape reduced stress. This result was further supported by the significantly lower stress values produced by case 4, the tapered elliptical branch compared with case 3, the tapered circular branch. Again the tapered branches had less stress than non-tapered branches.

For case 5, the effects of varying MOE values were analyzed. Case 5 was divided into three rings for this analysis, and MOE values were selected for each ring based on results from previous research (unpublished results). The percent increase in MOE values between the first and second rings was 113%, while the percent increase between the second and third rings was 25%. To study the effect of change of MOE in overall stress, the MOE of all three rings was increased or decreased by 5% until there was either an additional positive or negative 20% from the reference percent (Figure 15).

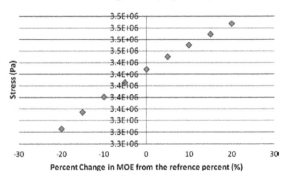

Figure 15. Percent change in MOE from reference percent versus maximum stress

As the percent change in MOE increased between concentric rings, stress increased. Thus, branches composed of materials with more uniform MOE throughout experienced less stress than branches with large differences of MOE (in the radial direction).

Case 6 accounted for the curviness of branches. Various dimensions were entered into the program to create branches with varying degrees of curves. For consistency all of the branches had the same base diameter and the same tip diameter. First, stress on a branch with no curves was compared to case 3. Both should produce similar results because they both had the same base diameter and were very narrow at the branch tips. However, there was a 33% difference between the two branches (Table 1).

Table 1. Comparison of the stresses in a non-curved branch as calculated by cases 3 and 6. The percent error calculation suggests that case 3 is more accurate

Case 3 Stress (Pa)	Case 6 Stress (Pa)	Percent Error
2.73×10^5	3.63×10^5	32.97%

The differences in stress values may result from the use of a best fit polynomial equation even when the branch may be more accurately represented by a straight line. Based on these results case 6 should only be used when the branch is curved, or when there is an abnormality in the branch which would shift the centroid of the branch away from the branch axis.

Figure 16 shows the shapes of two branches developed by the best fit polynomial equations and their corresponding maximum stress values.

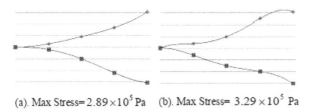

(a). Max Stress= 2.89×10^5 Pa   (b). Max Stress= 3.29×10^5 Pa

Figure 16. Branch Shapes generated by polynomial equations and their corresponding stress values

The tree branch generated in Figure 16a has small curves. The stress value calculated for this branch more closely aligns to the stress calculated using case 3 displayed in Table 1. The branch in Figure 16a has small curves so it resembles case 3. However, since curves were present, a polynomial a polynomial equation provided a better relationship than a linear equation. Figure 16b has more prominent curves, and produced a larger stress value. The Matlab program was written so that the polynomial fit equation will pass through all of the input points. Therefore, whenever a branch has a drastic change in height all changes should be entered. In addition, input points may be added at locations between changes in order to guide the path of the line. The FE analysis was also used to study the stress patterns on main branches with and without leaves. Figure 17 shows that while stresses increased when leaves were present, the overall stress pattern was the same as when leaves were not present.

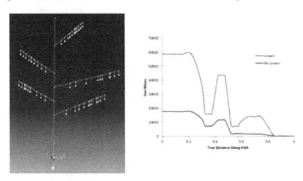

Figure 17. Stress pattern with and without leaves

Now that these tools were developed, the next step will be to simulate tree branches and calculate stress for a variety of tree species.

4. Conclusion

In this study we performed analytical stress analysis and a finite element simulation on tree branches for 6 theoretical cases. Next, using the developed methods, we analyzed the effects of several variables on branch stresses.

In addition, we developed a FE analysis to study 3D models of tree branches with leaves and fruits included. Both developed tools are operable by users with limited knowledge of Matlab and/or Abaqus.

The cases that had fixed cross sections produced stresses greater than those with tapered cross sections, since non-tapered branches had more mass. When both mass and angle of the branch were varied, an elliptical cross section proved to be advantageous in resisting stress. When variances of MOE in the radial direction were considered the results were very similar to the case with fixed circular cross section (case 1). Variances of MOE in the radial direction proved to be most beneficial when there are differences in MOE values between two consecutive, concentric wood rings. Considering curved branches with uniform material and a circular cross section were most beneficial for branches that have curves or abnormalities that shift the center of mass of the branch away from the branch axis. FE analysis showed that the stress pattern along the branch didn't change with and without leaves but stresses increased when leaves where present.

Future experiments may involve analysis of stress patterns on actual tree branches as they grow. Additional cases will be considered. The effects of added secondary branches will be determined. In most tree species, the biological center of the stem does not coincide with its geometric center. This causes the tree to develop irregularly shaped wood rings. In such cases, values of MOE will need to be adjusted [5]. To account for these wood properties additional programs will be written for compression and tensile woods with concomitant MOE values for the rings. In addition, additional FE models will be developed to study more geometrically complicated branches.

Acknowledgement

The authors are indebted to thank the Catherine and Robert Fenton Endowed Chair to LSE for financial support.

References

[1]   Evans, L., Kahn-Jetter, Z., Torres, J., Martinez, M., and Tarsia, P., "Mechanical stresses of primary branches: a survey of 40 woody tree and shrub species," *Trees*, Vol. 22, No. 3, 2008, pp. 283-289.

[2]   Sone, K., Noguchi, K., and Terashima, I., "Mechanical and ecophysiological significance of the form of a young acer rufinerve tree: vertical gradient in branch mechanical properties," *Tree Physiology*, Vol. 26, No. 12, 2006, pp. 1549-1558.

[3]   Mencuccini, M., Grace, J., and Fioravanti, M.,"Biomechanical and hydraulic determinants of tree structures: a Scots pine: anatomical characteristics," *Tree Physiol*, Vol. 17, No. 2, 1997, pp. 105-113.

[4]   Beer, F. P., Johnston, R. E. Jr., DeWolf, J. T., and Mazurek, D. F., *Mechanics of Materials*, 6th ed., McGraw Hill, New York, 2012, Chaps. 2, 4, 5.

[5]   Almeras, T., Thibaut, A., andGril, J., "Effect of circumferential heterogeneity of wood maturation strain, modulus of elasticity and radial growth on the regulation of stem orientation in trees," Vol. 19, No. 4, 2005, pp. 457-467.

[6]   Kaminski, A., Mysliwiec, S., Shahbazi, Z., and Evans, L., "Stress Analysis Along Tree Branches", ASME 2014 International Mechanical Engineering Congress and exposition, 2014, Montreal, Canada.

28

Design and Development of Mine Railcar Components

Pruthviraj Manne, Shibbir Ahmad*, James Waterman

Mechanical Engineering Department, Rowan University, Glassboro, New Jersey, USA
*Corresponding author: ahmadjerin@gmail.com

Abstract   In this paper, two of the vital mining railcar components have been designed. In this paper, a comparison has also made with current and proposed manufacturing process. Furthermore, Value stream mapping (VSM) has been analyzed through wagon and coupler manufacturing process and it has been shown that 15 % non-value adding time and 20 % labor can save though implementing the VSM methodology. Moreover, the proposed wagon body design has been proposed to save 30% unloading time. Additionally, proposed simplified coupler design can save 18% manufacturing cost.

Keywords: SM, coupling, retention time, cos effective, efficiency, laborintensive

1. Introduction

Mining industry is struggling for improving productivity. A gondola is a railcar used for transporting bulk materials, such as gravel or coal, at approximately 100 tons per railcar. These loads create a total railcar weight that can exceed 286,000 pounds [1]. As railcar weights increase to improve capacity efficiency, new methods and devices can be implemented to increase railcar performance and service life.

The body of the railcar sits on two wheel and axle suspension assemblies called trucks. The body of the railcar contacts each truck at a center bowl and two side bearings [2]. The center bowl, which takes approximately 90% of the cargo load and railcar self-weight, is a cylindrical bowl that contacts the center plate on a flat surface [3]. A center pin runs through the middle of the center bowl/center plate assembly and alignments the truck and railcar body during maintenance. The side bearings, which may or may not be in constant contact over various loading conditions, prevent excessive rocking of the railcar [2].

Currently, the Association of American Railroads (AAR) requires center bowl liners or other lubricants that reduce friction and wear between the center bowl/center plate bearing surfaces [4]. Over time, plastic or metallic center bowl liners wear down and must be replaced. Replacement decreases productivity and adds maintenance costs. Research shows that liners fail when center bowl/center plate contact shifts from flat, evenly distributed contact to point or edge contact [3].

In this research, illustrates the current manufacturing process and newly proposed manufacturing process of the mine car components. Figure 1 left side shows open top wagon which is generally used for the transportation for moving ore and materials produced in the process of traditional mining. Mine carts are seldom used in modern

operations, having largely been superseded in underground operations (Especially coal mines). On the other hand, This research also explain about the current manufacturing process and newly proposed manufacturing process of universal rail coupler in Figure 1 right side. It is universal coupling by the design made in Solid works and do the proper material selection for producing the coupler.

Figure 1. Multiple way unloading wagon body and coupler

2. Design

2.1. Design Concept

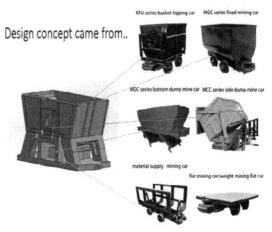

Figure 2. Design concept of multiple way unloading wagon body

By seeing the above Figure 2 present mine car are six types which are used for different purpose. One of the aims of the research to make one wagon which can do all the works by taking the unique components like bottom dump, side dump, material supply, fixed mining car. Furthermore, we will explain about material selection why we change the material, mechanism and cost of manufacturing the wagon.

2.1.1. Detail Explanation of the Wagon Body Proposed

- Under frame
- Centre Sill
- Body end
- End side
- Wheels
- Bogie
- Braking system.

2.1.2. Standard Feature of the 'MWUMC' WAGON

Table 1. Features of Wagon body

Sl.No	Features	Measurement
1	Length over head stock	3034mm
2	Length over couplers	3363mm
3	Length inside	2833mm
4	Width inside/Width Overall	1350/1450mm
5	Height inside/Height(max.) from RL	824/935 mm
6	Wheel dia. on tread (New/Worn)	350/650mm
7	Floor area (Sq.M)	13.56m2
8	Cubic Capacity (Cu.M)	1.54
9	No. of wagons per train	18
10	Coupler Buckeye coupler	
11	Bearing R.B.	
12	Maximum Speed (Loaded)	35kmph

2.1.3. Unloading of Material

The design of the steps in middle of the wagon is the key factor to unload the material. Once any door open the material drop immediately. In this paper, material unloading is other key factor and this material unloading can be done from three sides: 2 sideways and 1 bottom drop.

2.1.4. Sideways

The door has been designed in such a way that, once the pulley on the door touch the ramp where the material need to unload the door opens which is automatic no manual operation required which is **key advantage of the project.** The door can open up to 30° then the material comes out and it closes automatic because of the weight of the door.

2.1.5. Bottom Drop

It required manual operation. It has a mechanism to open the door where the material drops from the bottom without disturbing the base. The design is made using Francis turbine (draft tube concept) and support head on the bolster which is designed to protect the bolster while unloading. The design of the unloading can be seen from Figure 3.

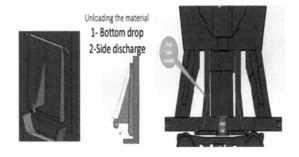

Figure 3. Unloading of material

2.2. Design of Coupler

A **coupling** (or a **coupler**) is a mechanism for connecting rolling stock in a cart. The design of the coupler is standard, and is almost as important as the track gauge, since flexibility and convenience are maximized if all rolling stock can be coupled together.

Reduce the components

Figure 4. Railcar coupler components

The core aim of the coupler is to reduce the components and provide the coupler with better material than the existing coupler. By reducing the components, it will be more flexible and absorb shocks. Moreover, resulting process would be easy, reduce the cost depend on the material with this design and easy to operate when it would be manually connected.

3. Material Selection

Material selection is the vital phase of any manufacturing environment. There are some of the important factors are involved behind proper material selection e.g product longevity, cost sensitivity etc. So 3cr12 stainless steel would be more productive than 301 stainless steel in the case of wagon body sheets production. Because density of 3cr12 is less, resulting in, the total weight of the body would be lower which shown in Table 2. Additionally, this material also could be able to generate savings in maintenance, improve productivity and longer service life. Table 3 depicts the material selection for producing body frame where Al5005 alloy has been proposed in lieu of 301 stainless steel because it can be machined by conventional method, density is pretty less compared to iron, and welding can done by standard techniques i.e TIG or MIG. Table 4 discuss the material selection for base plate which is very important component of the wagon body that should bare total weight of the coal and so Al5042 has proposed for this parts as it should have higher tensile strength. In addition, Al 5042 has more life time compared to Al5050H32. 3cr12 stainless steel material has been proposed for the door of the wagon body which shown in table 5. This material has selected for the wagon body because it has good corrosion resistance and low friction.

Table 6 shows that 303 stainless steel has been selected for locking pin of coupler because it can resist corrosion from environments, acids, moisture and at high or low temperature. From Table 7 and Table 8 shows that 301 stainless steel has proposed because it has high strength and excellent corrosion resistance than any other materials. It can further be used at low temperature regions.

Figure 5. Current wagon body materials practice

Figure 6. Proposed materials for wagon body

Table 2. Wagon body sheets material properties

301 Stainless Steel	3CR12 Stainless Steel
Current Material	Proposed Material
Density : 7880(kg/m³)	Density : 7740(kg/m³)
Elastic Modulus : 193 Gpa	Elastic Modulus : 200 Gpa
Yield Strength : 275 Mpa	Yield Strength : 280 Mpa
Hardness : 165	Hardness : 178
Ultimate TS: 758 Mpa	Ultimate TS: 650 Mpa

Table 3. Body frame material properties

301 Stainless Steel	Al 5005 Material
Current Material	Proposed Material
Density : 7880(kg/m³)	Density : 2700(kg/m³)
Elastic Modulus : 193 Gpa	Elastic Modulus : 68.9 Gpa
Yield Strength : 275 Mpa	Yield Strength : 131 Mpa
Hardness : 165	Hardness : 38
Ultimate TS: 758 Mpa	Ultimate TS: 200 Mpa

Table 4. Base plate material properties

Al5050H32	Al 5042 Material
Current Material	Proposed Material
Density : 2690(kg/m³)	Density : 2700(kg/m³)
Hardness : 46	Hardness : 96
Yield Strength : 145 Mpa	Yield Strength : 345 Mpa
Elastic Modulus : 68.9Gpa	Elastic Modulus : 70 Gpa
Ultimate TS: 172 Mpa	Ultimate TS: 360 Mpa

Table 5. Wagon body doors material properties

3cr12 Stainless Steel
Density : 7740(kg/m³)
Elastic Modulus : 200 Gpa
Yield Strength : 280 Mpa
Hardness : 178
Ultimate TS: 650 Mpa

Figure 7. Current railcar coupler materials practice

Figure 8. Proposed railcar coupler materials

Table 6. Coupler's locking pin material properties

Locking Pin	
Current Material (A396 ductile Iron)	Proposed Material(303 Stainless steel)
Density : 7113(kg/m³)	Density : 8027(kg/m³)
Hardness : 143	Hardness : 228
Yield Strength : 275 Mpa	Yield Strength : 415 Mpa
Elastic Modulus : 70Gpa	Elastic Modulus : 193 Gpa
Ultimate TS: 448 Mpa	Ultimate TS: 690 Mpa

Table 7. Coupler's knuckler material properties

Locking Pin	
Current Material (A396 ductile Iron)	Proposed Material(301 Stainless steel)
Density : 7113(kg/m3)	Density : 7880(kg/m3)
Hardness : 143	Hardness : 165
Yield Strength : 275 Mpa	Yield Strength : 275 Mpa
Elastic Modulus : 70Gpa	Elastic Modulus : 193 Gpa
Ultimate TS: 448 Mpa	Ultimate TS: 758 Mpa

Table 8. Coupler's jaw material properties

Jaw	
Current Material (A396 ductile Iron)	Proposed Material(301 Stainless steel)
Density : 7113(kg/m³)	Density : 7880(kg/m³)
Hardness : 143	Hardness : 165
Yield Strength : 275 Mpa	Yield Strength : 275 Mpa
Elastic Modulus : 70Gpa	Elastic Modulus : 193 Gpa
Ultimate TS: 448 Mpa	Ultimate TS: 758 Mpa

4. Process Flow Diagram

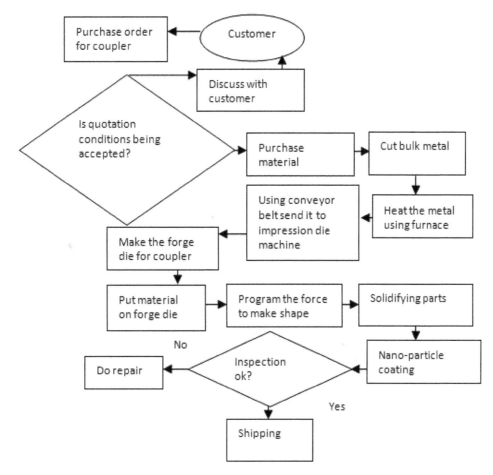

Figure 9. Current coupler manufacturing process flow

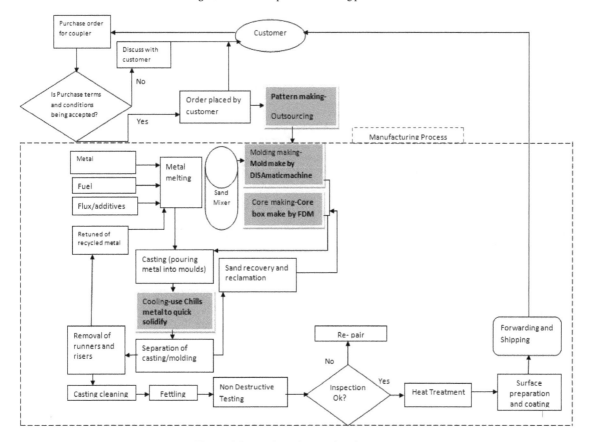

Figure 10. Proposed coupler manufacturing process

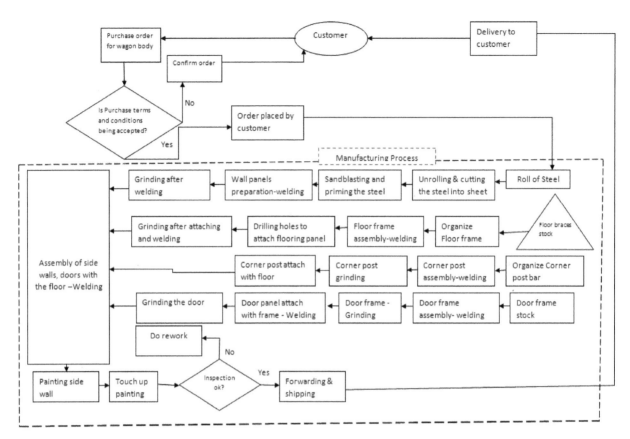

Figure 11. Current wagon body manufacturing process flow

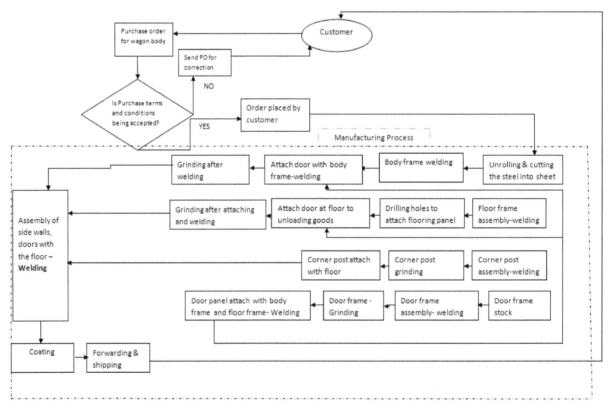

Figure 12. Proposed wagon body manufacturing process flow

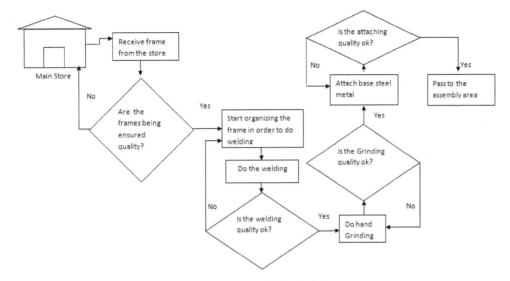

Figure 13. Framework of Poka Yoke

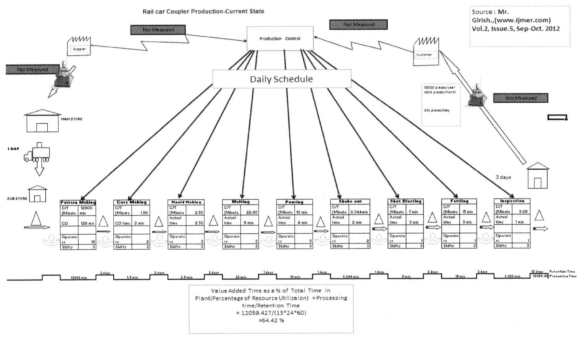

Figure 14. Current State of coupler's VSM

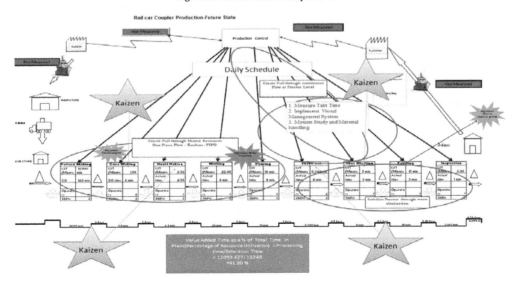

Figure 15. Future State of coupler's VSM

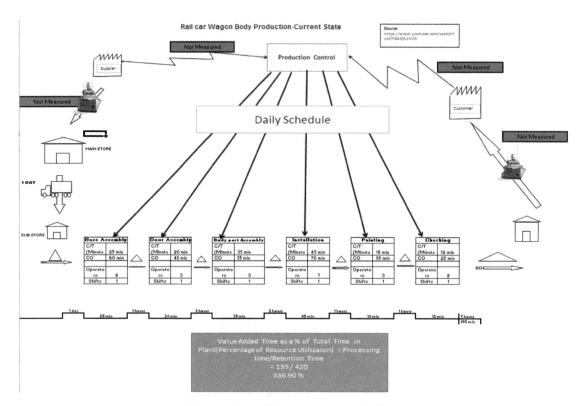

Figure 16. Current Value Stream Mapping for Wagon

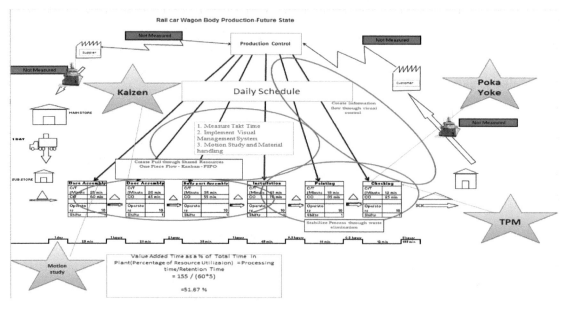

Figure 17. Future Value Stream Mapping for Wagon

Table 9. Cost –Benefit analysis for wagon body			Table 10. Cost –Benefit analysis for coupler		
Cost Benefit Analysis for wagon			**Cost benefit analysis for coupler**		

	Present Process	Our Project
Al,301 stainless steel		
Raw Material Cost/Ton	$3,800	$4,000
Material required(700Kg)	2,500	3,000
Transportation cost	100	100
Cost of Filler Metals	50	50
Stamping Cost	50	50
Painting Cost	50	50
Drilling Cost	25	25
Grinding Cost	25	25
Cost of rework	100	NA
Repairs and Maintenance	100	100
Labour Cost/Piece	100	75
Cost Of Inspection	20	20
Shipping cost	50	50
Miscellaneous Cost	50	50
Totals	$3,220	$3,595

3cr12 Stainless steel, 5005,5050h32 Al alloy
Sheets,body frame,base and doors
309 Filler material
5356 Filler material

	Present Process	Our Project
ductile iron		
Raw Material Cost/Ton	$2,800	$3,500
Material required (75kg)	210	265
Cost Of Sand Casting	50	50
Cost Of Shot Blasting	20	NA
Cost Of Fettling	20	20
Cost of cleaning the casting	5	5
Cost of heat treatment	15	15
Grinding Cost	15	15
Inspecting cost	10	5
Repairs and Maintenance	10	10
Labour Cost/Piece	10	10
Miscellaneous Cost	10	10
Totals	$375	$405

301,303 Stainless steel

4.1. Technologies to Improve Process Flow

4.1.1. Fused Deposition Moldering (FDM)

Fused Deposition Modeling (FDM) and PolyJet are two of the most advanced and effective additive manufacturing (AM) or 3D printing technologies available. They span the range from budget-friendly, desktop modeling devices to large-format, factory-floor equipment that draw from the capital expenditure budget, and can produce a range of output from precise, finely detailed models to durable production goods. While there is crossover in applications and advantages, these two technology platforms remaindistinct and bring different benefits. Understanding the differences is the baseline for selecting the right technology for your application, demands and constraints [5].

4.1.2. Disamatic Modeling

Currently, all Metal Technologies foundries utilize Disamatic Molding Machines to produce molds for our castings. Disamatics offer a highly efficient means of rapidly and automatically creating a string of flaskless molds. These molds are built for vertical casting and are created in a vertical molding environment. For an explanation of the process, read through the text below and click on the thumbnail images to see illustrations [6].

4.1.3. Rate of Solidification Time

Solidification of casting in permanent (metallic) mold is much faster than in sand mold due to its high thermal conductivity. The rate of heat transfer is controlled at the interface between the solidifying metal and the metallic mould where air gap quickly develops [7]. This is especially important in permanent mould casting where the heat transfer between the casting and the mould are primarily controlled by conditions at the mould-metal interface. At the metal-mould interface the casting tends to shrink as it solidifies, creating areas where gaps form between the casting and the mould surface [8]. Solidification involves extraction of heat from the molten metal thereby transforming it to solid state at the solid-liquid interface. The rate of solidification is determined mainly by rate of heat extraction through conduction and convection and can be represented using cooling curves [9]. Permanent mould casting is the casting of metals by continually using a reuseablemould, thus making casting time much shorter than in temporary sand casting. Permanent mould casting can be gravity die casting; centrifugal casting; and die casting [10]. In this research, chiller metal have been proposed to optimize solidification time which might expedite the whole casting process.

5. Results and Discussions

Manufacturing organization now is struggling for improving productivity. They look for cost effective method to produce and sell the product since the market is now highly competitive. The manufacturing of railcar wagon body is the labor intensive process. Though manufacturer launch automation in order to make body part in some of the process however there are still a plenty of opportunity to reduce the labor from that kind of manufacturing industry. Today, customer is not only looking for on time delivery but also they are emphasizing more on right first time. There are some manufacturing engineering tools and techniques which might mitigate the cost and improve the quality of the wagon body while manufacturing. In order to do faster production, designing the railcar components is the preliminary step. In Figure 1 has shown wagon body and coupler design. The concept of design has come from six different existing wagon bodies which portrayed in Figure 2. The core idea of design is to do unloading of mine coal faster than the existing wagon body does. To do that, there are three doors has inserted at the proposed design as if coal can be unloading through these three doors easily and quickly. There are two doors on the side wall and one door on the bottom of the wagon body. On the other hand, coupler has been designed by optimizing six components into three components by keeping the same functions of the coupler shown in Figure 4.

In **Traditional** Coupler manufacturing process, forging process has been utilized to manufacture a coupler which has shown in Figure 9. However this process is time consuming, and expensive. Customer looking for faster delivery with least cost. So Sand casting process would be more cheaper manufacturing process. Though sand casting is the antique manufacturing process, this processing is labor intensive too. There are still a lot of opportunities to make the sand casting process efficient. Currently, Sand casting patterns are commonly produced by using computer numerical control (CNC) machining, but problems like incorrect shrink compensation and design flaws can occur which in turn require the pattern be reworked – creating additional costs and lead time that manufacturers cannot afford to outlay while under increasing pressure to improve time, cost and production efficiencies. There are some suggestions for making casting process expediting has shown in Figure 10 which also described below. The suggestion for making pattern by using FDM technology. FDM technology can implement in order to make the pattern with the use of 3D printing which can save cost in labor, time and expedite the whole manufacturing process. FDM technology can not only reduce the pattern making time and cost but also there are some of the other development can be brought in the current casting process. FDM Technology can also be used for core box production. The molding is the most time consuming process in the coupler manufacturing by using the casting technology. So the extensive level of time and labor can be saved by using DISAMATIC sand molding machine which the has several advantages comparing to conventional molding processes. It does not use flasks, which avoids a need of their transporting, storing and maintaining. It is fully automatic and requires only one monitoring operator, which reduces labor costs. Molding sand consumption can be minimized due to variable mold thickness that can be adjusted to the necessary minimum. The efficiency of the sand casting process while manufacturing railcar coupler can be improved by optimal designing of Riser. Risers are generally used for putting off of shrinkage defects. However, they decrease the usage rate of metal and extend the cooling time of castings after solidification as well. Therefore, proper riser size needs to be designed to satisfy feeding with the smallest volume.

The coupling manufacturing through the sand casting process might be accelerating by doing optimal gating design too. The gating design and in gate position plays an important role in the quality and cost of a metal casting. Due to the lack of theoretical procedure to follow, the design process is normally carried out on a trial and error basis. The time saving at the time of cooling while manufacturing coupler is another vast time saving improvement. Using metals like chills is an object used to do rapid solidification. In this research, the suggestions for internal chills metal which can be melted. When the cavity is being formed, chill metal parts will melt and finally become the portion of the casting product and thus the chill must be the same material as the casting. The advantages of the internal chill can absorb both heat capacity and heat of fusion energy.

In Figure 11 discuss the current manufacturing flow of wagon body where non-value adding activities has been adopted. But there are four way wagon body manufacturing process has depicted in Figure 12. The side wall produces by welding process. But it does not need to do grinding after welding as the chemical treating process has been proposed which would be faster and smooth process. Similarly, chemical dipping proposed for each parts of wagon body manufacturing i.e base and support, body frame and coal unloading doors.

The cost benefit analysis for wagon body has been shown in table 9. The current wagon body manufacturing cost is $3200 but the proposed wagon body manufacturing cost would be approximately $3600 per body. This price is very competitive though many features have been added into wagon body. Table 10 shows the cost benefit analysis of the coupler. The current and proposed price almost same but the quality of the coupler would be higher than existing as well as coupler can be delivered to the customer very quickly since some manual process has been eliminated while manufacturing through sand casting process.

The total retention time or non-value adding time for coupler is 13 days and processing time is 12059.43 minutes from Figure 14. Calculated the resulting percentage of utilization or value added time as percentage which was 64.42 % for the current state in case of coupler while doing VSM analysis. This is meant that there are still approximately 35 % opportunities remaining for improving utilization that leads to improving efficiency of the company. But Figure 15 state that there are most important waste reduction tools and techniques have been proposed to implement while manufacturing coupler of railcar. From the current state, it was observed that unnecessary time taken in between the shakeout and shot blasting, fettling and shot blasting, shot blasting and checking. The unnecessary time from that operation could be reduce through introducing Pull production methodology instead of push production, FIFO method, Kaizen, Visual management system, motion study and right material handling techniques. The proposed improved has been shown in the future state and If the company would implement the above mentioned techniques, higher productivity can be achieved. The proposed calculated percentage of utilization or value added time 84 % which is competitive and serving in the manufacturing competitive environment.

The total retention time or non-value adding time for wagon body is 7 hours depicted at Figure 16 and processing time is 155 minutes. Calculated the resulting percentage of utilization or value added time as percentage which was 36.90 % for the current state in case of body part while doing VSM analysis. This is meant that there are still approximately 64 % opportunities remaining for improving utilization rate that leads to improve the efficiency of the company.

But in the future state which has shown in Figure 17 that there are most important waste reduction tools and techniques have been proposed to implement while manufacturing coupler of railcar. From the current state, it was observed that unnecessary time taken in between the base assembly, Installation, painting and checking. The unnecessary time from that operation could be reduce through introducing Pull production methodology instead of push production, Kaizen, Visual management system, motion study and right material handling techniques, Poka Yoke and TPM. The proposed improved has been shown in the future state improving and If the company implement the above mentioned techniques. The proposed calculated percentage of utilization or value added time as percentage which is53 %.

6. Conclusions

In this paper, a lot of new things have been adapted while developing design and process. Furthermore ,we have successfully compared the current process and proposed process then by using VSM techniques that canreduce labor and inspection time which made our process boost from the existing, machine used in this process are automated with high quality which improve the productivity. We made a continuous flow of material unloading which has eliminate dumping system. In this paper, moreover focus on design which made the expedition throughout the manufacturing process. Furthermore, In this research , a tremendous improvement has done by reducing labor from the manufacturing activities. Design of the railcar wheel mechanism as well as fabricating the wagon body and coupler as per design would be further recommendation for the future research.

Acknowledgement

An engineer with only theoretical knowledge is not a complete engineer, practical knowledge is very important to develop and apply engineering skills". It gives has a great pleasure to have an opportunity to acknowledge and to express gratitude to those who were associated with has during my course "Advanced Manufacturing-Concepts and applications" we are very great-full to for providing has anopportunity with this project. Furthermore, special thanks to **Prof. James Waterman** for his help, support and guidance.

We express my sincere thanks and gratitude to **Karl Dyler (Technician) and Engineering workshop** authorities for allowing has to 3d print parts. We will always remain indebted to them for their constant interest and excellent guidance in our progress to project, moreover for providing has with an opportunity to work and gain experience.

References

[1] Wolf, G. (2005b). "Vehicle Side Bearings: Function, Performance and Maintenance".

[2] Hay, W. W. (1982). Railroad Engineering, John Wiley & Sons, New York.

[3] Tournay, H.M., Lang, R., and Wolgram, T. (2006). "Performance History".

[4] Association of American Railroads (1982). Manual of Standards and Recommended.

[5] Fred Fischer, FDM and Poyjet 3D printing.

http://www.stratasys.com/resources/~/media/722E026DFF2F429C9DA6362BDD42ED2D.pdf.

[6] Altan Turkeli, Sand casting,MSE-432,Foudnry technology.

[7] K. C. Bala and R. H. Khan, Experimental Determination of the Effect of Mould Thickness on the Solidification Time of Aluminium Alloy (Al-Mn-Ni -Si) Casting in Rectangular Metallic Moulds, International Journal of Engineering Research & Technology, 2(3), 2013, 1-6.

[8] K. C. Bala, Simulation of Solidification of Aluminium Casting in Metallic Mould, doctoral diss., Mechanical Engineering Department, Federal University of Technology, Minna, Nigeria, 2012.

[9] R. E. P. DeGarmo, J. T. and R. A. Kohser, Materials and Processes in Manufacturing (New York, John Wiley & Sons, 2010).

[10] P. L. Jain, Principles of Foundry Technology (London: McGraw-Hill, 2003).

29

Design and Construction of a Magnetic Levitation System Using Programmable Logic Controller

Minhaj Ahmed[1,3,*], **Md. Fahad Hossen**[1,3], **Md. Emdadul Hoque**[1,3], **Omar Farrok**[2,4], **Mohammed Mynuddin**[2,5]

[1]Department of Mechanical Engineering
[2]Department of Electrical and Electronic Engineering
[3]Rajshahi University of Engineering & Technology, Rajshahi-6204, Bangladesh
[4]Ahsanullah University of Science and Technology, Dhaka-1208, Bangladesh
[5]Atish Dipankar University of Science and Technology, Dhaka-1230, Bangladesh
*Corresponding author: somo.ruet@gmail.com

Abstract The motivation of this paper is to design and fabrication a cost effective magnetic levitation (shortly called Maglev) system using PLC. For this purpose a stand, a 12 volt electromagnet, eddy current displacement sensor which input voltage is 24 volt and output voltage range is 0 to 10 volt for 0 to 10 mm displacement, Siemens Logo PLC setup board including CPU, cable, analog expansion module, an amplifier circuit & Logo soft software are used. Sensor senses the displacement of a target and gives corresponding signal in terms of voltage. This sensor output voltage is used as input in PLC input terminal. CPU works according to ladder diagram which is installed in program memory. The electromagnet is connected to the analog expansion output terminal. The current is controlled by PLC according to sensor signal which passes through the electromagnet. When distance between the sensor and target is increased the output sensor voltage is increased. PLC receives this voltage and decrease the current supply at output terminal. Thus the magnetic force is decreased and target is levitated at a desired position. When the distance between the sensor and the object is decreased, then current is increased in electromagnet which increases the magnetic force and the target returns to its levitate position. Finally the target is levitated. This paper will be helpful to simplify this control system and implement in different types of maglev study and industrial issues as well.

Keywords: *magnetic levitation, electromagnetic force, PLC, automatic control, mechatronics*

1. Introduction

Maglev is the technique to suspend an object in the air by manipulating magnetic force. This magnetic force is used to counter the gravitational force of object. Maglev is one of the best recent technology. It has no physical contact between the motion object and stable part of the system. So there is no friction and wear in such type of technique. Its application is rapidly increasing in various industry. A maglev device uses magnetic fields to suspend an object in a desired position. It was experimented by many researcher with superconductors, permanent magnets, diamagnetic materials and they comprehend that maglev is quite complex technique. When the gap between the object and magnetic source is too long, then the strength of magnetic field will inadequate to sustain the weight of the object. By placing near to the magnetic source, the strength of magnetic field becomes very powerful and the magnetic field can easily attract the object till it makes direct contact with the magnet. From the above, it is clear that a maglev device is an intrinsically unbalanced system. An electromagnet will have to be constructed to solve this uncertainty. This electromagnet is a simple device which is made up of a magnetic material wound with a current conducting coil. This coil permits to pass a varying current through it and hence generate a varying magnetic field or force. The generated magnetic force can attract any object in its domain. By controlling the amount of force exerted on the object and developing a stabilizing controller that can use measurements of position and velocity as feedback parameters in favor of balance the position of the object [1].

The objectives of this project are to design and fabricate the maglev system as well as to develop a ladder diagram and controlling system using LOGO PLC, to stabilize an object at a desired vertical position.

Figure 1. Magnetic Levitation [2]

2. Ease of Use

1) Maglev Train [3]. 2) Maglev Car [4,5]. 3) Maglev Stirrers [6]. 4) Bearing Less Centrifugal Pump [7].5) Magnetic Bearings [8]. 6) Maglev Wind Turbine[9].

3. Programmable Logic Controller (PLC)

3.1. Definition of PLC

PLC is a digital computer consists of integrated used for automation of electromechanical devices to execute control functions. According to NEMA PLC is a "Digital electronic devices that uses a programmable memory to store instructions and to implement specific functions such as logic, sequencing, timing, counting, and arithmetic to control machines and processes". It was created in favor of replace the sequential circuits which were commonly used for machine control. It is widely used for automation of electromechanical processes in many industries [10].

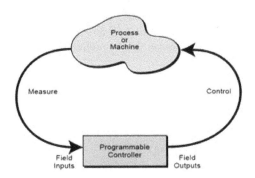

Figure 2. PLC conceptual application diagram

3.2. Basic Component

All programmable controllers contain a CPU, memory, power supply, I/O modules, and programmable devices. Basic parts of the PLC are as follows:

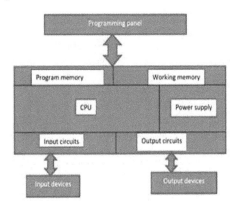

Figure 3. Basic components of PLC

4. Magnetic Levitation Techniques

History says that in the last few decades, maglev was experimented by many people using permanent magnets. Experiments were made to pursue the proper setting of permanent magnets to suspend another magnet (which size is must smaller than permanent magnet) or any

ferrous material object. Earnshaw's theorem [11] however, manifests that maglev using permanent magnet is not mathematically possible. There are many alternative method to generate magnetic fields other than permanent magnet those can be used for levitation. Electro-dynamic system is one of them. The operating principle of electro-dynamic system can explain from Lenz's law. Electro-dynamic levitation also results from an effect discovered in superconductors. This effect is known as Meissner effect [12].

Magnetic suspension is one of the simplest method which is used to levitate any object (ferrous material) electromechanically. An electromagnet is positioned just above the object which is to be levitated. The gravitational force of the object is opposed by the magnetic force of electromagnet. A position sensor which works as a feedback is placed just below the object at a certain distance. The senor senses the position of the object and delivers this information to the control. According to the information provided by the sensor, the control unit supplies current to the electromagnet.

There are many techniques for levitation. These techniques are as follows: (1) Levitation using permanent magnets [13,14]. (2) Levitation using diamagnetic materials [15]. (3) Levitation using superconductors [16]. (4) Levitation using induces eddy currents in a conducting body [17]. (5) Levitation using current passing conductor in a magnetic field. (6) Levitation using inductance, capacitance, resistance circuit and the electrostatic force of attraction [18]. (7) Levitation using inductance, capacitance, resistance circuit and electromagnetic force of attraction [19]. (8) Levitation using controlled DC electromagnet [20]. (9) Levitation using mixed μ (permeability of material) system [21].

5. Methodology

5.1. Basic Concept of Levitation

In this project current is controlled by PLC. This current passes through the electromagnet which creates magnetic force on the desired object (such as cylindrical ball, bar, ferrous materials, door hinge etc). If load is increased the sensor senses the displacement and delivers signal to the control unit (PLC). PLC receives the signal and increased current supplied to the electromagnet. Thus magnetic force is increased and object is levitated a stable position. If load is decreased vice-versa activity is occurred. The block diagram is given bellow:

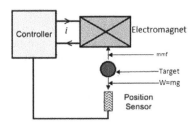

Figure 4. Block diagram of levitation concept with feedback system

5.2. Control and Regulate Basics [22]

In engineering field, quantities can be both controlled and regulated. At the time of controlling, a quantity is manipulated in such a way without being able to compensate for outside influences. Similarly to compensate for outside influences a quantity is maintained at a specific value at the time of regulating.

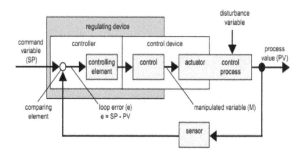

Figure 5. Basic concept of regulating with feedback control system [23]

In order to compare the command variable with the process value, the comparing element uses the sensor. If the command variables and process value differ from one another, this consequences a positive or negative loop error that in turn changes the process value.

5.3. Control Loop

By means of the regulating device, the process value x influences the manipulated variable M. Thus a close circuit is created which is also known as a control loop.

5.4. Loop Error

The difference between the command variable and the process value is the loop error. Again, loop error is the deviation of a process value (PV) from a set value (SP).

The loop error e brings about a change to the manipulated variable M. If with a desired temperature of 21 °C (= command value SP), the room temperature is 24 °C (= process value PV), then the loop error is:

$$e = SP - PV = 21°C - 24°C = -3 \ °C$$

In this case, the negative sign indicates a reversing action, the heat output is reduced.

There are mainly three kinds of controller: 1. P controller 2. I controller 3. D controller

5.5. P Controller

P controller stands for proportional-action controller. It alters the manipulated variable M proportional to the loop error. The P controller works instantly. P controller cannot minimize the loop error to zero.

5.6. I Controller

I controller stands for integral-action controller. It changes the manipulated variable M proportional to the loop error and to the time. I controller does not work instantly. It can completely minimize a loop error.

5.7. PI Controller

A PI controller minimizes the loop error instantly and will finally minimize the loop error to zero.

$$M_n = M_{Pn} + M_{In} = k_P \times e_n + k_I \times (T_S / T_I) \times e_n + M_{In-1}$$

M_n: Manipulated variable at the time n, M_{Pn}: Proportional part of the manipulated variable, M_{In}: Integral part of the manipulated variable, M_{In-1}: Manipulated variable of the I controller at the time n-1(also called integral sum), k_P: Gain of the P controller, k_I: Gain of the I controller, T_S: Sampling time, T_I: Integral time, e_n: Loop error at the time n.

The following picture shows a jump in process value and step response of the controller:

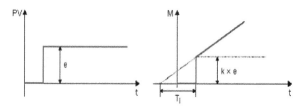

Figure 6. Response of PI- Controller Summary [23]

The features of PI controller is given below: 1.The P controller components instantly intercept an occurring loop error.2.The I controller components can minimize the rest loop error. 3. The controller components supplement each other so that the PI controller works instantly and accurately. For this reason this controller is used in this job.

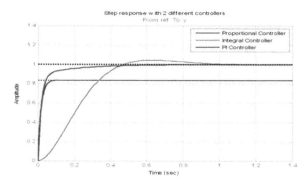

Figure 7. PI controller response

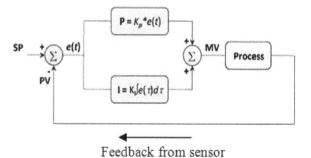

Feedback from sensor

Figure 8. Block diagram of maglev system with feedback control

6. Physical View of Maglev System

The physical structure as shown in Figure 9 consists of an electromagnet, a displacement sensor, a control unit (PLC), a limiter and an object or target. The target that is to be levitated between electromagnet and sensor without any physical contact. In general, when the current passes through the electromagnet, a magnetic force is produced which will withstand the weight of the levitated object. Since the electromagnetic force is very quick response, so any low amount of current that will tend the levitated object in unstable region. A controller is used for that case. The purpose of the controller is to control the amount of current to keep the levitated object in stable position.

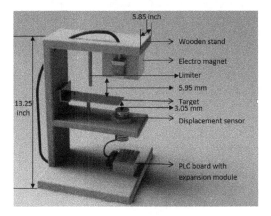

Figure 9. Structural drawing of the magnetic levitation

7. System Components

7.1. Actuator

A type of magnet in which the magnetic field is produced by the flow of electric current is shown in Figure 10. The magnetic field vanishes after the current is break off.

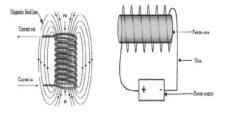

Figure 10. Basic construction of electromagnet [2]

The motivation of the magnetic actuator is to deliver a force. That magnetic force which will carry the load which is being supported. The suspended object must be capable to interact with magnetic forces because the bearing force is turning to be magnetic. This signifies the suspended load must be made of ferromagnetic material. The general block diagram of an electromagnet is shown in Figure 11.

Figure 11. Block diagram of actuator

7.2. The Sensor

The electromagnet helps to regulate the strength of the magnetic field by controlling the current flow through it, but it is important to know when and by how much to regulate the strength in a real application. The sensor gives this information. There are many different types of sensors [24] that detect the distance between objects.

1. Proximity Sensor 2. Hall Effect Sensor 3. Ultrasonic Sensor 4. Capacitive Displacement Sensor

7.3. Eddy Current Displacement Sensor

Eddy current displacement sensors are non-contact devices used for the measurement of noncontact position, displacement and proximity. It is capable to measure the position or change of position of any conductive target without any kind of physical contact with high accuracy. The other name of eddy-Current sensors are inductive sensors.

The basic operating principle of inductive proximity sensors is electrical inductance. A fluctuating current induces an electromotive force (emf) in a target object which is the basic phenomenon of inductance. Eddy circuits developed in the metallic object when a metal object moves into the inductive proximity sensor's detection field, by magnetically push back eventually minimize the inductive sensor's oscillation field. When the oscillator becomes decreased to an adequate level, the sensor's detection circuit monitors the oscillator's strength and activates an output from the output circuitry.

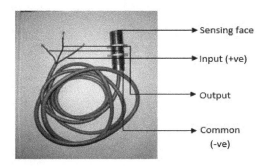

Figure 12. Physical view of a eddy current displacement sensor sensor

7.4. Control System

Analog Control, Digital Signal Processing (DSP), Microcontroller, PLC are some control system. Among those PLC is used. The sensor delivers a signal (voltage) that is proportional to the position of the target. When the object has displaced from its levitated position, the sensor provides a signal to the control unit (PLC). Then the PLC regulates the supply of current to electromagnet according to the information provide by sensor to bring back the disturbed object to its stable position.

8. Experimental Setup

8.1. Power Source

In this project 12 Volt & 24 Volt DC power supply are used which is obtained by 12V, 2A AC to DC Adapter & 24V, 6A AC to DC power supply. 12V Adapter & 24V power supply are needed to operate power amplifier with Electromagnet and Logo PLC with external module. The solar energy can also be used as power source of Electromagnet and Logo PLC. These systems are mounted together on a base plate to form the test bed. This configuration allows for portability of the system and rigid but adjustable positioning of the components. In order to design a user friendly controller for the maglev system, the system element must be modeled or characterized correctly.

8.2. Structural Elements

The elements whose are used to construct this system are given below: 1.Wooden frame2.Aluminums sheet 3.Iron bar 4.Plastic 5.Super enamel wire.

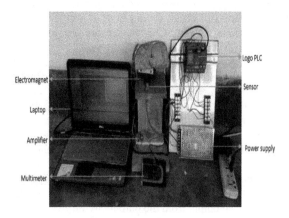

Figure 13. Experimental setup

8.3. Logo PLC CPU and Expansion Module Connection

In this project main controlling device is PLC. The analog output of Displacement sensor is the analog input of PLC I7 terminal act as analog input AI1 which range is 0-10V, 0-20 mA or 4-20 mA. Analog expansion card is attached with CPU and the analog output of AQ1 (current or voltage) is send to the power amplifier input. In online test mode the process is observed by the Computer monitor i.e the input and output analog voltage.

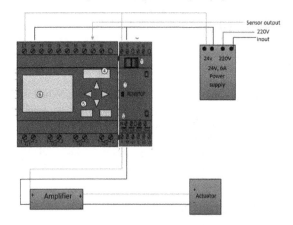

Figure 14. Wiring diagram of Logo to the system

Power amplifier Circuit consists of MJ11032 Darlington power transistor [25], potentiometer [26], resistor [27].

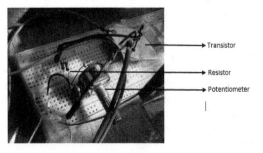

Figure 15. Power amplifier circuit

8.4. Actuator (Electromagnet) Design

An electromagnet is a type of magnet where a magnetic field is produced by electric current.The prime design criteria of an electromagnet is the lifting power. The lifting power of an electromagnet is depend on type of material and number of winding. E core shape is selected for design of electromagnet. Laminated iron core was selected to make the electromagnet because it has good permeability and hysteresis properties and the range of operating temperature is high. The core have the following dimension:

Length of the core, L_c : 39mm
Width of the core, D_c : 26mm
Height of the core, H_c : 33mm
Wire specification,18 AWG ; 1050 winding
Super enamel.
Air gap : 5.94mm
Weight of the object,
$W = mg = 24 \times 10^{-3} \times 9.8 = 0.235N$
Voltage rating, 12V($\pm$20%)
Current rating, 0.35 A
Power,$P = VI = 12 \times 0.35 = 4to4.5watt$

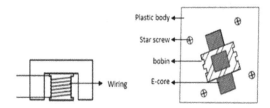

Figure 16. 2D top R.S view of actuator

8.5. Force Calculation

The strength of electromagnetic force depends on electromagnetic field. Again the strength of electromagnetic field depends on how much current will pass through the electromagnet. Electromagnetic field increases with increasing current. The basic characteristic of ferromagnetic material is that the saturated value of magnetic field is nearly 1.6 teslas that is for good permeability core steels. But when the magnetic field reaches at saturated value, then the magnetic field will not increase though the current still pass through the clectromagnet.

The equation of maximum electromagnetic force for I shape core electromagnet is:

$$F = \frac{B^2 A_p}{2\mu_0 \times 3}$$

where, F is the force (Newtons), B is the electromagnetic field (Teslas), A_p is the pole faces area = (57 × 17)= 0.969 × 10^{-3} m², μ_0 is the permeability of free space (air) = $4\pi \times 10^{-7}$ HM^{-1}

Since the E shape core has 3 poles, so for E-shape the force will be one-third of I shape core electromagnet.

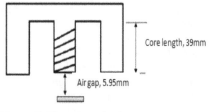

Figure 17. Dimensions of E-shaped electromagnet

Figure 17 depicts the dimensions of the E-shaped electromagnet.

The number of turns (N) for the electromagnet was chosen 1050 (18 AWG) and estimating the maximumu current (I) that could pass through the electromagnet is about 0.35 Amp. Again the magneto-motive force is the product of the current which will pass through the electromagnet and the number of turns of the wire across the electromagnet. Then, the magneto-motive force (mmf):

$$mmf = I \times N \ \text{AT} \left(\text{Ampere} - \text{Turns} \right) = 367.5 \ \text{AT}$$

Also, for the air-gap of 5.95mm mmf is:

$$AT = H \times L$$

$$H = \frac{AT}{L} = 61764.70 \ \text{T/m}$$

L is the air-gap length = 5.95mm.

In the air-gap the magnetizing force (H) is:

$$H = \frac{B}{\mu_0}$$

Therefore, $B = H \times \mu_0 = 0.0775 \ \text{wb/m}^2$.

As the ferromagnetic material is saturated at about 1.6t eslas, the flux in the air-gap and the total flux in the core will be same..

The total flux is obtained to be, $\Phi_c = B \times A_c$

Where, A_c is the core area $= 0.02972 \ \text{m}^2$

Hence,

$$\Phi_c = 2.3033 10^{-3} \ \text{wb}$$

Hence the electromagnetic force for the air-gap 5.95 mm is computed as

$$F = \frac{B^2 A_p}{2\mu_0 \times 3} = 0.772 \ \text{N}$$

But due to eddy current loss, poor insulation of the wire and lamination of the core this calculated lifting force must be multiplied by a factor (C) to obtain the actual lifting force.

The factor C is given by, $C = 0.1$

Then, the actual lifting force exerted by the electro-magnet is

$$F = 0.772 \text{N}.$$

9. Experimental Results

9.1. Sensitivity of the System

The sensitiveness of the system is how the system interacts at the time of functioning under harsh condition. The system depicts a high notch of sensitiveness with the variation of the distances of the hanging object. The output voltages remain in a consistent trend over the various set of data collected while operation is enduring.

Figure 18 represent that the curve shows almost linear behavior from 2.5 mm to 5 mm. So the minimum gap between the sensor and the target should be maintained within the above value. Therefore the reaction of the system when some ferromagnetic material is brought

closer to the sensing element is clearly evident by the curve.

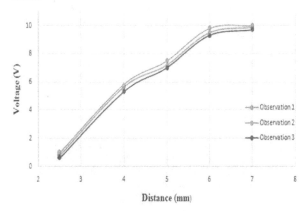

Figure 18. Sensitivity of the position sensor

9.2. Characteristics Curves of Maglev System

Identical trial works are performed to observe the attributes of the developed system. The below figures represent the different characteristics of the developed maglev system with varying different parameters (current, air gap, load). The motion of the existing setup can control only in the vertical direction. Therefore by aligning the object perfectly between the sensor and the electromagnet, the vibration of the object can be minimized. Figure 19 shows the current increases with increasing load because more current have to pass through the electromagnet coil to keep the object in stable position. Figure 20 illustrates the dynamic response of the developed maglev system. It is observed that with the increase in the air gap the control current also needs to increase to produce more magnetic force.

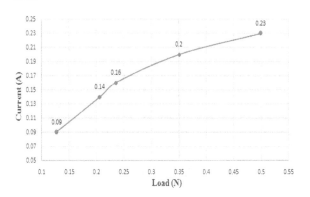

Figure 19. Variation of control current at different loads

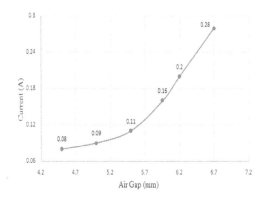

Figure 20. Variation of current at different air gap

9.3. System Realization

A stable maglev system is developed by logo PLC using a PI controller. At the same time, the system is suitably controlled by the controller. From the experiment, it is observed that the power consumption of the system is approximately 2 to 4 watt. It is considerably low regarding the stability of the system. It is seen from the figure that the system is stable using PLC. The levitation is shown in Figure 21 where the object is levitated at a stable position. The target is levitated at 6.56 V & 0.168 A.

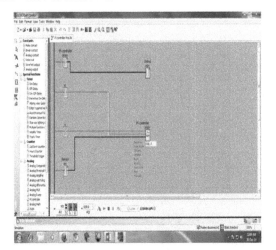

Figure 21. Realization of maglev system

9.4. Ladder Diagram

Ladder diagram of this project is quite simple and small. One digital make contact I2 and one brake contact I1, PI controller SF001, one analog input & one analog output A1 & A2 blocks are used to design the program. The parameter of PI controller is adjusted such as Controller amplification gain KC = 0.01, Integration Time (TI) = 1 second, Set value (SP) = 500 (5 volt), Manual output (Mq) = 0, Direction = Upwards (+), Sensor = 0-10 V, Sensor gain = 1.00 & offset = 0.

Figure 22. Ladder diagram online test simulation view

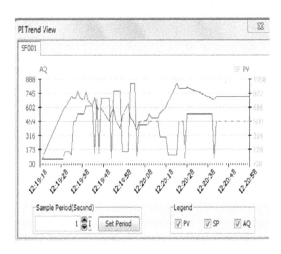

Figure 23. PI controller response and fluctuation diagram

In Figure 23 shows the PI response in online test mode. At starting time i.e 12h: 19min: 18s Process value (PV) was 100 (1V). After 1min 40s the process value is same as set point (SP) or reference value 500. During this time there are randomly changes the process value and output (AQ). When the process value is same as set value, the output is fixed at 637 (6.37 Volt) and the levitation is done by actuator.

10. Conclusions

This paper focuses on the arrangement of maglev system as well as design of a controller to keep the object in levitated position without any disturbances by controlling the amount of current through electromagnet.

Magnetic levitation system has been designed and fabricated. A ladder diagram has been developed for controlling system. Since the displacement sensor can detect the object to the amount of distance is limited to 1 to 10 mm, so when the distance range will increase then the sensor cannot deliver any signal (voltage) to the control unit (PLC). As the sensor response is not linear function, so the maglev system is intrinsically unbalanced system. However, this paper displays the control of single actuated, single axis maglev system. The single degree-of-freedom motion of the levitated object is controlled manually by tuning the controller gains. PI controller is used for its low cost, simplicity and stability. Furthermore, the energy consumption of the system is quite satisfactory. Finally, the object is levitated at a desired levitated position without friction and any kind of physical contact.

N.B: If anyone wants to see our project Video, Please check this link:-

https://www.youtube.com/watch?v=fjeD5chMfX4(low resolution)

https://www.youtube.com/watch?v=u_i9ByQ_LRM(high resolution)

Acknowledgement

A deep sense of gratitude to project supervisor Dr. Md. Emdadul Hoque, Professor and Head of the Department, Mechanical Engineering, Rajshahi University of Engineering and Technology, whose overall direction and guidance has been responsible for the successful completion of this project.

I also extend our sincere thanks to all honorable teachers of Mechanical Engineering Department, Rajshahi University of Engineering and Technology. Without their encouragement and valuable suggestion it would be impossible to bring project work to a success.

References

[1] S. Earnshaw, "On the Nature of the molecular forces which regulate the Constitution of the Luminiferous Ether", Transactions of the Cambridge Philosophical Society, Vol. 7, pp. 97-112, 1842.

[2] Google Maglev Image.

[3] Monika Yadav, Nivritti Mehta, Aman Gupta, Akshay Chaudhary, D. V. Mahindru, "Review of Magnetic Levitation (MAGLEV): A Technology to Propel Vehicles with Magnets", Global Journals Inc. (USA), Online ISSN: 2249-4596, Print ISSN: 0975-5861.

[4] Ken Watanabe, Hiroshi Yoshioka, Erimitu Suzuki, Takayuki Tohtake, Masao Nagai, "A Study of Vibration Control System for Superconducting Maglev Vehicles(Vibration Control of Lateral and Rolling Motions)", Transaction of the Japanese Society of Mechanical Engineers, Vol.71, No.701, C, pp.114-121, 1.(2005).

[5] Ken Watanabe, Hiroshi Yoshioka, Erimitu Suzuki, Takayuki Tohtake, Masao Nagai, "A Study of Vibration Control System for Superconducting Maglev Vehicles(Vibration Control of Vertical and Piching Motions)", Transaction of the Japanese Society of Mechanical Engineers, Vol.71,No.701,C,pp.122-128,1.(2005).

[6] http://www.google.com/patents/CN202893268U?cl=en.

[7] Thomas Gempp, " Centrifugal pump without bearings or seals" CHIMIA International Journal for Chemistry, Volume 57, Number 6, June 2003, pp. 331-333(3).

[8] J.C. Ji, Colin H. Hansen, Anthony C. Zander, "Nonlinear Dynamics of Magnetic Bearing Systems", School of Mechanical Engineering, The University of Adelaide, SA 5005, Australia.

[9] Minu John, Rohit John, Syamily P.S, Vyshak P.A, "Maglev Windmill", International Journal of Research in Engineering and Technology, Vol.03.

[10] Programmable Logic Controllers- fourth edition by Frank D. Petruzella.

[11] Philip Gibbs and Andre Geim, "magnetic levitation", March 1997.

[12] "Handbook of Superconducting Materials", Volume 1, edited by David A Cardwell and David S Ginley.

[13] Theraja B.L"A text book of Electrical Technology" S. Chand & Company Ltd.

[14] Ilpo Karjalainen, Teemu Sandelin, Riku Heikkila, R. Tuokko, "Using piezoelectric technology to improve servo gripper performance in mini and micro assembly, Assembly Automation, Vol. 25, No. 2, page 117-123, 2005.

[15] M. D. Simon, L. O. Heflinger, A. K. Geim, "Diamagnetically Stabilized Magnet levitation" American Association of Physics Teachers, Am. J. Phys., Vol. 69, No. 6, June 2001.

[16] A Patel, G Giunchi, A Figini Albisetti, Y Shi, S C Hopkins, R Palka, D A Cardwell, B A Glowacki : "High force magnetic levitation using magnetized superconducting bulks as a field source for bearing application", Physics Procedia 36 (2012) 937-942.

[17] M. T. Thompson, "Eddy current magnetic levitation. Models and experiments", IEEE Potentials, Vol. 19, Issue: 1, pages: 40-44.

[18] kutarr.lib.kochi-.tech.ac.jp/dspace/bitstream/10173/601/1/1118003601.pdf

[19] www.ece.ubc.ca/~tims/pubs/hollis93.pdf

[20] P. K. Biswas, S. Bannerjee, "Analysis of Multi-Magnet Based DC Electromagnetic Levitation using ANSYS Simulation Software", P.K. Biswasa et al. /IJESM Vol.2, No.2 (2012).

[21] www.ijeat.org/attachments/File/V1Issue4/D0356041412.pdf.

[22] "Introduction to programming a SIEMENS LOGO PLC" by Seyedreza Fattahzadeh.

[23] Siemens Logo soft manual (controlling basic & PI-control section)

[24] Sensors and Signal Conditioning- second edition by Ramon Pallas Areny, John G. Webster.

[25] www.researchgate.net/publication/228989498_Analog_and_LabView Based_Control_of_a_Maglev_System_With_NI-ELVIS.

[26] www.iasj.net/iasj?func=fulltext&aId=92151.

[27] www.iraj.in/journal/journal.../journal.../1-73-140688957336-42.pdf.

30

A Theoretical Analysis of Static Response in FG Rectangular Thick Plates with a Four-Parameter Power-Law Distribution

Fatemeh Farhatnia[*]

Mechanical Engineering department, Islamic Azad University Khomeinishahr Branch, Isfahan, Iran
*Corresponding author: farhatnia@iaukhsh.ac.ir

Abstract In this paper, we proposed a simple mathematical procedure to solve the differential equations governing the buckling and bending analysis of FG thick rectangular plates resting on two-parametric foundation based on Mindlin assumption. All edges are set on the simply supported conditions. Young modulus of the FG plate was assumed to vary according to a simple four-parameter power law across the thickness direction. For bending analysis, the plate was subjected to two kinds of loading: sinusoidal and uniform. For bucking analysis, two kinds of in-plane loading were applied to the plate: uniaxial and biaxial. Variations of FG material variation profile, thickness ratio, and foundation parameters on buckling critical load and out-plane displacement were examined. The distribution of axial and shear stress across the thickness, when the plate is exposed to uniform transverse loading, was further studied.

Keywords: *Mindlin rectangular plates, power law FG distribution, two parametric elastic foundations*

1. Introduction

Functionally graded materials (FGM) are the new composites that are microscopically heterogeneous. Mechanical properties of this kind of new composite change gradually from metal to ceramics in arbitrary directions. Ceramic can tolerate high temperature, and metal provides good machinery ability, high hardness, and flexibility [1]. Many structures can be modeled with rectangular FG plates in the aerospace engineering. Investigating the buckling critical loads of these structures is of great importance in achieving better performance. Plates with ratio of length to plate thickness less than10 times are known as the thick plates Since the classical plate theory underestimates the out-plane displacement and overestimates the buckling loads and frequencies of the thick plate due to ignoring the effect of shear deformation, many studies have focused on considering the behavior of thick plates with higher order shear deformation theory when subjected to static and dynamic loadings [2-8]. Since the theoretical analysis is carried out in this paper using First Shear Deformation Theory (FSDT), we introduce some papers related to it here. Levy, Reissner, and Mindlin were the first ones who tried to rectify this deficiency existing in classical plate theory [9]. In his first attempt in this area, Levy [10] achieved a solution for governing equations by employing the three-dimensional elasticity. Reissner [11] proposed the theory of thick plate by considering the influence of shear deformation. He performed his approach based on stress analysis for bending the elastic plates. Mindlin [12] assumed that the transverse shear stress is constant along the thickness, whereas this assumption makes to take the shear strain constant, too. To remove this deficiency, he introduced the shear factor in his formulation to predict the shear stress resultant. Levinson [13] refined the theory of Mindlin plate by omitting the shear correction factor from his approach based on the displacement theory. Lanhe [14] investigated the thermal buckling of FG rectangular moderately thick plates by first shear order theory using simply supported edge condition. Liew and Chen [15] proposed the buckling analysis of Mindlin rectangular plates subjected to partial in-plane loading, using radial point interpolation method. Shimpi et al. [16] proposed two new displacement-based, first-order shear deformation theories involving only two unknown functions to analyze static and dynamic problems.

Solving the governing equations based on first and higher-order shear deformation theories brings about computational complexity. In addition, while the neutral plane is not coincident to the middle one, there are extension-bending couplings in FG plates; therefore, to present an efficient and simply solving procedure in order to derive governing equations, are always attractive for researchers [17,18,19].

Solving the governing equations based on first and higher-order shear deformation theories brings about computational complexity. In addition, while the neutral plane is not coincident to the middle one, there are extension-bending couplings in plates made of functionally graded materials; therefore, to present an efficient and simply solving procedure, the derived governing equations always attract researchers [17,18,19].

Furthermore, considering plates resting on elastic foundation is of great importance in modern engineering structures, aerospace, biomechanics, petrochemical, construction, electronics, and nuclear and civil engineering [20]. Winkler simulated an elastic foundation by using a set of linear elastic springs which worked independently without taking into account the effects of shear coupling between them. It was known as the one-parameter model. Pasternak proposed the two-parameter model considering the influence of shear layer between springs [21]. Gupta et al. [22] investigated the buckling and vibration of orthotropic plate in Winkler elastic foundation. Rashed [23] studied the bending analysis of thick plates in two-parametric elastic foundation by exploiting boundary integral transformation. Additionally, in another paper Wen [24] proposed the Laplace transform for analysing moderately thick plate resting on two-parametric elastic foundation. Civalek [25] reported the discrete singular convolution method to solve the governing equation of bending, buckling, and vibration in Mindlin plates resting on two-parametric elastic foundation. Akhavan et al. [26] presented an exact solution for the buckling analysis of rectangular plates on two simply supported opposite edges resting on Pasternak foundation subjected to uniform and non-uniform in-plane loading. Hosseini-Hashemi et al. [27] studied the Hydro-elastic vibration and buckling rectangular plates resting on two-parametric elastic foundation for various edge conditions. In their research the plate was subjected to linear distributed in-plane loading. Bouderba et al. [28] presented thermo-mechanical bending analysis of FG rectangular plate resting on two-parametric elastic foundation. They exploited a developed trigonometric shear deformation theory. The benefit of their approach was dealing with four unknowns as against five in case of the other shear deformation theories. Zidi et al. [29] studied the bending behavior of FG plates, using a four-variable, refined plate theory. In another work, to reduce the number of unknowns in shear deformable theories Hamidi et al. [30] employed a sinusoidal plate theory to study the thermo-mechanical bending of sandwich plates. They dealt with 5 unknowns in the governing equations of this kind of plate. Bennoun et al. [31] proposed a simple method for considering the vibration analysis of FG sandwich plates. They divided the displacement into three parts. The unknowns were diminished to five, as opposed to six or more in the other shear deformation theories.

With respect to this important issue that the governing equation of Mindlin plates are included in three displacement components in three axes of rectangular Cartesian system and two rotations in in-plane directions, solving the governing equations simultaneously to obtain the exact solution is a complicated task. Therefore, in this research in the first stage, we managed to reduce three differential equilibrium equations of plates to one equation in terms of lateral displacement. For this purpose, we rewrote the equations based on the neutral plane of FG plates. Consequently, the stress resultants of FG plates were formulated as isotropic, homogenous ones. In the next stage, by exploiting some algebraic operations, these equations were reduced to one in terms of lateral

displacement. Following this method, the influences of elastic foundation parameters, thickness ratio, loading factor, and various FG power indices were further investigated on deflection, normal and shear stresses, and critical buckling load.

The rest of the paper is organized as follows: In Section 1, the governing equilibrium equations are derived based on Mindlin theory and neutral plane. In Section 2, the solving procedure is devoted to decreasing the three differential equilibrium equations from three to one. By employing this procedure, the quantities of unknowns are diminished from three (rotation in two directions and lateral displacement) to one. Section 3 presents some numerical examples to illustrate the accuracy and precision of the present approach by comparing it with other related approaches in the literature.

2. Displacement Field and Constitutive Equation

Consider the rectangular plate of length a, width b and thickness h, defined in xyz rectangular coordinate system. Its center is located in the middle plane of plate, whereas z is representative for distance of any arbitrary points in the thickness of the plate with respect to the middle plane. Here, we consider a relatively thick plate. In this category of plates, the validation of hypothesis of straight normal is not established. This is due to the existence of shear deformation effects that cannot be ignored. In this paper, we assumed the shear deformation behavior is based on Mindlin model; therefore it can be described by first shear deformation theory (FSDT). The displacement field can be written as follows:

$$
\begin{aligned}
u_\alpha &= u_\alpha\left(x, y, z\right) = (u_0)_\alpha + z\psi_\alpha \\
w &= w_0\left(x, y\right) \qquad \alpha, \beta = x, y
\end{aligned}
\tag{1}
$$

Where u_α and w are the displacement components in x, y and z directions, $(u_0)_\alpha$ and w_0 are mid-plane displacement and are rotation components in in-plane directions. By appropriate differentiation with respect to the coordinate axes, the strain components of Mindlin plate in any arbitrary point in state of the infinitesimal deformation are obtained as follows:

$$
\begin{aligned}
\varepsilon_{\alpha\beta} &= \varepsilon_{\alpha\beta}^0 + 0.5z(\psi_{\alpha,\beta} + \psi_{\beta,\alpha}) \\
\gamma_{\alpha z} &= w_{,\alpha} + \psi_\alpha
\end{aligned}
\tag{2}
$$

In the above relations is middle plane strain. In two-dimensional stress state, the stress-strain relations of Hook's law are defined as follows:

$$
\sigma_{\alpha\beta} = \tilde{E}\left(z\right)\left((1-\mu)(\varepsilon_{\alpha\beta} + \delta_{\alpha\beta}\mu\varepsilon_{\gamma\gamma})\right)
\tag{3.1}
$$

$$
\sigma_{\alpha z} = \tilde{E}\left(z\right)(1-\mu)\varepsilon_{\alpha z}, \quad \tilde{E}\left(z\right) = \frac{E\left(z\right)}{1-\mu^2}
\tag{3.2}
$$

In above equation: $\alpha, \beta, \gamma = x, y$. Based on equations (2) and (3), the stress components can be obtained in terms of strains of the middle plans and rotation in two directions x and y, as follows:

$$\sigma_{\alpha\beta} = \left\{ \begin{array}{l} \tilde{E}(z)((1-\mu)\varepsilon_{\alpha\beta}^0 + \delta_{\alpha\beta}\mu\varepsilon_{\gamma\gamma}^0 \\ +z((0.5(1-\mu)(\psi_{\alpha,\beta}+\psi_{\beta,\alpha})+\delta_{\alpha\beta}\mu\psi_{\gamma,\gamma})) \end{array} \right\} \quad (4)$$

where is Young modulus of the FG plate that assumed to vary according to a simple power law, across the thickness coordinate z. It can be expressed as:

$$\tilde{E}(z) = \tilde{E}_m + (\tilde{E}_c - \tilde{E}_m)V_c \quad (5)$$

Vc, Em and Ec represent ceramic volume fraction, Young modulus of metal and ceramics, respectively. In this study, it is assumed ceramic volume fraction of functionally graded material follows two simple four-parameter power-law distributions. Therefore, Young modulus is defined as follows[32]:

model 1:

$$\tilde{E}(z)\big|_{1(a^*/b^*/c/p)}$$
$$= \tilde{E}_m + \tilde{E}_{cm}(1-a^*(\frac{1}{2}+\frac{z}{h})+b^*(\frac{1}{2}+\frac{z}{h})^c)^p \quad (6\text{-}1)$$

model 2:

$$\tilde{E}(z)\big|_{2(a^*/b^*/c/p)}$$
$$= \tilde{E}_m + (\tilde{E}_{cm})(1-a^*(\frac{1}{2}-\frac{z}{h})+b^*(\frac{1}{2}-\frac{z}{h})^c)^p \quad (6\text{-}2)$$
$$\tilde{E}_{cm} = \tilde{E}_c - \tilde{E}_m$$

In above equation, p is functionally graded power index. When p equals to zero and infinity, it corresponds to pure metal and pure ceramic plate, respectively. The parameters a^*, b^*, c corresponds for the material variation profile through the thickness of functionally graded plate.

3. Stress Resultants

The stress resultants can be expressed as follows:

$$(N_{\alpha\beta}, M_{\alpha\beta}) = \int_{-h/2}^{h/2} \sigma_{\alpha\beta}(1,z)dz \quad \alpha,\beta = x,y$$
$$Q_{\alpha z} = \int_{-h/2}^{h/2} \kappa^2 \sigma_{\alpha z} dz \quad \kappa^2 = \frac{\pi^2}{12} \quad (7)$$

κ^2 is the transverse shear correction coefficient, applied to the transverse shear forces based on the parabolic distribution of shear transverse strains across thickness. By inserting relations (4) into (7) and integration across the thickness, the stress resultants are evaluated, as follows:

$$N_{\alpha\beta} = A\left((1-\mu)\varepsilon^0{}_{\alpha\beta} + \delta_{\alpha\beta}\mu\varepsilon^0{}_{\gamma\gamma}\right)$$
$$+B\left(0.5(1-\mu)(\psi_{\alpha,\beta}+\psi_{\beta,\alpha})+\delta_{\alpha\beta}\mu\psi_{\gamma,\gamma}\right) \quad (8a)$$

$$M_{\alpha\beta} = A\left((1-\mu)\varepsilon^0{}_{\alpha\beta} + \delta_{\alpha\beta}\mu\varepsilon^0{}_{\gamma\gamma}\right)$$
$$+B\left(0.5(1-\mu)(\psi_{\alpha,\beta}+\psi_{\beta,\alpha})+\delta_{\alpha\beta}\mu\psi_{\gamma,\gamma}\right) \quad (8b)$$

where,

$$(A,B,D) = \int_{-\frac{h}{2}}^{\frac{h}{2}} (1,z,z^2)\tilde{E}(z)dz \quad (9)$$

It is clear that the extension-bending coupling existing in the above relations makes the solution procedure get more complicated. For anisotropic plates, the neutral plane doesn't coincide with the middle one. For an isotropic plate or a composite one with symmetrical mechanical properties with respect to mid-plane of the plate, B gets value of zero and consequently the displacement of the middle plane equals zero, too. Abrate [18] showed that instead of expressing the stress resultant relations in base of middle plane, writing them with respect to the neutral one, the governing equations are derived as those of ones for homogenous materials. To achieve the aim, we introduced z_0 as the position of the neutral plane, as follows [33]:

$$z_0 = \int_{-h/2}^{h/2} z\tilde{E}(z)dz \Big/ \int_{-h/2}^{h/2} \tilde{E}(z)dz = B/A \quad (10)$$

It is obvious that for an isotropic or symmetric laminated composite plate, the neutral plane is coincident with the middle surface. After substituting into relations (9) and by employing relation:

$$(\bar{A},\bar{B},\bar{D}) = \int_{-h/2}^{h/2} \tilde{E}(z)(1,\bar{z},\bar{z}^2)dz \quad \bar{z} = z - z_0 \quad (11)$$

$\bar{B}$ equals to zero. Consequently the bending-extension coupling is removed.

$$\bar{A} = A, \quad \bar{B} = 0, \quad \bar{D} = z_0^2 A - 2z_0 B + D$$

Therefore, the stress resultants are expressed as:

$$N_{\alpha\beta} = A\left((1-\mu)\varepsilon_{\alpha\beta}^0 + \delta_{\alpha\beta}\mu\varepsilon_{\gamma\gamma}^0\right)$$
$$M_{\alpha\beta} = \bar{D}\left(0.5(1-\mu)(\psi_{\alpha,\beta}+\psi_{\beta,\alpha})+\delta_{\alpha\beta}\mu\psi_{\gamma,\gamma}\right) \quad (12)$$
$$Q_{\alpha z} = A\kappa^2(1-\mu)(w_{,\alpha}+\psi_\alpha),$$

4. Governing equation of Mindlin Plates

The governing equation can be derived through principal of virtual work. It can be stated as follows:

$$\iiint_V \sigma_{\alpha\beta}\delta\varepsilon_{\alpha\beta}dV + \int_A (F_k + q)\delta w = 0 \quad (13)$$

In above equation F_k and q are elastic foundation reaction force and external loading, respectively. Substituting equation (2) and (4) into equation (13) and employing this principle leads to the following governing equations of Mindlin plate, as follow:

$$N_{\alpha\beta,\beta} = 0 \quad \alpha,\beta = x,y \quad (14a)$$

$$M_{x,xx} + 2M_{xy,xy} + M_{y,yy} + q + k_1 w_0 - k_2\nabla^2 w = 0 \quad (14b)$$

Where k_1 and k_2 are Pasternak elastic foundation parameters. By assumption to infinitesimal deformation and writing the equilibrium equations based on the neutral surface, the in-plane displacement u_0, v_0 are eliminated. Then by substituting the relations (12) into equilibrium equations (14), the governing equations for a Mindlin plate are expressed as:

$$D_1 \nabla^2 \psi_\alpha + D_2 \psi_{,\alpha} + A\left(w_{,\alpha} - \psi_\alpha\right) = 0 \qquad (15a)$$

$$A\left(\nabla^2 w - \psi\right) + q^* = 0 \qquad (15b)$$
$$D_1 = \bar{D}(1-\mu)/2, \quad D_2 = \bar{D}(1+\mu)/2$$

Where $\psi = \psi_{\alpha,\alpha}$, q^* in the equation (15b) is defined as follow:

$$q^* = q - k_1 w + k_2 \nabla^2 w \qquad (16)$$

where q, k_1, k_2 are the transverse loading, elastic foundation reaction parameters of Pasternak model, respectively. Now By differentiation of equation (15a) with respect to α and rearranging it, then:

$$\left(D_1 + D_2\right)\nabla^2 \psi + A\left(\nabla^2 w - \psi\right) = 0 \qquad (17)$$

In state of buckling analysis, q is expressed as:

$$q = N_{\alpha\beta} w_{,\alpha\beta} \qquad \alpha,\beta = x,y \qquad (18)$$

In above equation, $N_{\alpha\beta}$ denotes to in-plane loading. By eliminating of rotation parameters from governing equation (17) by substituting from equation (15b) into it, bi-harmonic resultant equation is obtained as:

$$\bar{D}\nabla^4 w = \hat{q}, \quad \hat{q} = -s\nabla^2 q^* + q^* \quad s = \frac{\bar{D}}{A} \qquad (19)$$

As observed, the above equation assembles the same one that employed for thin plate with classical plate theory (CPT). By introducing the non-dimensional parameters and operator L, as follows:

$$\xi = \frac{x}{a}, \zeta = \frac{y}{b} \quad , W = \frac{w}{a}$$

$$\delta = \frac{a}{b} \quad , \frac{k_1 a^4}{\bar{D}} = \bar{k}_1 \quad , \frac{k_2 a^2}{\bar{D}} = \bar{k}_2 , \bar{q} = \frac{qa^3}{\bar{D}}$$

$$L = \frac{\partial^2}{\partial \xi^2} + \delta^2 \frac{\partial^2}{\partial \zeta^2}, \quad L(\ ) = a^2 \nabla^2(\ )$$

Using in equation (20), the governing equation of Mindlin plate resting on Pasternak elastic foundation exposed on transverse loading, can be written in non-dimensional form as follow:

$$c_1 L^2\left(W\right) + c_2 L\left(W\right) + c_3 W = \hat{q}^* \qquad (20)$$

where,

$$c_1 = 1 + \bar{k}_2 \frac{s}{a^2}, \quad c_2 = -\bar{k}_1 \frac{s}{a^2} - \bar{k}_2 \quad , c_3 = \bar{k}_1,$$

$$\hat{q}^* = \bar{q} - \frac{s}{a^2} L\left(\bar{q}\right)$$

In state of buckling analysis, the governing differential equation is derived in non-dimensional form as:

$$c_1 L^2\left(W\right) + c_2\left(W\right) + c_3 W + c_4 L^*\left(L(W) + W\right) = 0 \qquad (21)$$

where,

$$L^* = \frac{\partial^2}{\partial \xi^2} + r\delta^2 \frac{\partial^2}{\partial \zeta^2}, \qquad \lambda = \frac{N_x a^2}{\bar{D}}$$

$$\frac{N_y}{N_x} = r, \qquad c_4 = \lambda(\frac{s}{a^2} - 1).$$

When all edge set in simply supported condition for a thick rectangular plate, the out-plane displacement component w and rotations ψ_x and ψ_y can be defined in the form of sinus Fourier double series as follows:

$$W = \sum_n \sum_m W_{mn} \sin \alpha_m \xi \sin \beta_n \zeta \qquad (22a)$$

$$\psi_\xi = \sum_n \sum_m E_{mn} \cos \alpha_m \xi \sin \beta_n \zeta \qquad (22b)$$

$$\psi_\zeta = \sum_n \sum_m F_{mn} \sin \alpha_m \xi \cos \beta_n \zeta \qquad (22c)$$

where $\alpha_m = m\pi$, $\beta_n = n\pi$

4.1. Bending Analysis

By considering the transverse loading in two form of sinusoidal and uniform distribution, as:

$$q = \begin{cases} q_0 \sin(\pi\xi)\sin(\pi\zeta) & type\ 1 \\ q_0 & type\ 2 \end{cases} \qquad (23)$$

By utilizing aforementioned function and equation (22) in governing equation (20), the coefficient of W_{mn} is determined, as follows:
-type 1

$$W_0 = \frac{\bar{q}_0(1 + \frac{s}{a^2}\left(1+\delta^2\right)\pi^2)}{c_3 - c_2 \pi^2\left(1+\delta^2\right) + c_1 \pi^4\left(1+\delta^2\right)^2} \qquad (24a)$$

-type 2

$$W_{mn} = \frac{16\frac{\bar{q}_0}{mn\pi^2}(1 + \frac{s}{a^2}\left(m^2 + \delta^2 n^2\right)\pi^2)}{c_3 - c_2 \pi^2\left(m^2 + \delta^2 n^2\right) + c_1 \pi^4\left(m^2 + \delta^2 n^2\right)^2} \qquad (24b)$$

To find the rotation in direction x, firstly ψ can be determined based on equation (15b) and then by substituting into equation (15a), ψ_x may be derived from the following equation:

$$D_1 \nabla^2 \psi_x - A\psi_x + D_2(\nabla^2 w + q^* / A)_{,x} + Aw_{,x} = 0 \quad (25)$$

The non-dimensional form of equation (26) is expressed as follows:

$$L(\psi_\xi) + p_1 \psi_\xi + p_2 L(W_{,\xi}) + p_3 W_{,\xi} + p_4 \bar{q}_{,\xi} = 0 \quad (26)$$

where,

$$p_1 = -\frac{2a^2}{s(1-\mu)}, \qquad p_2 = \mu^*(1 + \frac{s}{a^2}\bar{k}_2)$$

$$p_3 = \frac{2a^2}{s(1-\mu)} - \mu^*\bar{k}_1 \frac{s}{a^2}, \quad p_4 = \frac{s}{a^2}\mu^*, \qquad \mu^* = \frac{1+\mu}{1-\mu}$$

By substituting the equation (22a) into aforementioned differential equation, ψ_ξ may be expressed in terms of out-plane displacement component as follows:
-type 1

$$E_0 = \frac{W_0(\pi^3 p_2(1+\delta^2) - \pi p_3) - \pi p_4 \bar{q}_0}{-\pi^2(1+\delta^2) + p_1} \quad (27a)$$

-type 2

$$E_{mn} = \frac{W_{mn}(\pi^3 m p_2(m^2 + \delta^2 n^2) - \pi m p_3) - 16 p_4 \dfrac{\bar{q}_0}{n\pi}}{-\pi^2(m^2 + \delta^2 n^2) + p_1} \quad (27b)$$

By carrying out the similar steps to determine ψ_η, this parameter can be obtained, too.
-type 1

$$F_0 = -\frac{\pi E_0 + a\psi_0}{\delta\pi} \quad (28a)$$

$$a\psi_0 = q_0\frac{s}{a^2} + W_0\left(c_3\frac{s}{a^2} - c_1\pi^2(1+\delta^2)\right)$$

-type 2

$$F_{mn} = -\frac{\pi m E_{mn} + a\psi_{mn}}{\delta\pi n} \quad (28b)$$

$$a\psi_{mn} = \frac{16 q_0}{\pi^2 mn}\frac{s}{a^2} + W_{mn}\left(c_3\frac{s}{a^2} - c_1\pi^2(m^2 + \delta^2 n^2)\right)$$

The distribution of shear stresses $\sigma_{\alpha z}$ through thickness can be obtained from equilibrium equations as follows:

$$\frac{\partial\sigma_{\alpha\beta}}{\partial x_\beta} + \frac{\partial\sigma_{\alpha z}}{\partial z} = 0 \rightarrow$$

$$\sigma_{\alpha z} = -\int_{-h/2}^{z}\frac{\partial\sigma_{\alpha\beta}}{\partial x_\beta}d\varsigma \quad -h/2 < \varsigma < z \quad (29)$$

Upon substituting of the in-plane stress equation (4) based on neutral plane, into equation (29):

$$\sigma_{\alpha\beta} = \frac{z - z_0}{h}\tilde{E}(z)\begin{Bmatrix} 0.5(1-\mu)(\psi_{\alpha,\beta} + \psi_{\beta,\alpha}) \\ +\delta_{\alpha\beta}\mu\psi_{\gamma,\gamma} \end{Bmatrix}. \quad (30)$$

The distribution of shear stress σ_{xz} and σ_{yz} can be obtained as:

$$\sigma_{\alpha z} = B^*(z)(\mu^+\nabla^2\psi_\alpha + \mu^-\psi_{,\alpha}) \quad (31)$$

where $B^*(z)$, $\delta_{\alpha\beta}$ is Kronecker delta and μ^+ and μ^- are defined as:

$$B^*(z) = \int_{-h/2}^{z} z_1\tilde{E}(z_1)dz \quad -h/2 < z_1 < z$$

$$\mu^+ = (1+\mu)/2 \quad , \quad \mu^- = (1-\mu)/2.$$

4.2. Buckling Analysis

In bucking problem, an analytical solution for critical bucking loads of FGM rectangular plate with four-side simply supported boundaries can be obtained by utilizing equation (20a) into governing equation (21), as follows:
- Biaxial in-plane loading (r=1):

$$\lambda_{cr} = \frac{\begin{bmatrix} (1+\bar{k}_2\dfrac{s}{a^2})\pi^2(m^2 + \delta^2 n^2) \\ +(\bar{k}_1\dfrac{s}{a^2} + \bar{k}_2) + \dfrac{\bar{k}_1}{\pi^2(m^2 + \delta^2 n^2)} \end{bmatrix}}{(1-\dfrac{s}{a^2})(\pi^2(m^2 + \delta^2 n^2) - 1)} \quad m,n = 1,3,5,\dots \quad (32)$$

- Uniaxial in-plane loading (r=0):

$$\lambda_{cr} = \frac{\begin{bmatrix} (1+\bar{k}_2\dfrac{s}{a^2})\pi^2(m^2 + \delta^2 n^2) \\ +(\bar{k}_1\dfrac{s}{a^2} + \bar{k}_2) + \dfrac{\bar{k}_1}{\pi^2(m^2 + \delta^2 n^2)} \end{bmatrix}}{(1-\dfrac{s}{a^2})(\pi^2 m^2 - \dfrac{m^2}{(m^2 + \delta^2 n^2)})} \quad (33)$$

$$m,n = 1,3,5,\dots$$

5. Numerical Results

In this section, the author represented some numerical results. A computer program has been prepared in MATLAB. These results are categorized in two subsections, one for bending analysis and the other for buckling analysis. In both, the following is assumed:

$$a/b = 1, \quad a/h = 10, \quad a = 1\ m, \quad \nu = 0.3,$$
$$E_m = 68 \quad GPa, \quad E_c = 168 \quad GPa$$

E_m and E_c are Aluminum (metal) and Alumina (ceramic) Young modulus, respectively. The simply supported FG plate resting on Pasternak foundation parameters: $k_1 = 0.1 q_0$, $k_2 = 0.1 k_1$. In all numerical results, a shear correction factor of $\pi^2/12$ is used. Depending on the FG model in equations (6), three categorized cases can be considered: case1 ($a^* = 1, b^* = 0$), case2 ($a^* = b^* = 1$), case3 ($a^* = 1, b^* = 0.5$). In all cases, c=2 is kept as constant. The FG power index p is chosen: (0.125 0.25 0.5 1 2 5 10 15 20 50 100).

As observed in Figure 1- Figure 3, the FG material distribution obeys asymmetric profile in cases 1 and 3, whereas in case 2, the material distribution through the thickness is symmetric through the thickness. In cases 1 and 3, the distributions of metal and ceramic as constitution are characterized by the fact that bottom surface and top surface is ceramic rich in model and model 2, respectively. However, there is a mixture of two constituents through the thickness on the one surface when ceramic rich is obtained in one of two surfaces.

As observed in Figure 1a - Figure 1f, the FG material distribution obeys asymmetric profile in cases 1 and 3, whereas in case 2, the material distribution through the thickness is symmetric through the thickness. In cases 1 and 3, the distributions of metal and ceramic as constitution are characterized by the fact that bottom surface and top surface is ceramic rich in model and model 2, respectively. However, there is a mixture of two constituents through the thickness on the one surface when ceramic rich is obtained in one of two surfaces.

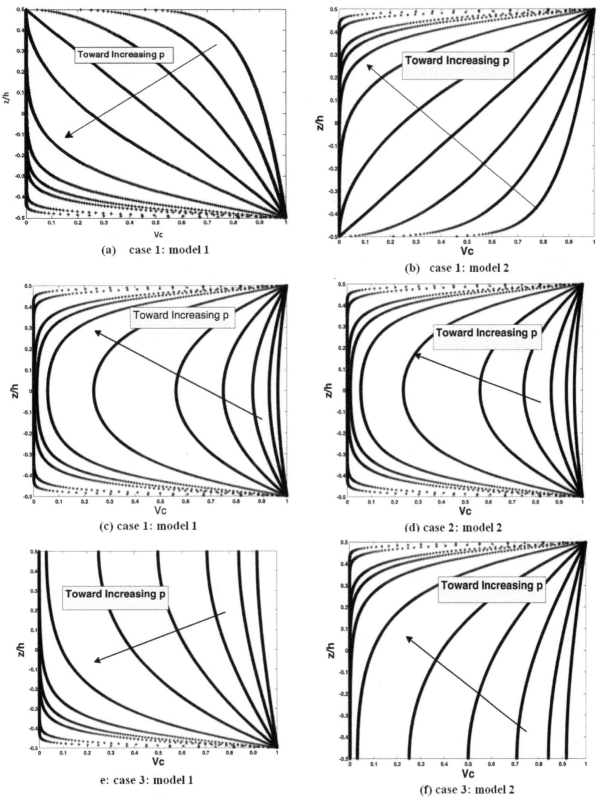

(a)   case 1: model 1

(b)   case 1: model 2

(c) case 1: model 1

(d) case 2: model 2

e: case 3: model 1

(f) case 3: model 2

Figure 1. Variations of the ceramic volume fraction V_C through the thickness for cases 1,2,3

5.1. Bending Analysis

The plate is subjected to two forms of transverse loading: Sinusoidal and uniformly distributed model. The variation of maximum deflection (at the center of the plate) in non-dimensional form for the aforementioned cases is tabulated in Table 1 and Table 2. The following non-dimensional parameters are used in presenting the numerical results in figures and tables:

$$\bar{w} = w \, \frac{E_m h^3}{q_0 l^4}, \qquad \bar{\tau}_{xz} = \frac{\tau_{xz} h^2}{q_0 l^2}, \qquad \bar{\sigma}_{xx} = \frac{\sigma_{xx} h^2}{q_0 l^2}$$

Table 1. variation of non-dimensional displacement $\bar{w}$ **with respect to variation of FG law index for uniform loading**

case	(k1,k2)	p					
		0	0.5	1	5	10	20
1	(0,0)	0.22287	0.29286	0.33339	0.40159	0.43012	0.46270
	(0.1,0.01) q_0	0.18466	0.23019	0.25447	0.29224	0.30701	0.32320
2	(0,0)	0.22287	0.22986	0.23672	0.28632	0.33416	0.39394
	(0.1,0.01) q_0	0.18466	0.18943	0.19406	0.22613	0.25490	0.28818
3	(0,0)	0.22287	0.24846	0.27307	0.38819	0.42673	0.46079
	(0.1,0.01) q_0	0.18466	0.20188	0.21781	0.28510	0.30529	0.32227

Table 2. variation of non-dimensional displacement $\bar{w}$ **with respect to variation of FG index for sinusoidal loading, h=100 mm**

case	(k₁,k₂)	p					
		0	0.5	1	5	10	20
1	(0,0)	0.14086	0.18510	0.21072	0.25383	0.27187	0.29246
	(0.1,0.01)q_0	0.11727	0.14639	0.16196	0.18628	0.19582	0.20628
2	(0,0)	0.14086	0.14528	0.14962	0.18098	0.21122	0.24900
	(0.1,0.01)q_0	0.11727	0.12031	0.12327	0.14380	0.16226	0.18367
3	(0,0)	0.14086	0.15704	0.17260	0.24536	0.26972	0.29125
	(0.1,0.01)q_0	0.11727	0.12827	0.13846	0.18168	0.19470	0.20567

As observed, in case of symmetric volume fraction profile, the deflection has less value than that of other two cases. Also, by raising p, the volume fraction of ceramic increased, thereby decreasing the values of deflection. The validity of present results was verified for the non-dimensional deflection ($\bar{w}$), as shown in Table 3. It can be concluded that the present yields very good results compared to that presented by Civalek [24].

Table 3. comparison $\bar{w} = 100w\bar{D}/q_0 l^4$ **in center of square SSSS plate**

h/a	Source	
	Civalek(2009)	Present study
0.01	0.4062	0.4062
0.1	0.4274	0.4273
0.2	0.4906	0.4905
0.3	0.5956	0.5957

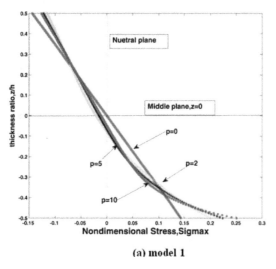

(a) model 1

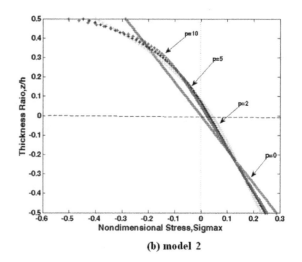

(b) model 2

Figure 2. Distribution of non-dimensional stress σ_{xx} across the thickness for case 1, $k_1=k_2=0$

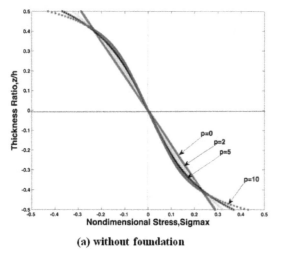

(a) without foundation

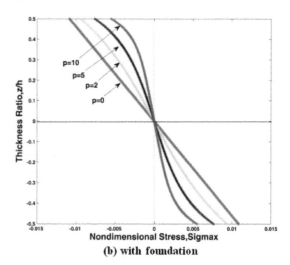

(b) with foundation

Figure 3. Distribution of non-dimensional stress σ_{xx} across the thickness for case 2, model 2

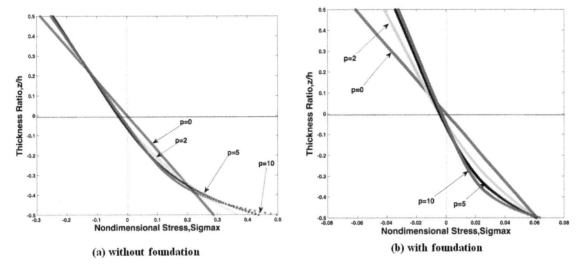

(a) without foundation   **(b) with foundation**

Figure 4. Distribution of nondimensional stress σ_{xx} across thickness for case 3, model 1

In Figure 2 – Figure 4 to show the capability of the presented method, distribution of bending stress component σ_x across the thickness with and without influence of elastic foundation under simply supported boundary condition, are illustrated for cases (1-3). As observed, stress component does not take value of zero in the middle surface for a FG plate. In addition, the amount of bending stress decreases effectively in the presence of elastic foundation.

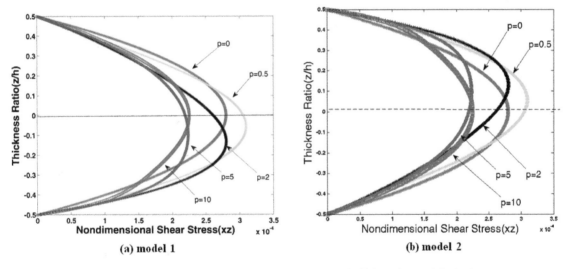

(a) model 1   **(b) model 2**

Figure 5. Distribution of nondimensional shear stress τ_{xz} across the thickness for case 1, $k_1=k_2=0$

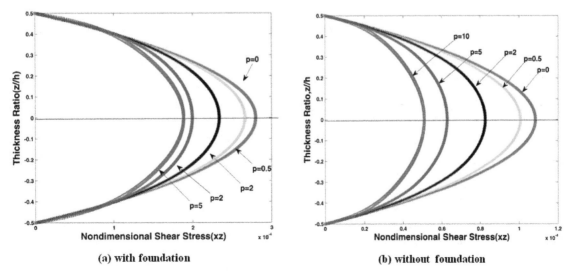

(a) with foundation   **(b) without foundation**

Figure 6. Distribution of nondimensional shear stress τ_{xz} across the thickness for case 2, model 1

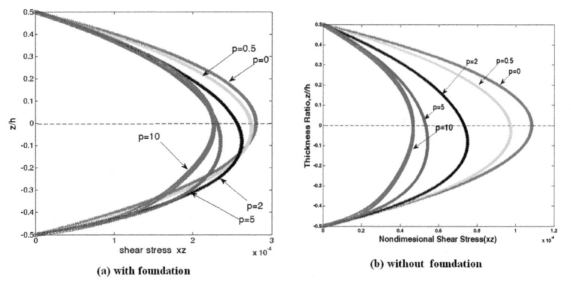

Figure 7. Distribution of nondimensional shear stress τ_{xz} across the thickness for case 3, model 1

In Figure 5-Figure 7, the distribution of shear stress τ_{xz} across the thickness is depicted. As observed, the maximum of shear stress doesn't occur at the middle surface for cases 1 and 3, due to be not symmetric distribution of mixture of ceramic and metal throughout the thickness. On the other hand, as shown in Figure 6, in case 2, this distribution seems to be symmetric as profile of ceramic volume fraction has symmetric pattern, too.

5.2. Buckling Analysis

The buckling analysis of FG rectangular thick plate is carried out for simply supported boundary condition. The results are given for case 1, n=1 as tabulated (Table 4). As observed, increasing the FG power law decreases the critical buckling load. This is due to the fact that increasing n raises the volumetric percentage of metal in FG plate, and as a result decreasing the bending rigidity of FG plate. Thus, when the plate is exposed to biaxial compression in-plane loads, the decline in the resistance against buckling is more diminished in comparison with the state of uniaxial compression loading. Also, the results are compared with those have been reported to be obtained by Thai and Kim [34].

Table 4. Influence of FG power index, elastic foundation parameters and thickness ratio (l/h) on λ_{cr}

r	(k1,k2)	l/h	p					
			0	0.5	1	2	5	10
0	(0,0)	10	19.4801	12.8540	9.3530	7.3312	6.5330	5.1005
		20	20.1748	12.9017	9.7784	7.5214	6.8948	5.3600
		100	20.9414	13.3521	10.550	7.9971	6.9140	5.9981
	(100,10)	10	21.8509	13.9880	12.014	10.5784	9.9947	8.7808
		20	23.0011	15.7456	12.5941	10.7015	9.1041	8.6721
		100	22.9870	16.1540	12.4432	10.6668	9.4516	8.4511
1	(0,0)	10	8.9841	6.6664	5.001	4.0011	2.8745	3.1880
		20	9.1245	6.8780	5.1145	4.3120	3.0050	3.0101
		100	9.3332	6.9956	5.3830	4.2879	3.1241	3.2241
	(100,10)	10	10.1987	7.5790	6.8540	5.1133	3.2183	4.2154
		20	10.6547	7.6666	6.9780	5.1402	4.1521	4.2634
		100	10.9914	7.7780	7.5333	5.2321	4.3316	4.5250

Table 5. Comparison the critical buckling load in state of biaxial compression loading with Thai and Kim [34] for case 1, $k_1=k_2=0$

l/h	p									
	0		1		2		5		10	
	Present work	Thai and Kim [34]	Present work	Thai and Kim [34]	Present work	Thai and Kim [34]	Present work	Thai and Kim [34]	Present work	Thai and Kim [34]
10	8.9841	9.2893	5.001	4.6695	4.0011	3.6315	2.8745	3.0177	3.1880	2.7264
20	9.1245	9.6764	5.1145	4.8337	4.3120	3.7686	3.0050	3.1724	3.2241	2.8834

6. Conclusions

In this paper, the buckling and bending responses of rectangular thick plates made of functionally graded materials resting on two-parametric foundation were

investigated. Since in FSDT theory the governing equations involve five unknown functions including three displacements and two rotations, the author aimed to represent a simple and effective procedure to reduce the complexity of the solution. As a result, three coupled governing equations were diminished to one ordinary differential equation in terms of out-plane displacement. The influences of power law index, thickness ratio, and foundation parameters on the critical buckling load were investigated in two cases of uniaxial and biaxial in-plane loadings of FG plates. The distribution of bending and shear stresses across the thickness direction was studied as well as out-plane displacement. The result showed that in the case of symmetric volume fraction profile, the deflection had less value than those of the other two cases. It is suggested that this approach should be used in vibration and post buckling of Mindlin FG plates.

Acknowledgments

The research described in this paper was financially supported by Islamic Azad University, Khomeinishahr branch.

References

[1] Shenshen H., Functionally Graded Materials Nonlinear Analysis of Plates and Shells, CRC Press, Taylor and Francis Group, USA , 2009.

[2] Tounsi, A., Houari, M.S.A., Benyoucef, S., Adda Bedia, E.A., "A refined trigonometric shear deformation theory for thermoelastic bending of functionally graded sandwich plates", Aerosp Sci Technol, 24.209-220. 2013.

[3] Belabed, Z., Houari, M.S.A., Tounsi, A., Mahmoud, S.R., Anwar Bég, O., "An efficient and simple higher order shear and normal deformation theory for functionally graded material (FGM) plates", Compos Part- B, 60. 274-283. 2014.

[4] Meziane, A. A., Abdelaziz, H.H., Tounsi, A., "An efficient and simple refined theory for buckling and free vibration of exponentially graded sandwich plates under various boundary conditions", J. Sandwich Struct. Mater. 16(3). 293-318. 2014.

[5] Hebali, H., Tounsi, A., Hourai, M.S.A., Bessian, A., "A new quasi-3D hyperbolic shear deformation theory for the static and free vibration analysis of functionally graded plates", ASCE J. Eng. Mech., 140.374-383. 2014.

[6] Mahi, A., Adda Bedia, E.A., Tounsi, A. (2015), "A new hyperbolic shear deformation theory for bending and free vibration analysis of isotropic, functionally graded, sandwich and laminated composite plates", Appl. Math. Modell, 39. 2489-2508.

[7] Ait Yahia, S., Ait Atmane, H., Houari, M.S.A., Tounsi, A., "Wave propagation in functionally graded plates with porosities using various higher-order shear deformation plate theories", Struct. Eng. Mech., 53(6). 1143-1165. 2015.

[8] Bourada, M., Kaci, A., Houari, M.S.A., Tounsi, A., "A new simple shear and normal deformations theory for functionally graded beams", Steel Compos. Struct. 18(2), 409-423. 2015.

[9] Szilard, R., Theories and Applications of Plate Analysis, John Wiley and Sons, New Jersy, USA, 2004.

[10] Lévy, M., "Mémoire sur la théorie des plaques élastiques ", J. Math. Pure Appl., 30. 219-306. 1877.

[11] Reissner, E., "On the theory of bending of elastic plates", J. Math Phys. 23, 184-191, 1944.

[12] Mindlin, R.D., "Influence of rotary inertia on flexural motions of isotropic elastic plates", J. Appl. Mech., 18.31-38. 1951.

[13] Levinson, M.,"An accurate, simple theory of the statics and dynamics of elastic plates", Mech. Res. Commun., 7(6). 343-350.1980.

[14] Lanhe, W., "Thermal buckling of a simply supported moderately thick rectangular FGM plate", Compos. Struct., 64. 211-218.2004.

[15] Liew, K.M.,Chen X.L., "Buckling of rectangular Mindlin plates subjected to partial in-plane edge loads using the radial point interpolation method", Int. J. Solids Struct., 41(5-6). 1677-1695.2004.

[16] Shimpi, R. P. ,Patel, H. G. , Arya, H. , New first-order shear deformation plate theories, J. Appl. Mech., 74. 523-533. 2007.

[17] Morimoto T., Tanigawa, Y.,"Linear buckling analysis of orthotropic inhomogeneous rectangular plates under uniform in-plane compression", Acta Mech., 187(1), 219-229. 2006.

[18] Abrate, S., "Functionally graded plates behave like homogeneous plates", Compos. Part B: Eng. 39 (1). 151-158. 2008.

[19] Bousahla, A.A., Houari, M.S.A., Tounsi, A., Adda Bedia, E.A., "A novel higher order shear and normal deformation theory based on neutral surface position for bending analysis of advanced composite plates", Int. J. Comput. Methods, 11(6), 1350082, (2014).

[20] Abbasi, S., Farhatnia, F., Jazi, S. R., "A semi-analytical solution on static analysis of circular plate exposed to non-uniform axisymmetric transverse loading resting on Winkler elastic foundation", Arch. Civil Mech. Eng. (ACME), 14. 476-488. 2014.

[21] Birman, V., Plates Structures, Springer, New York, NY, USA, 2011.

[22] Gupta, U.S., Ansari, A.H., Sharma, S. "Buckling and vibration of polar orthotropic circular plate resting on Winkler foundation", J. Sound Vib., 297. 457-476. 2006.

[23] Rashed, Y.F., "A boundary integral transformation for ending analysis of thick plates resting on Bi-parameter foundation, Adv. Struct. Eng., 5(1). 13-22.2009.

[24] Wen, P.H., "The fundamental solution of Mindlin plates resting on an elastic foundation in the Laplace domain and its applications", Int. J. Solids Struct., 45. 1032-1050. 2008.

[25] Civalek, O., "Three-dimensional vibration, buckling and bending analyses of thick rectangular plates based on discrete singular convolution method", Int. J. Mech. Sci., 49. 752-765. 2008.

[26] Akhavan, H., Hosseini-Hashemi, S., Damavandi-Taher, HR, Alibeigloo A., Vahabi, S., "Exact solutions for rectangular Mindlin plates under in-plane loads resting on Pasternak elastic foundation, Part I: buckling analysis", Comput. Mater. Sci., 44(3). 968-978. 2009.

[27] Hosseini-Hashemi, Sh., Karimi, M., Hossein Rokni, D.T., "Hydroelastic vibration and buckling of rectangular Mindlin plates on Pasternak foundations under linearly varying in-plane loads", Soil Dyn. Earthq. Eng., 30, 1487-1499. 2010.

[28] Bouderba, B., Tounsi, A., Hourai, M.S.A., "Thermomechanical bending response of FGM thick plates resting on Winkler–Pasternak elastic foundations", Steel Compos. Struct, 14(1). 85-104. 2013.

[29] Zidi, M., Tounsi, A., Houari, M.S.A., Adda Bedia, E.A., Anwar Bég, O. (2014), "Bending analysis of FGM plates under hygro-thermo-mechanical loading using a four variable refined plate theory", Aerosp. Sci. Techno., 34. 24-34. 2014.

[30] Hamidi, A., Houari, M.S.A., Mahmoud, S.R., Tounsi, A.,"A sinusoidal plate theory with 5-unknowns and stretching effect for thermomechanical bending of functionally graded sandwich plates", Steel Compos. Struct.,, 18(1), 235-253. 2015.

[31] Bennoun, M., Houari, M.S.A., Tounsi, A., "A novel five variable refined plate theory for vibration analysis of functionally graded sandwich plates", Mech. Adv. Mat. Struct., 23(4). 423-431. 2016.

[32] Tornabene F.,"Free vibration analysis of functionally graded conical, cylindrical shell and annular plate structures with a four-parameter power-law distribution", Comput. Methods Appl. Mech. Eng., 198, 2911-2935. 2009.

[33] Latifi, M., Farhatnia F. and Kadkhodaei, M. "Buckling analysis of rectangular functionally graded plates under various edge conditions using Fourier series expansion", Eur. J. Mech.-A/Solids, 41. 16-27. 2013.

[34] Thai, H., Kim, S., "Closed-form solution for buckling analysis of thick functionally graded plates on elastic foundation", Int. J. Mech. Sci., 5. 34-44. 2015.

Permissions

All chapters in this book were first published in AJME, by Science and Education Publishing; hereby published with permission under the Creative Commons Attribution License or equivalent. Every chapter published in this book has been scrutinized by our experts. Their significance has been extensively debated. The topics covered herein carry significant findings which will fuel the growth of the discipline. They may even be implemented as practical applications or may be referred to as a beginning point for another development.

The contributors of this book come from diverse backgrounds, making this book a truly international effort. This book will bring forth new frontiers with its revolutionizing research information and detailed analysis of the nascent developments around the world.

We would like to thank all the contributing authors for lending their expertise to make the book truly unique. They have played a crucial role in the development of this book. Without their invaluable contributions this book wouldn't have been possible. They have made vital efforts to compile up to date information on the varied aspects of this subject to make this book a valuable addition to the collection of many professionals and students.

This book was conceptualized with the vision of imparting up-to-date information and advanced data in this field. To ensure the same, a matchless editorial board was set up. Every individual on the board went through rigorous rounds of assessment to prove their worth. After which they invested a large part of their time researching and compiling the most relevant data for our readers.

The editorial board has been involved in producing this book since its inception. They have spent rigorous hours researching and exploring the diverse topics which have resulted in the successful publishing of this book. They have passed on their knowledge of decades through this book. To expedite this challenging task, the publisher supported the team at every step. A small team of assistant editors was also appointed to further simplify the editing procedure and attain best results for the readers.

Apart from the editorial board, the designing team has also invested a significant amount of their time in understanding the subject and creating the most relevant covers. They scrutinized every image to scout for the most suitable representation of the subject and create an appropriate cover for the book.

The publishing team has been an ardent support to the editorial, designing and production team. Their endless efforts to recruit the best for this project, has resulted in the accomplishment of this book. They are a veteran in the field of academics and their pool of knowledge is as vast as their experience in printing. Their expertise and guidance has proved useful at every step. Their uncompromising quality standards have made this book an exceptional effort. Their encouragement from time to time has been an inspiration for everyone.

The publisher and the editorial board hope that this book will prove to be a valuable piece of knowledge for researchers, students, practitioners and scholars across the globe.

List of Contributors

Ján Kostka, Peter Frankovský, František Trebuňa, Miroslav Pástor and František Šimčák
Technical University of Košice, Faculty of Mechanical Engineering, Košice, Slovakia

Marian Dudziak and Michał Śledziński
Poznan University of Technology, Chair of Basics of Machine Design, 60-365 Poznań, ul. Piotrowo 3, Poland

Andrzej Lewandowski
Jan Sehn Institute of Forensic Research in Kraków, 31-033 Kraków, ul. Westerplatte 9, Poland

H. Volkan Ersoy
Department of Mechanical Engineering, Yildiz Technical University, Istanbul, Turkey

Róbert Huňady, Martin Hagara and František Trebuňa
Technical University of Košice, Faculty of Mechanical Engineering, Košice, Slovakia

Hirave S.H. and Jadhav S.M.
Department of Mechanical Design Engineering, NBN Sinhgad School of Engineering, Pune

Joshi M.M. and Gajjal S.Y.
Department of Mechanical Engineering, NBN Sinhgad School of Engineering, Pune

O.A. Olugboji, M.S. Abolarin, I.E. Ohiemi and K.C. Ajani
Department of Mechanical Engineering, School of Engineering and Engineering Technology, Federal University of Technology, Minna, Nigeria

Ekuase Austin, Aduloju Sunday Christopher, Ogenekaro Peter, Ebhota Williams Saturday and Dania David E.
National Engineering Design Development Institute, Nnewi, Nigeria

Aysar A. Alamery, Zainab F. Mahdi and Hussein A. Jawad
Institute of laser for postgraduate studies, University of Baghdad, Iraq

Sobhi FRIKHA, Zied DRISS and Mohamed Aymen Hagui
Laboratory of Electro-Mechanic Systems (LASEM), National School of Engineers of Sfax (ENIS), University of Sfax (US), B.P. 1173, Road Soukra km 3.5, 3038 Sfax, TUNISIA

A.O. Osayi, E.A.P. Egbe and S.A. Lawal
Department of Mechanical Engineering, School of Engineering and Engineering Technology, Federal University of Technology, PMB 65 Minna, Nigeria

Mounir Muhammad Koura
Prof. Dr. in Faculty of Engineering, Ain Shams University, Egypt

Muhammad Lotfy Zamzam
Asst. Prof. in Faculty of Engineering, Ain Shams University, Egypt

Amr Ahmed Sayed Shaaban
PhD. Student in Faculty of Engineering, Ain Shams University, Egypt

A. Lebied and B. Necib
Mechanical Engineering Department, Faculty of Technology Sciences, University Constantine 1, Algeria

M. SAHLI
Departement de Mécanique Appliquée, ENSMM, 24 chemin de l'Epitaphe, 25030 Besançon, France

Khudhayer J. Jadee
Technical Engineering College-Baghdad, Middle Technical University, Baghdad, Iraq

A.R. Othman
School of Mechanical Engineering, Universiti Sains Malaysia, Malaysia

Peizheng Ma and Lin-Shu Wang
Department of Mechanical Engineering, Stony Brook University, Stony Brook, United States

Nianhua Guo
Department of Asian and Asian American Studies, Stony Brook University, Stony Brook, United States

Miqdam Tariq Chaichan
Mechanical Engineering Department, University of Technology, Baghdad, Iraq

Peter Frankovský
Department of Mechatronics, Faculty of Mechanical Engineering, Technical University of Košice, Košice, Slovakia

František Trebuňa, Ján Kostka, František Šimčák and Oskar Ostertag
Department of Applied Mechanics and Mechatronics, Faculty of Mechanical Engineering, Technical University of Košice, Košice, Slovakia

Władysław Papacz
Instytut Budowy i Eksploatacji Maszyn, Uniwersytet Zielonogórski, Zielona Góra, Poland

Michał Śledziński
Chair of Machine Design Fundamentals, Poznan University of Technology, Poznań, Poland

Omar Monir Koura
Mechanical Department, Faculty of Engineering, Modern University for Technology & Information, Egypt

Tamer Hassan Sayed
Design & Prod. Eng. Department, Faculty of Engineering, Ain Shams University, Egypt

Hassan Ghassemi, Manouchehr Fadavie and Daniel Nematy
Department of Ocean Engineering, Amirkabir University of Technology, Tehran, Iran

Darina Hroncová, Peter Frankovský, Ivan Virgala and Ingrid Delyová
Technical University of Košice, Faculty of Mechanical Engineering, Letná 9, 042 00 Košice, Slovakia

D. Lalmi
Faculty of exact sciences, natural sciences and life, Larbi Ben Mhidi University, Oeb Algeria

R. Hadef
Faculty of sciences and applid sciences, Larbi Ben Mhidi university, Oeb Algeria

Slah Driss, Zied Driss and Imen Kallel Kammoun
Laboratory of Electro-Mechanic Systems (LASEM), National School of Engineers of Sfax (ENIS), University of Sfax (US), B.P. 1173, Road Soukra km 3.5, 3038 Sfax, TUNISIA

Zahra Shahbazi
Department of Mechanical Engineering, Manhattan College, NY

Faisal Kader, Abdullah Hil Baky, Muhammad Nazmul Hassan Khan and Habibullah Amin Chowdhury
Islamic University of Technology, Gazipur, Bangladesh

LAHIOUEL Yasmina
Laboratoire d'Analyse Industrielle et de Génie des Matériaux (LAIGM), Faculté des sciences et de la Technologie, Université de Guelma, B.P. 401, Guelma

LAHIOUEL Rachid
Laboratoire de Physique de Guelma (GPL), Université de Guelma, B.P. 401, Guelma, Algérie

Zied Driss, Olfa Mlayeh, Slah Driss, Makram Maaloul and Mohamed Salah Abid
Laboratory of Electro-Mechanic Systems (LASEM), National School of Engineers of Sfax (ENIS), University of Sfax (US), B.P. 1173, Road Soukra km 3.5, 3038 Sfax, TUNISIA

Zahra Shahbazi and Allison Kaminski
Manhattan College, Mechanical engineering department, Riverdale, NY, United States

Lance Evans
Manhattan College, Biology department, Riverdale, NY, United States

Pruthviraj Manne, Shibbir Ahmad and James Waterman
Mechanical Engineering Department, Rowan University, Glassboro, New Jersey, USA

Minhaj Ahmed, Md. Fahad Hossen and Md. Emdadul Hoque
Department of Mechanical Engineering Rajshahi University of Engineering & Technology, Rajshahi-6204, Bangladesh

Omar Farrok
Department of Electrical and Electronic Engineering Ahsanullah University of Science and Technology, Dhaka-1208, Bangladesh

Mohammed Mynuddin
Department of Electrical and Electronic Engineering Atish Dipankar University of Science and Technology, Dhaka-1230, Bangladesh

Fatemeh Farhatnia
Mechanical Engineering department, Islamic Azad University Khomeinishahr Branch, Isfahan, Iran

Index

Printed in the USA
CPSIA information can be obtained
at www.ICGtesting.com
JSHW051439221024
72173JS00006B/1524